DAUGHTERS OF ALCHEMY

*Women and Scientific Culture
in Early Modern Italy*

Meredith K. Ray

HARVARD UNIVERSITY PRESS
Cambridge, Massachusetts
London, England
2015

Second printing

Library of Congress Cataloging-in-Publication Data
Ray, Meredith K., 1969–
Daughters of alchemy : women and scientific culture
in early modern Italy / Meredith K. Ray.
pages cm
Includes bibliographical references and index.
ISBN 978-0-674-50423-3 (alk. paper)
1. Women in science—Italy—History. 2. Women scientists—Italy—Biography.
3. Science—Italy—History. I. Title.
Q130.R37 2015
509.2'5220945—dc23   2014035683

For Owen

Contents

DAUGHTERS OF ALCHEMY

Introduction

In 1581, the Venetian writer Moderata Fonte, borrowing the imagery of metallurgy, wrote eloquently of women's untapped potential in the liberal arts as "buried gold," needing only to be unearthed and given the same attention as that afforded to men: "Gold which stays hidden in the mines / is no less gold, though buried, / and when it is drawn out and worked / it is as rich and beautiful as other gold."[1] *Daughters of Alchemy* undertakes a similar task, mining the rich and complex landscape of scientific culture in sixteenth- and seventeenth-century Italy to excavate the integral, but often overlooked, contributions of women. This book rethinks existing paradigms of early modern science by expanding the parameters of the investigation and reintroducing the work of women: as practitioners and patrons, authors and readers. While canonical figures such as Galileo loom large in most narratives of this dynamic intellectual and historical moment, scientific culture was woven from an evolving plurality of ideas, and encompassed the participation of people from diverse social and cultural contexts. Less visible in historical accounts than Galileo, but no less essential to the processes of cultural and epistemological change, women contributed in crucial ways to the production of knowledge on the cusp of what has traditionally been known as the Scientific Revolution.[2]

Women were, in many respects, at the vanguard of the new science with which Galileo is so closely associated. They were deeply invested in empirical culture through their experiments in medicine and practical alchemy—conducted at home, at court, and through collaborative networks of male and female practitioners—as well as in the study of natural philosophy, on display in their literary production. They could be found in the thick of public debate over astronomy and meteorology. Scholars have begun to restore vital aspects of women's scientific activity to the historical record, particularly with regard to medicine: work on nun apothecaries in Renaissance Italy, noblewomen healers and female practitioners in early modern Germany and France, and women's recipe books in seventeenth-century England has furnished important insight into the participation of women in medical and pharmaceutical culture.[3] Studies of female alchemists in England and northern Europe have expanded further our understanding of women's engagement with science in domestic and courtly contexts.[4] The influence of those activities on the work of early modern women writers, however, has received less attention. As Virginia Cox has observed, natural science "features more significantly in women's writing of this than any earlier period."[5] *Daughters of Alchemy* seeks to understand not only how women's participation in scientific activity helped to shape new directions in natural inquiry, but also how it influenced literary discourse.

My approach to the study of women in early modern scientific culture is both social and literary, seeing practice and representation as deeply interconnected. My focus is on Italy, not because women did not "do" science elsewhere in Europe—they clearly did—but in order to examine the interplay between these contexts among women facing a more similar (if still diverse) set of cultural, social, and religious circumstances, in what was a crucible for the emerging new science epitomized by Galileo. The material examined here encompasses the period from approximately 1500 to 1623, as the advent of print began to transform the dissemination of scientific knowledge throughout Europe. Medical and alchemical texts of previous centuries now found an unprecedented breadth of circulation, shaping the thought and practice of new generations, while the increased accessibility of print allowed for greater access to literary channels, including by women. My study begins in the Romagna, with Caterina Sforza's experiments with alchemical recipes, moves on to examine the vogue for printed "books of secrets" in sixteenth-century Italy, then travels to Venice to consider literary works by Moderata Fonte and Lucrezia

Marinella. It concludes in Padua, with Camilla Erculiani's letters on natural philosophy and, finally, Rome, where Margherita Sarrocchi demonstrated her prowess in astronomy by defending Galileo's discovery of the "Medicean" stars.

As historians have increasingly stressed, early modern science was not studied and practiced only in universities, laboratories, anatomy theaters, or other public spaces. Rather, natural inquiry unfolded in a variety of other contexts as well, many of which were more hospitable to the participation of women. As I show in *Daughters of Alchemy,* women engaged with science in the home, where they turned to medicinal and alchemical techniques to care for their families and run their households; in courts, where science played a crucial role in dynamics of patronage and power; in the pages of vernacular literature, in which women displayed their erudition and made science a new category of analysis in the ongoing *querelle des femmes,* or debate over women; and in academies, salons, and epistolary correspondence. Not all women who practiced science left a written record; not all women who wrote about science practiced it in a hands-on way. Nonetheless, the interaction of women with scientific culture is unmistakable in both realms. There were precedents for such activity: one of the most important gynecological works of the Middle Ages, known as the *Trotula,* was attributed to a woman physician, Trota of Salerno, and addressed treatments for women's ailments as well as recipes for women's cosmetics.[6] Like Trota herself, the learned *mulieres salernitane* named in the *Trotula* were also acclaimed for their medical skill. In northern Europe, the Benedictine abbess Hildegard of Bingen authored several works on medicine in the twelfth century: although these had a limited manuscript circulation, they were published together with the *Trotula* in a sixteenth-century edition that deliberately linked the texts' female authorship to the editor's own "empirical enterprise," establishing a connection between gender and empirical authority that would continue throughout the sixteenth and seventeenth centuries.[7]

The increasing emphasis on empiricism and the corresponding preoccupation with uncovering the "secrets of nature," in their many permutations and uses, fostered women's continued engagement with scientific culture and expanded the range of their activity. Women acquired scientific knowledge in different ways, depending on their circumstance: some had access to a degree of formal, humanist education, while others cultivated it through experience—in the workplace, the apothecary shop, or other arenas. Some women looked to past models to reach conclusions about matters of medicine, alchemy, and astronomy, studying the authoritative texts of Antiquity and the Middle Ages; others were influenced by the popularizations of Aristotelian thought that flourished in Renaissance Italy. Still others imagined innovative, sometimes transgressive, new formulations for explaining nature's operations. By the sixteenth century, literary, epistolary, and personal networks revolving around the exchange of scientific ideas stretched across Italy and abroad, connecting men and women throughout Europe more than ever before. As Carol Pal notes in a study of the Republic of Letters in the seventeenth century, the marginalization of women from our understanding of such intellectual networks stems from "our concept of the margins themselves."[8] It is not women who are missing from the picture: it is our lens that must be adjusted to perceive them. A similar widening of focus, I argue—one that encompasses women's hands-on scientific activity as well as the reflection and representation of that activity in literary discourse—allows us to see the real influence of women upon early modern scientific culture.

Just as the traditional conception of a monolithic Scientific Revolution has been called into question, so too the very term "science" has been problematized. As scholars including Paula Findlen, Katharine Park, and Lorraine Daston point out, "science" is a term used, imprecisely, to indicate what were in reality numerous strands of inquiry: philosophy, medicine, mathematics, alchemy, grammar, theology, even poetry. In sixteenth-century books of secrets, for example, alchemy, popular medicine, and the manufacture of cosmetics are all closely and inextricably intertwined; while the literary production of women writers in the later sixteenth and early seventeenth centuries collapses boundaries between natural history, astronomy, and astrology, and between the "old" Aristotelian science and the "new" empiricism.[9] Pamela H. Smith notes that historians of science are increasingly interested in "the uses made of natural knowledge more generally"; these new approaches to early modern science incorporate both the social context in which scientific knowledge was produced as well as its material and technical aspects. While Smith suggests creating more inclusive and accurate terms such as "technoscience," or "techno-medico-science," she and others continue to use the term "science" to indicate broadly the investigation of nature in the early modern period.[10] Mirroring the heterogeneous and fluid nature of early modern scientific inquiry, my own approach applies both a social and literary perspective to the role of women in scientific culture, encompassing women's activity in areas including alchemy, medicine, the production of cosmetics, astrology, meteorology, and astronomy.

Despite the evident problems with the term "science" to indicate this wealth of pursuits, like Smith I employ this term throughout for its usefulness in drawing connections to modern science, in addition to using "natural philosophy" to indicate a range of knowledge areas including medicine, alchemy, and astronomy.[11]

The Chemical Wedding

Reconsiderations of early modern scientific culture have led, among other things, to a reevaluation of the impact of empirical objectives and values on the development of the new philosophy. Work by Allen Debus, William Eamon, Pamela O. Long, Bruce Moran, Lawrence M. Principe, and others has helped resuscitate alchemy, in particular, as an area of serious scientific inquiry that contributed in fundamental ways to the development of new laboratory techniques and approaches to scientific observation and experiment. As Tara Nummedal notes in her study of Anna Maria Zieglerin, a court alchemist in sixteenth-century Germany, women played an active role in the "patronage, theory and practice of alchemy" in early modern Europe.[12] Not only did women engage in the quest to produce alchemical gold, often in court contexts, but they also incorporated alchemical practice into their quotidian lives, most especially with regard to the management of the household and the care of their own bodies and those of their families. As an examination of sixteenth-century manuscripts and printed books of secrets reveals, alchemical ingredients and processes commonly underlay instructions for the preparation of medicines and cosmetics and for the preservation of food and wine, areas closely associated with women's duties and activities. Women practiced science on a daily basis: as Lynette Hunter observes, the "non-formal" nature of this practice and its location within the domestic sphere has traditionally led scholars to undervalue or ignore it, but the kinds of work carried out in the home—cooking, gardening, tending the sick—involve aspects of physical and organic chemistry as well as "all aspects of preventive medicine and pharmacy."[13]

The centrality of women to alchemical experiment is strikingly evident in the gendered imagery of alchemy itself, which, in addition to establishing female origins through figures such as Maria the Prophetess, is often depicted as "women's work" (for example, cooking or washing) or as a gestational process that takes place in a womblike alchemical vessel.[14] The emblems in Michael Maier's *Atalanta fugiens* (1617), which contains detailed explanations of

Plumbo habito candido fac opus mulierum, hoc est, COQUE:

EPIGRAMMA XXII.

Q Uisquis amas facili multum præstare labore,
    Saturni in faciem (quæ nigra) sparge nives:
Et dabitur tibi materies albissima plumbi,
    Post quod, fœmineum nil nisi restat opus.
Tum COQUE, ceu mulier, quæ collocat ignibus ollas,
    Fac sed ut in propriis TRUTA liquescat aquis,

                                                    N   QUEM-

Emblem XXII, Michael Maier, *Atalanta fugiens, hoc est, emblemata nova de secretis naturae chymica . . .* With engravings by Theodore de Bry. Oppenheim: Hieronymi Galleri, 1618

alchemical images, draw on a long tradition of such associations. A particularly evocative example is emblem XXII, which depicts a woman tending a fire with a bellows, a bowl containing a pair of fish at her feet.

The motto reads: "When you have obtained the white lead, then do women's work, that is to say, cook."[15] The allegorical meanings of the emblem are complex and multivalent, but two major themes are evident. First, the alchemical operations performed on the philosophical subject are akin to those performed in the kitchen: it must be softened and distilled, through heat, just as the fish is cooked over the fire by the woman. Second, the two fish represent the spirit and the soul, or, alternatively, Mercury and Sulfur, which must be joined as one in a philosophical union or "chemical wedding" of male and female elements.[16] Emblem III, depicting a woman washing sheets, draws a similar parallel between alchemy and women's work. Again, the motto urges the alchemist to observe the woman's example: "Go to the Woman who washes the sheets and do as she does." The accompanying epigram explains that the alchemist should observe the world around him, taking from the mundane inspiration for the secret:

> Let not he who loves scrutinizing secret dogmas
> Neglect to take as an example everything that can help him.
> Don't you see how a woman is accustomed to cleaning dirty laundry
> By pouring hot water over it?
> Follow her example, so that you will not fail in your art.[17]

Transmutation, the ultimate goal of alchemy, is often depicted as occurring within the female womb. Emblem II, for instance, illustrates Earth as a pregnant woman carrying in her belly the philosophical child: the motto reads, "Its Nurse is the Earth."[18]

Such metaphorical language and symbolic imagery has been the subject of a wide range of interpretations, with some reading it as a Jungian elaboration of the collective unconscious, others as reflective of a masculinist attempt to seize reproductive capability from women (an analysis akin to feminist critiques of science such as that presented by Carolyn Merchant).[19] More recently, scholars such as Principe have tried to reground such gendered and erotic language and imagery in historical context, arguing that it results from the alchemists' "opposing desires" to reveal and conceal the secrets of their work.[20] Others focus on the complex and intersexual nature of such instances: Kathleen P. Long, for example, sees these examples as evidence of a (potentially

Vade ad mulierem lavantem pannos, tu fac similiter.

EPIGRAMMA III.

Abdita quisquis amas scrutari dogmata, ne sis
   Deses, in exemplum, quod juvet, omne trahas:
Anné vides, mulier maculis abstergere pannos
  Ut soleat calidis, quas superaddit, aquis?
Hanc imitare, tuá nec sic frustraberis arte,
  Námque nigri fœcem corporis unda lavat.

C 3   Si

Emblem III, Michael Maier, *Atalanta fugiens, hoc est, emblemata nova de secretis naturae chymica* . . . With engravings by Theodore de Bry. Oppenheim: Hieronymi Galleri, 1618

liberating) blurring of boundaries between masculine and feminine realms and attributes, a true "chemical wedding" that encompasses and values the female as well as the male in the resulting hermaphroditic product.[21] Such a reading may serve as way into women's participation in scientific activity more generally. I do not read women's activity in alchemy, medicine, and other scientific areas as existing in isolation from that of men. Rather, I take into account the networks of intellectual exchange and collaboration among circles of women and men of varying social status to consider how ideas circulated, traveled, and developed, whether in practice or on the printed page. Like alchemical operations, literary representations of science, too, were a kind of experiment— both involved a union of disparate elements that had to be integrated and dissolved into one another to form something new and valuable.

Scientific Culture in Practice and Print

Daughters of Alchemy moves from a consideration of women's practical engagement with science to an analysis of women's literary engagement with natural philosophy. While the practical and the literary are always enmeshed in some ways, this framework allows us to focus on the contexts in which science was practiced—as empirical work undertaken in spaces like the laboratory or apothecary shop, in the workplace, the court, or the home—as well as on the integration of scientific knowledge and the imagery of new discoveries into literary outlets such as treatises, letters, and dialogues.[22] It also allows us to see the long reach of alchemy's influence, from the court of Caterina Sforza to Margherita Sarrocchi's involvement with Galileo—whose early patron, Antonio de' Medici, was a descendant of Caterina and inherited her interest in scientific experiment. In their efforts to understand and unravel the secrets of nature, the women studied here are truly daughters of alchemy.

The first part of the book examines the fluid contexts of women's engagement with the interconnected areas of alchemy and medicine, the ways in which they established networks of scientific knowledge, and the literary reflection of such activity in the pages of sixteenth-century *libri di segreti,* or books of secrets. Chapter 1 presents the case of Caterina Sforza (1463–1509), progenitrix of the Medici dynasty of grand dukes. Immortalized for her military exploits in Machiavelli's *The Prince,* Sforza had an abiding interest in medicine and alchemy. Like noblewomen in courts throughout Europe, Sforza sought out alchemical, pharmaceutical, and cosmetic secrets, maintaining her own apothecary in Forlì and acquiring from other cities ingredients and instructions for her preparations. She compiled her discoveries in a manuscript titled *Experimenti,* or *Experiments,* which, despite being deemed by her nineteenth-century biographer Pier Desiderio Pasolini the "most complete and important" alchemical-medical document of the late fifteenth and early sixteenth centuries, has received remarkably limited scholarly attention.[23] The *Experiments* documents Sforza's pursuit of recipes for curing headache, fever, syphilis, and epilepsy; lightening the hair or improving the skin; treating infertility; making poisons and panaceas; and producing alchemical gems and gold. The manuscript offers valuable evidence of the ways in which Renaissance women incorporated scientific experimentation into their daily lives, trading "secrets" and applying the processes of the alchemical laboratory not just to the transmutation of metals, but also to the management of the household, the pursuit of beauty, and the preservation of health. Sforza's *Experiments,* along with her letters to her apothecary, other alchemists, and the nuns of the convent of Le Murate in Florence (known for its commercial pharmacy), offers a compelling picture of the avenues through which women pursued scientific knowledge, the uses to which it was put, and the manner in which it was preserved and circulated. Not only does Sforza's example help contextualize a broader manuscript tradition of women's recipe books in early modern Europe and England, it also lays the groundwork for understanding the emerging print tradition of books of secrets in sixteenth-century Italy, while at the same time elaborating a genealogy for the Medici interest in alchemy.

In Chapter 2, I argue that many books of secrets specifically targeted female readers by focusing on issues such as feminine beauty, domestic management, and women's health. There were precedents for such compilations of recipes, from the pseudo-Aristotelian *Secretum secretorum,* which circulated in Latin and various translations from the twelfth century, to the *Trotula* noted earlier.[24] In their sixteenth-century incarnation, many of these works circulated in the vernacular, rendering them more readily accessible to a reading public that included women. The influential *De' secreti del reverendo donno Alessio Piemontese* (*Secrets of Don Alessio of Piedmont,* 1555), for example, which had over one hundred editions, addresses problems with infertility, pregnancy, and lactation alongside general remedies for skin rashes and stomachache; and includes recipes for face creams, lotions, and rouges along with methods for preserving food and wine and removing stains. Drawing connections between books of secrets and the activities of female practitioners of science and medicine, I contend that these texts fit into a growing interest in describing and defining women's experience that played out in debates over women's social role and intellectual equality to men. I pay particular attention to the *Secreti della signora Isabella Cortese* (Secrets of Signora Isabella Cortese, 1561), dedicated to a female audience and one of few such texts attributed to a female author. Cortese's recipes for beauty waters, soaps, and perfumes, along with others for calcinated mercury and potable gold, are deemed useful to "every noble lady" and meant to reflect the range of uses that Renaissance women made of scientific experiment. My discussion of Cortese's text leads to a consideration of the complex intertextual web created between books of secrets, alchemical texts, medical treatises, herbals, and manuscript records, as well as the circles of scientific knowledge that underlay the compilation of such works (Cortese, for example, claims connections to a scientific community in Ragusa, while Girolamo Ruscelli, founder of the Accademia Segreta in Naples, was likely the author of the *Secrets of Don Alessio*).

Chapter 2 also considers the influence of books of secrets on other early modern literary genres, including vernacular treatises, dialogues, and epistolary collections in which such knowledge is often sharply satirized—especially as it pertains to women. As scholarship by David Gentilcore, M. A. Katritzky, and Tara Nummedal has demonstrated, the figure of the charlatan—whether medical or alchemical—was an increasing locus of early modern anxiety.[25] I take this line of inquiry further, investigating the gendered connotations of secrecy as a critical aspect of such anxiety.[26] Fraudulent, vanity-producing, a distraction from the proper pursuits of respectable women: the flip side of secrets is moral and spiritual corruption.

Moving from the compilation of secrets to broader meditations on natural philosophy and the newest scientific discoveries, scientific discourse comes to play an increasingly important role in the literature of the *querelle des femmes* that flourished in the sixteenth and early seventeenth centuries. Stretching back to Boccaccio's *Famous Women* (ca. 1360) and Christine de Pizan's *Book of the City of Ladies* (1405), the debate over women was a constantly evolving cultural conversation that occupied the attention of writers throughout Europe. Among the best-known female contributors to the debate are Moderata Fonte (1555–1592) and Lucrezia Marinella (1571–1653), considered in Chapter 3. Both Fonte and Marinella stressed women's capacity for scientific observation in their writings, framing it as an argument for women's intellectual equality (or, indeed, superiority) to men. The female characters in Marinella's epic poem

L'Enrico, overo Bizantio acquistato (*Enrico; or, Byzantium Conquered,* 1635) and her pastoral work *Arcadia felice* (Happy Arcadia, 1605) display a grasp of natural philosophy that is integral to the narrative structure of these works; while in Fonte's dialogue, *Il merito delle donne* (*The Worth of Women,* 1600), women become the custodians of an encyclopedic array of scientific and medical knowledge that synthesizes the "old" and "new" science. As I argue in this chapter, Fonte, in particular, places science at the service of her pro-woman argument, using it to bolster her claims for women's innate capacities and rights to equal status with men. The works of both Marinella and Fonte tap into a curiosity about female experience engendered in large part by the ongoing debate over women, and demonstrate a deep engagement with the kinds of medical and scientific discourse that were gaining traction by the end of the sixteenth century.

In my final chapter, I posit that by the early seventeenth century women had begun to participate in scientific discourse in more deliberate, formal ways: publishing treatises on natural philosophy, allying themselves with scientific academies, and entering into correspondence with major scientific figures. The two cases examined in Chapter 4 illustrate the flow of scientific knowledge among networks of both men and women that stretched throughout Italy and abroad. Camilla Erculiani (died post-1584), whose *Lettere di philosophia naturale* (Letters on natural philosophy, 1584) identifies her as an apothecary in Padua, had connections to scientific communities as far away as Poland (indeed, the publication of her work in Kraków suggests both the reach of her web of contacts as well as the obstacles she faced in publishing her work in Italy, given the unorthodox nature of her arguments). Erculiani places gender at the forefront of her scientific investigations, framing her *Letters* with a defense of women's aptitude in the sciences. The experience of Erculiani, who was questioned by the Inquisition on suspicion of heresy, underscores not only the links between gender and science, but also the ways in which these tensions subtended the conflict between science and faith later confronted by Galileo. Margherita Sarrocchi (c. 1560–1617), best known for her epic poem *La Scanderbeide,* hosted a celebrated salon in Rome that included Galileo and the mathematician Luca Valerio and was recognized throughout Italy for her learning in the sciences. In exchange for Galileo's advice on revisions to her poem, Sarrocchi offered her opinion of his works and, for a time, championed his discoveries in the world of the Roman academies. However, her participation in such intellectual communities eventually declined as her influence

with members of the Accademia dei Lincei and her orthodox poetic sensibili-
ties earned her the hostility of former allies. The complicated position of
women such as Sarrocchi vis-à-vis the formal institutions of science was reflected elsewhere in Europe: for example, in Germany, where the astronomer
Maria Winkelman was refused membership in the Academy of Sciences, marginalized by the increasing professionalization of intellectual culture.[27] In
Italy, attitudes toward women writers became increasingly more restrictive in
the Counter-Reformation climate of seventeenth-century Italy.[28] Sarrocchi's
example, as I demonstrate in this chapter, while indicative of the influence of
women on the circulation of scientific knowledge, also reflects such shifts in
literary and scientific culture.

Daughters of Alchemy seeks to reconstruct a broader picture of women's interactions with scientific culture, one that integrates the practice of medicine
and alchemy by women with the efforts of women writers to incorporate the
study of natural philosophy into their literary works. It is a study of the many
facets of women's participation in scientific culture, in arenas both private and
public, as well as an examination of the ways in which that participation was
absorbed and reflected in literary discourse, not just in discussions of science
and medicine but with regard to gender and women's intellectual potential.

I

Caterina Sforza's Experiments with Alchemy

Talc is the star of the earth and has gleaming scales; it is found on the isle of Cyprus and its color is similar to citrine; in a mass it looks green, and dissolved in air it looks crystalline; and it has the following virtues, not to mention others not noted in this book, which will be the alchemist's desire to discover: First, to make women beautiful and remove all spots or marks from the face, such that a woman of sixty will appear to be twenty. . . . Also . . . mixed with white wine, its powder will cure one who is poisoned; and he who drinks the powder in white wine will be protected that day from poison and all disease or plague. . . . Also . . . this water turns silver to gold, and makes false jewels perfect and fine.[1]

In a manuscript titled *Experimenti (Experiments),* Caterina Sforza (1463–1509), regent of Forlì and Imola in Italy's Romagna region, recorded this recipe for water of talc, a mineral-based solution produced through distillation and used by early modern alchemists for a variety of purposes. The recipe, which claims to restore youth, provide an antidote to poison and plague, and transform silver into gold, is the first of over four hundred prescriptions for medicinal remedies, cosmetics, and alchemical procedures compiled by Caterina Sforza over the course of her lifetime. The *Experiments* includes instructions for treating ailments ranging from fever, cough, and intestinal worms to sciatica, epilepsy, and cancer; creating lip colors, lotions, and hair dye; and—most valuable of all—producing the transmutatory philosopher's stone and quintessence: the elixir thought to cure all illness, protect against disease, and prolong youth (perhaps indefinitely).[2] Many recipes accomplish several of these things at once.

The *Experiments,* which has been characterized as a "foundational text in the history of pharmacology," not to mention that of alchemy, offers valuable insight into a little-studied aspect of this important Renaissance figure, the progenitrix of the Medici granducal dynasty.[3] Admired by her contemporaries for her political leadership and fearlessness in battle and immortalized in sixteenth-century works including Machiavelli's *Discourses* and *The Prince,* Caterina Sforza—like many noblewomen in early modern Europe—had a keen interest in scientific experiment.[4] As her *Experiments* demonstrates, this interest was wide-ranging and incorporated the pursuit of both practical and esoteric knowledge, assembled from a variety of sources: learned texts and popular tradition, direct experience and the accounts of others. Letters attesting to the enduring business relationship between Caterina and her apothecary in Forlì, along with others addressed to her that provide or request assistance with alchemical, medicinal, and cosmetic recipes, confirm that Caterina actively collected the kinds of prescriptions contained in the *Experiments* and support the proposition that she was compiling this work up until her death in 1509.

Caterina's compilation was passed down to her youngest son, the condottiere Giovanni dalle Bande Nere, father to the first Medici grand duke, Cosimo I. It therefore stands at the origins of what would be a long and profound engagement with alchemy and medicine on the part of the Medici family. From the first foundry established by Cosimo I in Palazzo Vecchio, to the Casino erected by his son, Francesco I, in San Marco, the Medici princes were well known for their interest in experiments. In addition to these laboratory spaces, the Medici were also responsible for installing the first botanical gardens at Florence and Pisa, which would have supplied primary ingredients for such activities.[5] Suggesting a lineage of alchemical activity that stretched into the past as well as the future, Caterina's manuscript contains a recipe for making counterfeit gold attributed to Cosimo the Elder (Cosimo di Giovanni de' Medici).[6] A medicinal secret for a potent antipyretic is also attributed to Cosimo.[7]

The example of Caterina Sforza offers an opportunity to explore some of the ways in which women—and men—engaged with scientific culture on the cusp of the Scientific Revolution, using it to take control of their physical health, enhance their appearance, and govern their households. Like the print tradition of "books of secrets" that flourished in the mid-sixteenth century, Caterina's *Experiments* reflect the empirical character of early modern scientific culture: indeed, the very term "experiment" has a complex history that speaks to its position between direct experience, observation, and application.[8] The collection also mirrors the fascination with secrets (valuable or unusual recipes relating to alchemy, medicine, cosmetics, perfumery, and metallurgy) that enthralled courts throughout early modern Europe, giving rise to a lively market for recipes transmitted via letters, manuscript collections, and printed works as well as by word of mouth.

The collection and circulation of practical secrets—that is, medical and alchemical knowledge turned to the needs and demands of daily life—was by the sixteenth century a common pastime for women as well as men.[9] Italian archives brim with such compilations: the Fondo Magliabechiano in Florence alone contains dozens of such works.[10] In these collections, recipes for beauty waters, oils, and lotions are often attributed to well-known noblewomen in order to spark the interest of female readers, in particular, and underscore their value and authenticity. A sixteenth-century *Gallant Recipe-Book (Ricettario galante)* credits distillations of roses and lemons to Elisabetta Gonzaga (1471–1526), duchess of Urbino, and alchemical recipes based on mercury and alum to Isabella d'Aragona (1470–1524), daughter of Alfonso II of Naples; while an anonymous Florentine manuscript of the same period describes a hand lotion made from pulverized bone and a delicate powder of roses said to originate with Ippolita Sforza of Calabria (1446–1484).[11] Echoing widely circulating medieval predecessors such as the gynecological texts of *The Trotula,* medicinal remedies included in such works typically address women's complaints ranging from problems with menstruation to pregnancy, childbirth, and nursing, in addition to a host of ailments common to both men and women. The compilation of recipe books was widespread throughout Europe, as well as in England, where they flourished well into the seventeenth century, in both print and manuscript form. The cookery books that also proliferated in this period commonly included medicinal recipes, with food serving multiple purposes for health and beauty as well as sustenance.[12]

The collection of recipes was not a solely textual pursuit; on the contrary, it found expression in the quotidian practice of both women and men, and in a variety of intellectual contexts. Like men, women sought out medicinal, alchemical, and, especially, cosmetic "secrets," experimenting with them in court spaces, where they could be used to establish status and reciprocity among aristocratic networks; and in domestic contexts, where they served the needs of the family and the household. Like Caterina Sforza, other early modern women associated with prominent courts collected and circulated secrets: Catherine de' Medici (1519–1589), regent of France, was known for her interest in cosmetics as well as alchemy and medicine (and, it was rumored, poisons); Marie de' Medici (1575–1642), also a queen of France (and a direct descendant of Caterina), established her own alchemical laboratory in order to conduct experiments. Bianca Cappello (1548–1587), wife of Caterina's great-grandson, Franceso I de' Medici, likely participated with her husband in alchemical experiments. Isabella d'Este (1474–1539), marchioness of Mantua, produced her own perfumes to give as gifts.[13]

Although the collection of recipes was by no means confined to court culture, the pursuit and exchange of secrets found particularly fertile ground in court settings. This valuable knowledge functioned as a form of currency, a tool through which to establish social and intellectual position, and a means to cement networks of communication with like-minded collectors across boundaries of gender as well as geography. Daniel Jütte refers to an "entire economy of secrecy" that characterized European court culture in the early modern period.[14] Courts were important sites for introducing new ideas and technologies, spurred by the same spirit of competition that galvanized courtiers in other areas of court life.[15] Princes patronized scientists and alchemists who supplied novel and valuable ideas that could benefit or enhance their power. The courtly association with secrets was epitomized in later decades by figures such as Giambattista della Porta, whose explorations of the secrets of nature in works such as *Magia naturalis* (1589) were directed at the intellectual curiosity of the learned prince. William Eamon stresses the exchange value of secrets at court, noting, "Secrets, paradoxes, allegories, and other forms of privileged information, whose hidden meaning presented a challenge and promised to surprise and delight the discoverer, were among the most appropriate kinds of gifts a client could offer a patron."[16] Yet it was not only clients who engaged in the pursuit and exchange of such information. As the example of Caterina Sforza illustrates, *signori*—and indeed *signore*—were equally fascinated by experiments and also engaged in them directly. The *Experiments* demonstrates Sforza's interest in new scientific technologies and techniques as a tool for shaping and maintaining political power (by producing alchemical gold, counterfeit coins, and even poisons and their antidotes); but also that her involvement with science had a personal and familiar component (sending and reciprocating gifts, tending to health and hygiene, managing the household). Along with her *Experiments,* Caterina's correspondence with her apothecary in Forlì and other men and women with whom she exchanged recipes reveals the extent to which she was invested in what has been termed "court experimentalism": that is, courtly enthusiasm for experimental knowledge including alchemy, medicine, and the production of cosmetics.[17] Caterina's example amply reflects the complex contours of knowledge and practice in early modern scientific culture, in which medicine and alchemy were deeply intertwined (while cosmetics, produced through alchemical processes, had medicinal—and sometimes moral—effects). Her pursuit of secrets blurs lines not just between alchemy and medicine, but also between domestic and commercial, private and public, male and female. Turned to both political and personal purpose, Caterina's recipes are amassed from an amalgam of learned and popular sources; and from men and women of varying social status, including kings, noblewomen, courtiers, nuns, and Jews.[18] After looking closely at Caterina's volume of *Experiments* and what it can tell us about the fluid arenas in which women engaged with science and medicine as well as the material conditions in which such experiments were undertaken, this chapter will then consider Caterina's participation in networks of scientific knowledge beyond the pages of her compilation. In doing so, it examines her interactions with practitioners of alchemy and medicine in other Renaissance courts as well as in other contexts—particularly the convent, which played a key role in the development and commercialization of pharmaceutical medicine in sixteenth-century Italy. It also considers the political and financial implications of Caterina's collection of secrets and the importance of her legacy with respect to the scientific pursuits of her Medici descendants.

Caterina at Court

The natural daughter of Duke Galeazzo Maria Sforza and Lucrezia Landriani, Caterina was raised and educated in her father's household in Milan. Like most aristocratic women, she likely received some form of humanist education similar to that of her brothers. Caterina had a close relationship with her father's second wife, Bona Maria di Savoia, who arrived at the Sforza court in 1468. Bona's apothecary, Cristoforo de Brugora, kept a medicinal garden, and it has been suggested that it was through Bona and her apothecary that Caterina was first introduced to the world of botanical pharmaceuticals she would later explore in her experiments.[19] After her first marriage, Caterina designed medicinal gardens of her own in Imola and Forlì, likely used for the cultivation of simples, the basic botanical ingredients employed to make medicines and cosmetics. A half-century later, Caterina's grandson Cosimo I would establish Europe's first public botanical garden in Pisa, as well as the *Giardino de' semplici* in Florence, both designed by Luca Ghini.[20]

In 1473 Caterina was married to Girolamo Riario, a nephew of Pope Sixtus IV, as part of a papal plan to regain control of the Romagna, now under Sforza authority.[21] At age fourteen, she left Milan to celebrate her wedding and join her husband in Rome. There she participated in the city's lively court culture for several years, until civil unrest forced the pair to leave Rome and take up residence in the Romagna, where Girolamo had been invested with the signoria of Imola and Forlì.

Caterina was regarded by her contemporaries with both admiration and trepidation for her occupation of the Castel Sant'Angelo in Rome in 1484 following the death of Sixtus IV as well as for her tenacious governance of Imola and Forlì after Girolamo's assassination in 1488.[22] Especially legendary is the story of Caterina's defense of the Rocca di Ravaldino, the main fortress of Forlì, in the wake of the coup against Girolamo, when she successfully outwitted her enemies by overturning their expectations of her as a female ruler. Machiavelli, who had occasion to treat with Caterina on behalf of Florence, famously relates the story in his *Discourses,* describing how Caterina, in response to her foes' threats to kill her hostage children, lifted her skirts, pointed to her genitalia, and retorted that she could make others.[23] While Machiavelli's account of the incident is rather more theatrical than the description offered by contemporary chroniclers, it cemented her enduring public image as a virago who had defiantly preferred losing her children to losing her state.[24] In *The Prince,* Machiavelli further immortalized Caterina as an audacious leader, celebrating her courage even as he noted her defeat by Cesare Borgia and his forces.[25]

In the early years of their tenure as *signori* of Imola and Forlì, Caterina and Girolamo funded a variety of public works and commissioned architectural and artistic projects in an effort to establish their court. In Imola, this included fortifying the city's defenses, creating a large central piazza lined by merchants' shops, and building a personal residence and lodgings for important visitors. In Forlì, the principal seat of the signoria, they likewise made a series of military fortifications and renovated another residence next to the Rocca di Ravaldino fortress, which they called "Paradiso."[26] In both towns, Caterina oversaw projects for new gardens and park spaces: the enclosed garden at the Rocca di Ravaldino was a walled area situated between the Paradiso and the fortress, where she likely grew the vegetables, flowers, and medicinal herbs used in her recipes. The chronicler Andrea Bernardi of Forlì tells us that she also built a larger park along the southern border of the Rocca, where she could "keep wild animals of every type and plant fruit-bearing trees."[27] In addition to satisfying Caterina's passion for hunting, these spaces would have provided additional materials for her experiments, which sometimes included fruits and meats. It is likely that it was during this largely peaceful period, between her arrival in Forlì in 1484 and the assassination of Girolamo Riario four years later, that Caterina began experimenting in earnest.

Compared to other cultural centers of the time, the Riario-Sforza court was smaller and more provincial. Agriculture remained the principal activity of the Forlivesi, and the aristocratic class was composed of only a handful of noble families, together with a burgeoning professional elite, which included doctors, lawyers, and, especially, apothecaries—of whose services Caterina availed herself. In residence at Forlì were a court master, seneschal, servants, several slaves, and a *buffona,* or jester, as was fashionable in other courts, including that of Isabella d'Este in Mantua; there were also tutors for the children and secretaries for political and personal matters.[28] While it may have been more difficult to attract and retain the artists and intellectuals who flocked to richer venues such as Urbino or Ferrara, Girolamo and Caterina worked to create an image of princely magnificence, participating actively in public festivities within as well as outside of their territories as part of a "larger scheme of public presentation and diplomacy."[29] They also surrounded themselves with luxury items for their household and their personal adornment, from fine textiles and silver to brocade garments and satin cloaks.[30]

Following the assassination of Girolamo, Caterina served as regent for her eldest son Ottaviano (1479–1533), continuing to solidify her power through a combination of force, diplomacy, and public works. She had relationships with (and is thought to have secretly wed) Giacomo Feo, the brother of her castellan Tommaso Feo and, after Feo's death (also by assassination), Giovanni di Pierfrancesco de' Medici of Florence, called "il Popolano." This final union resulted in a son who would become the renowned condottiere Giovanni dalle Bande Nere, father to Cosimo I.

In 1499, Caterina was besieged in the Rocca of Ravaldino by Cesare Borgia and the forces of Louis XII. Vastly outnumbered, Caterina and her forces held out for weeks before surrendering, earning the admiration of her contemporaries in Italy and abroad: as Isabella d'Este wrote, "if the French condemn the cowardice of [Italian] men, they must at least praise the daring and valor of Italian women."[31] Caterina was taken to Rome, where, after refusing to re-

Title page, *Experimenti de la ex.ma s.r. Caterina da Furli . . .*

nounce her claim to her state, she was charged with attempting to poison Pope Alexander VI, Cesare Borgia's father, and imprisoned. After a year of solitary confinement in the same Castel Sant'Angelo she had occupied so defiantly years earlier, Caterina was finally compelled to give up her titles to Imola and Forlì and permitted to leave Rome for Florence.[32] There, she would enter into a lengthy legal battle with her Medici in-laws over the custody and patrimony of her son Giovanni, while working at the same time to secure the safety and future prospects of her older children. She nurtured connections with the Dominican convent of Annalena and, especially, the Benedictine convent of Le Murate, where she spent time as a boarder.[33] Convent chronicles recount that she resided in a cell next to the main *sala,* and that she often expressed her wish to retire among the sisters of Le Murate.[34] Archival records show that Caterina's patronage of this convent, which had begun in the 1490s, continued until her death from pneumonia in 1509; that she maintained a cordial correspondence with the abbess, Elena Bini; and that she sent gifts to the nuns.[35] Caterina remained in Florence until her death, continuously but unsuccessfully seeking to negotiate a return to power in Imola and Forlì.[36] She was buried, according to her wishes, in the convent church of Le Murate.[37]

The *Experiments*

In her will, Caterina distributed what wealth and possessions she had among her sons, also making provisions for two granddaughters.[38] Her collection of *Experiments* she passed down to her son from her Medici marriage. Only one copy of the *Experiments* is extant today: a transcription made for Giovanni dalle Bande Nere by Lucantonio Cuppano, a colonel who served under him, now held in a private archive.[39] Numbering 553 pages in total and bound in leather, the manuscript bears a note made in a later hand stating, "This parchment was procured by Gio[vanni] de' Medici . . . as inheritance from his mother Caterina Sforza Riario, 1514."[40] The frontispiece also contains a code to the work's encrypted passages, apparently supplied by the transcriber or copied from the original.[41] That Caterina bequeathed the *Experiments* to her son is also suggested by a letter written by Giovanni dalle Bande Nere in 1525, well after Caterina's death, to Don Francesco Suasio of Trebbio, where the family had territory, which seems to refer to the gift. Indicating that a manuscript collection of recipes has been mislaid, Giovanni makes clear that he wants the precious volume returned to him: "We find missing from the strongboxes in Rome a handwritten book of recipes for many and various things: we must find it, because, one way or another, we want it."[42]

Cuppano's copy of the *Experiments* was published in the nineteenth century by Caterina's biographer Pier Desiderio Pasolini, who termed it the "most complete and important known document on medicine and perfume" of the early sixteenth century.[43] Cuppano's preface refers repeatedly to Caterina as the work's author, and to himself as the transcriber of her original, handwritten instructions. He takes care to underscore that he works from an autograph copy of Caterina's manuscript, emphasizing her direct engagement with the volume as both author and medical/alchemical practitioner: "In the name of God, in this book you will find some experiments taken from the original by the most illustrious madonna Caterina of Forlì, mother of the most illustrious signor Giovanni de Medici my lord and patron; and since the original was written in said madonna's own hand . . . I will not mind the fatigue I undergo in copying them."[44]

Pasolini devotes some attention to the question of transcription in his edition of the *Experiments,* noting that the kinds of errors that appear in the manuscript demonstrate that it was copied from a written source rather than from dictation.[45] Because the original manuscript is not extant, we cannot compare Cuppano's copy to Caterina's own. We must rely on Cuppano's testimony as to the work's authorship in his preface and at several points in the body of the text where Caterina's ownership and mastery of these recipes is reiterated: for example, in a reminder to the reader that a recipe for calcinated talc must be carefully followed, "just as Madame of Furlì used to do it."[46] The letters Caterina exchanged with her *speziale* and other collectors of secrets offer further evidence of her authorship of the manuscript. A valuable cultural artifact, the *Experiments* suggests that alchemy, in its many configurations, was a central and continuing pursuit for Caterina, one that occupied her attention even amid her most pressing political battles and up until the end of her life.[47]

Caterina's collection reflects the interest in experimental methodology that was characteristic of scientific engagement in this period. Cuppano's foreword underscores that these are recipes Caterina personally tried and tested; therefore "one must assume [all the recipes] to be effective, for they have been proven so by this great lady."[48] Their efficacy is practical, not theoretical, and the recipes do not seek to explain why they work, only to show that they do work. Accordingly, the *Experiments* is plain in style and largely devoid of narrative framework or flourish, with an authorial voice making only rare appearances: for example, to direct a reader's attention to an especially valuable prescription. The recipes' instructions are often general in nature, with measurements frequently left unspecified, assuming some prior knowledge or familiarity on the part of the user.[49] The extensive markings in the Cuppano transcription (notes in the margins, asterisks, or illustrated index fingers directing the reader to particularly important points) suggest that this was a well-used text, not a presentation copy or a work meant only for display.

The recipes in the *Experiments* are presented as tools for navigating daily life and daily problems, whether relating to basic preventive medicine or cures for more serious ailments (remedies for epilepsy and venereal disease appear with particular regularity); attending to personal appearance; or making one's domestic environment more healthful and pleasant through the production of perfumes and pomanders. Some of the more overtly alchemical recipes likely derive from Caterina's political concerns as regent of a territory in constant need of financial backing: for example, alchemical recipes for coloring or adding weight to metal so that it appears to be gold. Occasional asides inserted into the recipes further underscore the collection's practical function. One such note (whether Cuppano's or Caterina's own is unclear) follows the recipe for water of talc cited earlier. It explains that extra pages have deliberately been bound with the manuscript so that anyone who discovers additional uses for this or other mixtures can record them for the benefit of future readers: "I will leave space in this volume so that if anyone should acquire additional knowledge, they may deign to communicate it."[50] Indeed, the note elaborates, it would be "a mortal sin to keep such treasure hidden."[51] The tension revealed here between the secrecy and obscurity traditionally associated with alchemical experiment and the practical imperative to disclose and share information, especially (but not only) when it relates to health, is characteristic of scientific culture in the early sixteenth century. The conflict that arises from maintaining the privileged nature of the recipes even as they are revealed to the reader requires a constant negotiation, one that becomes still more fraught once the kinds of "secrets" collected by Caterina in manuscript form find a wider diffusion through print, as in the published books of secrets that proliferate in the mid-sixteenth century.[52]

Further contributing to an aura of secrecy is a handful of Caterina's recipes that are offered either partially or wholly in Latin rather than Italian. These recipes are for the most part related to the transmutation of metals, the Latin intended to signal their particular gravity and importance, both in terms of their alchemical content and their significance as a tool for the creation of currency. In some instances, the Latin may indicate a recipe's provenance from another, authoritative source. Several alchemical recipes—particularly those that describe experimental techniques rather than ingredients or uses—are partially presented in code, the key to which is provided in the manuscript. Interestingly, a handful of recipes relating to treatments for impotence or lack of libido are also given in partial code, suggesting a reticence around the discussion of sexual function that is curious given the marked and matter-of-fact attention paid to this subject at other points in the manuscript.[53] Again reflecting the tension between secrecy and divulgence, the code is simple and easy to decipher with the key provided; its primary function to mark a recipe as notably unique, demanding, or valuable.

Of the 454 recipes recorded in Caterina's *Experiments,* the majority are primarily medicinal in nature; the others are divided among cosmetics and alchemical recipes. However, it is difficult to draw real distinctions between these types of recipes, as cosmetic and medical prescriptions rely heavily on alchemical ingredients and processes, blurring the boundaries between areas of knowledge and production. Among Caterina's medicinal recipes are a number of distilled waters, unguents, and elixirs produced through alchemical procedures such as multiplication, a kind of progressive distillation whereby a substance assumes greater and more diverse powers during the course of preparation. A "marvelous secret unguent" *(una untione mirabile segreta),* good for gout, sciatica, and nervous tension, requires boiling meat and salad leaves in wine for eight hours, then draining the mixture and gradually adding more ingredients such as rosemary and other herbs. To make a truly marvelous unguent, the mixture should be left outside for a month, so that it may gain additional force from the heat of the sun.[54] A recipe for a potent "marvelous water" *(acqua mirabile)* capable of reversing infection, aiding liver function, and curing tumors and paralysis, is given in Latin and combines nutmeg, cardamom, cinnamon, and antimony (a metal used in alchemy and, in medicine, as an emetic): "Finely pulverize all these ingredients and combine with antimony and as much white wine as you deem fit, then distill in an alembic over low heat and you will have a most powerful water."[55] Other general panaceas described by Caterina include an *elisir vitae,* an elixir capable of conferring perpetual health and youth—a principal goal of therapeutic alchemy. This elixir is so powerful, it can raise the dead: "it causes a person to regain his youth and brings the dead back to life, and if someone were so ill as to have been abandoned by his physicians as an incurable case, it will restore him to health."[56] To create it, Caterina advises combining spices, flowers, herbs, fruit, and sugar in a mortar, then distilling the mixture with water in a tightly sealed glass alembic.[57] After two days it is stirred over low heat until the water changes color from clear to white. Her instructions recall the principal stages of the alchemical process as described in many formal alchemical texts, in which the phases of transformation are marked by corresponding changes in color (rendering the use of a glass vessel of central importance to the alchemist).[58] Similarly, a "marvelous and divine water" *(acqua mirabile et divina)* also produced from a mixture of spices, flowers, wine, and other ingredients improves memory, treats toothache, earache, and melancholy, and can even cure leprosy, paralysis, and other grave illnesses. Again distilled over a period of several days, this recipe, like the elixir, results in three separate stages of liquid, each more powerful than the last: here, the third stage produces the deep red color most prized by alchemists and, again, capable of bringing a patient back from the brink of death: "the second [liquid] is better than the first but the third is the color of blood, and better than all the rest."[59]

Caterina includes many other medicinal recipes to treat conditions from lice and colic to wounds, tumors, and epilepsy, and even one for a powerful, opium-based surgical anesthetic that places the patient in a drug-induced sleep.[60] Skin conditions receive much attention, and prescriptions for dealing with fever and plague abound. As befitting a ruler in a Renaissance court, Caterina also records recipes for poisons and their antidotes, including one attributed to Pope Paul II (1464–1471), which takes a distillation made from scorpions as its central ingredient (similarly born of the realities of court intrigue are recipes for concealing and protecting written communication with invisible or disappearing ink).[61] Like most medicinal preparations of the time, Caterina's recipes rely extensively on familiar herbs, spices, and foodstuffs, but also on more exotic ingredients thought to contain magical and curative properties, such as minerals, precious gems, and metals.[62] In this respect they are evocative of Marsilio Ficino's instructions for good health in his widely disseminated *De vita libri tres* (1480–1489), which extols the benefits of ingredients ranging from cheese, almonds, and apples to infusions of gold and silver leaf.[63] Many of Caterina's ingredients suggest that she subscribed to the law of similars, according to which a substance is thought to share a common characteristic with the ailment it is used to treat (such as using goat's spleen to treat diseases of the spleen) or, alternatively, to impart its characteristics to the patient (for example, egg whites to whiten the complexion). Caterina's faith in similars is akin to the "doctrine of signatures" that would characterize Paracelsus's influential approach to medicine a few decades later and formed a central element of Della Porta's theory of natural magic.[64] Such correlations are particularly common among recipes for improving sexual function: several of Caterina's aphrodisiacal prescriptions rely on *testiculisvulpis,* an orchid bulb whose suggestive shape made it precious in "preparing for the exertions of Venus."[65] Astrology, considered of crucial relevance for matters of science and medicine in the early modern period, plays a role in some recipes, with the time of day or year adding potency to the results. Instructions for a preparation of thistle *(cardo benedetto)* cures headache when eaten in salad and body aches when taken with red wine; but, when distilled into a liquid during the months of May or August, it rids the body of all "bad humors" *(cattivi humori).*[66]

Caterina's correspondence indicates that she depended on her apothecary, Lodovico Albertini, based in Forlì, to supply her with the herbs and other ingredients she needed for her recipes. When Caterina died in 1509 she owed Albertini more than 587 florins. Caterina's third husband, Giovanni de' Medici, also died owing Albertini money, suggesting that he probably participated in his wife's experiments as well. In a letter written by Albertini to Francesco Fortunati, Caterina's administrator and confessor, shortly after her death (and suggesting their closeness during her life), the apothecary writes that he couldn't bring himself to speak of the debt when he visited Caterina during her final illness: "it didn't seem appropriate to mention it, but the illustrious Lady—bless her memory—owed me 587 florins, and more, for materials I provided her with in Forlì, as my account books clearly show. Likewise the magnificent Giovanni di Medici—bless his memory—owes me for the materials I provided him from my shop."[67]

The Renaissance *speziale* played an important role in early modern medical and scientific culture. Occupying a rung on the social ladder below the physician but well above the herbalist or itinerant salesman of remedies, the *speziale* furnished his clients not just with ingredients for medical prescriptions, but also with foods, sugared candies, oils, pigments, papers, and ink. Apothecary's shops also served increasingly as meeting places for a variety of clients and intellectuals, making them spaces for the exchange of political and cultural information and knowledge.[68] In addition to providing Caterina with the necessary ingredients for her experiments and with other kinds of goods, as well as the benefit of his professional expertise, Albertini would likely have had his ear to the ground in Forlì, serving an additional role by keeping Caterina apprised of the mood among her subjects. Caterina also maintained her own walled garden at the Rocca di Ravaldino, where she could produce many of the herbs, plants, and flowers required by her recipes. Caterina's residence within the Rocca (the renovation of which she undertook after Girolamo Riario's assassination in the family's central palazzo in Forlì) included elegantly decorated living quarters and, perhaps, an *officina* where she could carry out her experiments.[69]

Caterina's correspondence with the abbess of the Florentine convent of Le Murate, Elena Bini, indicates that she also received gifts of flowers and fruits from the nuns. She reciprocated their generosity with gifts of her own, generally donations of money or grain.[70] In a letter sent from Forlì on July 18, 1499, Caterina thanks Elena for a basket of flowers, although she begs her to refrain from such expense in the future. In a letter dated a few months later, Caterina expresses gratitude for Elena's correspondence and for the prayers offered by the nuns on her behalf; she also mentions a gift of pomegranates *(pome granate)* and other fruits from the convent's garden: "I read your letters with the greatest pleasure in seeing that you have not forgotten me, and that you always remember me in your prayers. . . . The pomegranates and other fruits from your garden were most welcome, for many reasons."[71] Caterina's clarification that the gift was welcome "for many reasons" is significant, since pomegranates served a variety of medicinal purposes in the early modern period, including as a vermifuge, and were also used to make red dye. It is plausible that Caterina used some of these gifts from the convent of Le Murate—which, in addition to maintaining an orchard and garden, would soon establish a commercial pharmacy—in her own medicinal and cosmetic remedies. Indeed, we might speculate that Caterina benefited from the pharmaceutical knowledge of the Murate nuns, who were well versed—whether through formal learning or direct experience—in illness and the ingredients needed to treat it. As Sharon Strocchia has shown, female convents turned increasingly to the business of pharmacy in the sixteenth century in an effort to combine the spiritual imperative to aid the sick with the need for revenue that could ensure a convent's survival. They produced a range of medicinal remedies, including pills, unguents, liquors, elixirs, and possibly cosmetics, overlapping in many respects with traditional commercial apothecaries. By the mid-sixteenth century Florence was home to six convent-run apothecaries.[72] Given the time Caterina spent as a boarder at Le Murate, her enduring connection to its nuns, and her interest in medicinal recipes, it is likely that at least some of her prescriptions originated there. Many of the prescriptions for preventive care and common

ailments that appear in her *Experiments* are similar to the kinds of preparations that would have been available in convent-run apothecaries, which in
many respects functioned as an organized extension of the kinds of scientific
and medical knowledge women like Sforza had long collected. The convent
of Le Murate, moreover, was famed for its scriptorium. While the nuns concentrated primarily on liturgical works, perhaps these dual interests in manuscript production and medicine had occasion to converge in the collection of
prescriptions and recipes.[73]

Although it is not clear whether the purview of the apothecary nuns at Le
Murate extended to cosmetics, Caterina's own certainly did. Caterina had a
keen interest in cosmetic secrets and her *Experiments* includes a number of
beauty waters, skin lotions, hair dyes, and face and lip colors.[74] Like her medicinal remedies, Caterina's cosmetic recipes make use of herbs and plants distilled in alchemical vessels, in addition to minerals, precious stones, and metals.
In keeping with Renaissance ideals of feminine beauty (pale skin, blonde or
reddish hair, small breasts), many recipes focus on smoothing and lightening
the skin or removing freckles, bleaching dark hair, and reshaping the female
form. One recipe to protect the complexion against unsightly spots calls for
the distillation of fennel, betony, endive, roses, and white wine. Over a period of days, the recipe will—in accordance with the alchemical theory that
prolonged distillation produces increasingly powerful results—generate, in
turn, "rose water," "silver water," and, finally, "golden water" or "balsam," a
precious panacea.[75] Caterina's instructions (again keyed to the alchemical theory
of color stages) recall similar recipes included in other collections, including
one dating to the early sixteenth century that likewise offers a recipe for a water
distilled in three stages (here, the first water, a cosmetic remedy, removes
freckles; the second cures fistulas; the third, the strongest, cuts through iron).[76]
Some of Caterina's cosmetic recipes are simple and straightforward, relying
on egg whites or fava flowers distilled with water to whiten the skin.[77] More
complex entries, however, revolve around alchemical ingredients and procedures to produce multifaceted remedies. One recipe for removing blemishes
from the face is based on pulverized silver litharge dissolved in vinegar, a process that involves heating and distilling.[78] A "most perfect beauty water" *(acqua
perfettissima per far bella)* requires mixing "argento vivo," or mercury—a central alchemical ingredient—in a rounded flask with sage and then pulverizing
it in a stone mortar with a walnut pestle, followed by various stages of boiling.[79]

A number of other recipes describe beauty waters likewise based on mercury, silver sublimate, camphor, borax, alum, and salt—all part of the alchemical laboratory.[80]

Caterina's numerous recipes for the hair seek to lighten it in order to achieve what Agnolo Firenzuola, author of a sixteenth-century dialogue on female beauty, called "the proper and true color of hair."[81] Many of these recipes also require the distillation of ingredients such as cinnabar (a common sulfide ore of mercury), white alum, and sulfur (and indeed are said to produce an alchemical sort of result, that is, "hair as blond as gold"); but some are composed of more easily acquired herbal ingredients such as rhubarb and nettle. One could choose, therefore, to conduct a more complicated experiment, requiring mineral substances, such as this one: "Take cinnabar, saffron, and sulfur and distill these things in an alembic, and when you have washed your hair, comb it in the sun and wet the comb frequently with this distilled water and when you have dried it in the sun your hair will be as beautiful as gold."[82] Or, one might prefer the more straightforward advice to "take nettle seeds and boil them in lye made from leftover ashes and wash [your hair] and it will come out beautifully."[83]

Neglecting no aspect of female beauty, Caterina's *Experiments* also includes advice for achieving the small, firm breasts idealized by Firenzuola and evident in many Renaissance portraits. In *On the Beauty of Women,* Firenzuola's interlocutors agree that the bosom should swell "in such a way as to be hardly noticeably by the eye," and "must be well set . . . and small."[84] Likewise, Caterina's recipes include instructions to "make the breasts small and hard so that they do not grow any larger." Interestingly, this is one of very few points in the volume where an authorial voice is evident and addresses a readership directly, one composed solely of women. One such case specifies: "Take hemlock juice and use every day, and even if [the breasts] are large they will become small; and know this, young women: if you are still girls, and have not reached maturity, if you use this on your breasts every day they will not grow any larger and they will remain beautiful and firm."[85] Unlike Caterina's other cosmetic recipes, these prescriptions for the breasts do not require any kind of alchemical preparation or instruments. On the contrary, they can be produced and applied easily, even by young girls.

Some of Caterina's most elaborate beauty recipes, however, require real expertise, and demonstrate how tightly intertwined are cosmetics, medicine, and alchemy for Caterina. The talc water described earlier, when calcinated and then slowly heated, can make a woman of sixty appear to be only twenty. The substance not only confers youth and beauty, but in powder form and mixed with white wine it serves as an antidote to plague or poison, while the water itself can cause inferior pearls to increase their luster and even their size, and turn silver to gold.[86] If this recipe focuses on the myriad uses of talc water, another explains how to produce the liquid itself: "To extract water from talc: Take calcinated talc, prepared as described earlier, and place in an alembic; in this way you will make a little open stove and place a copper vessel inside . . . and seal it well so it cannot evaporate, then heat it slowly over charcoal."[87] That this recipe is one of those partially encoded again suggests a deeper level of secrecy associated with the actual techniques of alchemy.

Caterina herself was widely cited by her contemporaries as an exemplar of female beauty. Although few certain likenesses of Caterina survive, her fifteenth-century biographer, Jacopo Filippo Foresti da Bergamo described her as "one of the most beautiful women of our century, of elegant appearance and blessed with a marvelous figure."[88] A portrait by Lorenzo di Credi (1459–1537), thought to depict Caterina, presents a slender, composed woman with reddish-blonde hair and pale skin. A fortress in the background evokes the Rocca di Ravaldino, while the little pot of jasmine resting in the sitter's hand signals Caterina's skilled use of botanicals in her experiments.[89]

As Joyce de Vries notes, physical adornment and cosmetic enhancement functioned as part of Caterina's broader strategy of self-fashioning (an effort that also included actions from charitable works to the casting of portrait medals in her image). Archival documents, including descriptions of the engagement and marriage presents she received from Girolamo Riario, record Caterina's acquisition of brocade and velvet gowns, gold-trimmed purses, and rings set with precious gems.[90] Just as the surviving inventories of Caterina's luxurious wardrobe and jewelry suggest her concern to dress the part of power, so do her cosmetic recipes attest to a corresponding interest in achieving and maintaining the ideal of feminine beauty.[91] Caterina was an active participant in her own self-fashioning. Her pursuit of secrets—cosmetic, medicinal, alchemical—added an important facet to the authority of her public image as collector and practitioner.

In addition to medical and cosmetic recipes that clearly draw on elements of alchemical practice, the *Experiments* also includes thirty recipes that are specifically alchemical in nature: instructions for coloring metals, dissolving precious stones, calcifying mercury or converting it to water, and preparing

Partially encoded recipe for talc water, *Experimenti de la ex.ma s.r. Caterina da Furli . . .* c. 6

Lorenzo di Credi, *Ritratto di giovane donna o Dama dei gelsomini*

Portrait of Young Woman or the Lady of the Jasmine Flowers, c. 1481–1483 (oil on panel), Credi, Lorenzo di (1459–1537) / De Agostini Picture Library / Bridgeman Images

tinctures of gold. Alchemy had a visible presence in courts across early modern Europe: the Hapsburg emperor Rudolf II was intrigued by it and invited alchemists to the imperial court in Prague, while Italian alchemists were welcomed at the court of Philip II of Spain.[92] Caterina shared alchemical secrets with Maximilian I, the husband of her sister Bianca Maria Sforza and Holy Roman Emperor from 1508; and her descendants—notably the grand dukes of Tuscany Cosimo I and Franceso I de' Medici—would construct *studioli* and laboratories in which to conduct experiments.[93] The practical aspects of alchemy were particularly attractive to princes: in the fifteenth century the Medici held a monopoly on the sale of papal alum (used in glassmaking, tanning, and the textile industry); while the production of saltpeter (used in gunpowder) was of great interest to early modern rulers.[94] Saltpeter (salt niter) appears with regularity as an ingredient in Caterina's *Experiments;* and indeed the kinds of alchemical recipes recorded in her compilation reflect such a practical mentality. These prescriptions waste no time on discussions of the philosopher's stone, or on any allegorical significance that might be associated with alchemy, focusing instead on the production of materials that have the potential for concrete use and tangible political or financial return. In this sense, alchemical recipes could be considered a kind of "state secret": if successful, they had the potential to upend the existing socioeconomic order by disrupting, diluting, or corrupting markets.[95]

In some cases, alchemical recipes are recorded partially or wholly in Latin, underscoring the value of the instruction, as in a water that promises to "dissolve iron and any other metal and congeal mercury" as well as serve a medicinal use as balsam.[96] Several others for turning tin to silver or silver to gold are also given in Latin; one particularly valuable recipe adds a reddish tint to any metal, giving it the appearance of twenty-four-carat gold.[97] Such pragmatic instructions for making metals and lesser coins appear to be more valuable than they are numerous in the *Experiments*. One recipe explains how to prepare a "[w]ater that produces new silver coins that look old";[98] another, for "adding greater weight to a scudo or golden ducat," attributed to Cosimo the Elder, calls for saltpeter distilled in water with "filings of Saturn," or lead *(limatura de Saturno)*, then calcinated and distilled once again in an alembic. To add weight to the coin, it must merely be dipped in the mixture and weighed; and dipped again until it reaches the desired weight.[99] With a note of self-consciousness, this recipe is specified as one that allows the user to commit the fraud without pangs of conscience *(senza carigo de conscientia)*. The dis-

tinction is important, suggesting an awareness not only of any moral implications of such an action, but also the penalties that were associated with such
fraud, which at the most extreme could include death.[100] Other chemical recipes for lending color to ducats and coloring copper and pewter do not contain such caveats.

These alchemical recipes underscore that Caterina saw potential in the application of alchemy to the realities of her own life and political situation, in
which the need for money was a perpetual burden. Like other aristocratic men
and women, Caterina regularly pawned her belongings or used her jewels as
collateral for loans. Even before Girolamo Riario's death, the financial reality
was difficult; after his assassination in 1488, Caterina faced the challenge of
maintaining the appearance of courtly magnificence while keeping her actual
expenditures down.[101] Especially after 1500, when she had been divested of her
claims on Imola and Forlì and found herself a widow in Florence, Caterina
felt the pressure of financial necessity—not only for herself, but on behalf of
her children, who wrote her constantly to ask for money and whom she felt
did not do enough to look after their mother's financial welfare. In a letter to
her son Ottaviano, Caterina lamented, "I have twenty-four mouths to feed,
five horses and three mules; I have to pay for them all and I am penniless; and
I have found no relief here at all, nor anyone who wants to offer me so much
as a glass of water."[102] Her circumstances improved somewhat after she won
her legal battle over her son Giovanni's custody (and patrimony) in 1505. Nonetheless, when she died in 1509, Caterina retained only a fraction of the precious jewels and valuable objects she had once possessed.[103] As we will see,
Caterina's interest in alchemy as a means to manufacture additional, tangible
resources is attested to not just by her *Experiments,* but by letters she exchanged
with others seeking ingredients and information necessary to undertake these
activities.

Alchemy and Epistolary Networks

Caterina's personal correspondence corroborates that she sought out and circulated recipes similar to those included in her *Experiments.* Caterina cast a
wide epistolary web: hundreds of documents survive detailing her exchanges
with highly placed interlocutors such as Lorenzo de' Medici in Florence, Federico Gonzaga in Ferrara, and her uncle Lodovico Sforza (called "il Moro")
in Milan, as well as with her stepmother, sisters, children, and her agents.[104]

A number of letters to and from Caterina show that the pursuit of secrets established her within a network of others, women and men, of varying social status, who called upon her for the same, and that this activity was an important part of Caterina's public identity. The exchange of recipes was common in Renaissance courts and is attested to in the letters of other early modern *signore:* Caterina's contemporary, Isabella d'Este, for example, procured face powder from Venice and had a preferred recipe for tooth powder.[105] Recipes performed different functions, depending on the circumstance. By supplying a valuable secret to someone in a higher social position, one might earn status or financial reward. Those already established among the highest levels of society circulated recipes among themselves to reciprocate favors or gifts, establish goodwill and maintain networks of communication, and cement their own position and reputation through access to valuable and even clandestine knowledge. Isabella d'Este was famous throughout Europe for the "compositioni" she created: perfumes and scented oils sent as gifts to recipients including Anne de Bretagne, the queen of France, and her ladies. As important as the perfumes themselves were the elegant boxes *(bossoli)* that contained them, ornamented with jewels or ornately carved.[106] The significance of such gifts in bolstering Isabella's reputation is clear in the letters that accompany them. In one instance, she refers to herself as "unsurpassed by any perfumer in the world."[107] In another, a devoted friend recounts how he defended Isabella before a group of professional perfumers: "I argued with every perfumer in this city and with every lady, Spanish or Italian, [to convince them] that *Vostra Excellenza* makes and uses the best *compositione* in the world."[108] Well known for her delight in new acquisitions, Isabella sought out recipes in addition to creating them, whether for the cosmetic powders described earlier or for the perfumed gloves she avidly collected.[109] Isabella had no qualms about associating herself with the manual labor involved in creating her compositions: indeed, she begs the recipient of one of her gifts not to "change [apothecary] shop" *(non cambiare la bottega).*[110] For Isabella, as for Caterina, it was desirable to cultivate a public persona as both purveyor of secrets and practitioner.

Like Isabella, Caterina was known both for her interest in acquiring new cosmetic, medicinal, or alchemical recipes, and for her skill in preparing them. Letters exchanged with her contacts throughout Italy, from Rome to Forlì to, indeed, the Gonzaga court in Mantua, reflect this side of Caterina's engagement with empirical culture. A letter written to Caterina in Florence in 1502 by Luigi Ciochi makes reference to a secret recipe, compounded from egg and saffron, gleaned from the court of Isabella d'Este. It is too precious to describe on paper, but Ciochi promises to reveal it to Caterina in person: "I don't want to send [by letter] the best recipe, [composed of] eggs and saffron, because it is too marvelous and of real value."[111] Ciochi (sometimes written Cioca, Ciocca, or Ciocha) was involved in political negotiations on Caterina's behalf and wrote her frequently from Mantua to keep her and her sons informed of doings there and abroad.[112] As this and other letters from Ciochi to Caterina demonstrate, he was accustomed to seeking out recipes for his patron—often from the same source, for example, the "Madona [sic] Costanza" named here (likely a lady at Isabella's court), from whom he has also obtained an unguent for the face and hands ("el vero oncto . . . per el volto et per le mane")—and that he had an established interest in performing experiments with Caterina. Insisting that he wishes to be present when Caterina tests the recipe, Ciochi describes himself as an indispensable accomplice, possessed of all the requisite qualities of the alchemical adept:

> I assure your Excellency that I will bring you the recipe myself because I wish to be present for so great an experiment and so great a satisfaction . . . and besides, your Excellency would never find another man like me, because courage is required (that is, he can't fear spirits); faith (that is, he must believe); secrecy (that is, he can't disclose anything to anyone); and [he must] have the instruments necessary for such a great experiment and there are none in the laboratories of Bologna, nor Ferrara, nor Paris, nor Rome, similar to mine.[113]

The qualities Ciochi praises in himself—courage, faith, discretion—are among those traditionally prescribed for the alchemist. A printed book of secrets attributed to Isabella Cortese, published in 1561, stresses these very attributes when describing the ideal alchemical assistant. He should be, according to Cortese, "a servant who is faithful, discreet, and courageous in spirit."[114] More important, Ciochi has access to a well-stocked alchemical laboratory.

A second letter from Ciochi to Caterina similarly underscores his role as a procurer of recipes—this time a potent medical remedy—and a participant in the experiments performed by the Lady of Forlì. Like the other, this letter reveals the links forged by Caterina with other noblewomen through her pursuit of marvelous secrets, as Ciochi passes on the same Madonna Costanza's request for certain perfumes and "powder of cypress" (the same "polvere di Cipri" that Isabella d'Este was known to purchase from Venice on occasion).

Ciochi urges Caterina to comply, so that Costanza may later have occasion to reciprocate (as, in fact, she does): "Madonna Costanza prays Your Excellency send her some perfume and powder of Cyprus, and I beg you to satisfy her because she has some nice things *(de le gentileze)* with which to reciprocate."[115] Once again underscoring the empirical underpinnings of these exchanges of secrets, Ciochi signs himself Caterina's "servant and faithful supporter in undertaking all the experiments in the world."[116] A letter from Ciochi to Caterina's son Ottaviano, finally, filled with political news, also makes reference to a medical remedy procured at the Mantuan court—pills for an unspecified ailment, the efficacy of which Ciochi is curious to know: "twice the marchese asked me whether the pills had helped you . . . please advise me whether they helped or not."[117]

Caterina's other correspondents likewise provide her with a variety of cosmetic and medicinal remedies. A certain "Anna Hebrea [Anna the Jewess] of Rome" sends, at Caterina's request, a black unguent for the complexion, capable of removing age spots and softening the skin—just the kind of cosmetic recipe that appears with great frequency in Caterina's own *Experiments*. Anna gives detailed instructions on how the cream is to be applied (at night, and rinsed off in the morning), along with a price list for her various products, suggesting both that Anna was practiced in selling cosmetics to noblewomen, and that she sought to establish a commercial relationship with this client in particular:

> The [black] unguent should be applied in the evening and worn until morning, when you will rinse it off with pure water from the river. Next, wash your face with the water marked "Smoothing water" [*Acqua da Canicare*], then use a little of this white unguent. Then take some of the sublimate (not even the size of a chickpea) and dilute it with the water marked "Gentle water" [*Acqua dolce*], and use it on your face—and it is better to use just a very little bit of everything. The black unguent costs four carlini per ounce, and the *acqua da canicare* costs four carlini per leaf. The waxy substance, that is to say the white unguent, costs eight carlini per ounce; the sublimate costs one gold ducat per ounce, and the gentle water one gold ducat per leaf. If you use [these products], I will make certain to continue to send them to you.[118]

Jews were known in Italy for their production of cosmetics, and sometimes maligned for it (as in Ariosto's *Satira V,* which launches an anti-Semitic condemnation of cosmetics for women).[119] As Jütte notes, the early modern interest in clandestine knowledge furnished opportunities for people from many parts of society—including the lower classes, women, and Jews—to participate and communicate with one another in the "marketplace of secrets." For Jews, this resulted in increased and safe "zones of contact" and patronage opportunities at court.[120] Anna's letter to Caterina demonstrates the value of certain kinds of secrets and the willingness of noble clients to pay for them.

Caterina's contacts also supplied her with medicines of various sorts. Frate Bernardino di Gariboldi sends Caterina a medical remedy: three small flasks of "acqua celeste," a distilled liquid capable of curing a range of maladies. As Bernardino explains, the first flask is for treating headache and stomachache; the second for the liver; and the third to protect against plague:

> I am sending you three little flasks of *acqua celeste* by way of my companion Don Piero, with whom I am staying in Firenzuola, because he is a trustworthy messenger for these flasks. One is for headache and stomachache. The other is for the liver, but you must mix it with an ounce of *diarodon abbatis* [a kind of rosewater]. The other is to protect against plague: if you take a little in the morning, you will be protected from plague for the day. This is all explained in a note attached to the neck of each one.[121]

Bernardino further promises to provide an oil, which he has not yet finished producing.[122] Remedies for all these complaints are included in Caterina's *Experiments,* often in numerous variations; although which of these, if any, corresponds directly to the remedy furnished by Bernardino is impossible to tell.

Caterina did not just collect recipes: she also provided prescriptions and remedies to others. A letter from her daughter Bianca asks Caterina to prepare "two flasks of pine water," commonly used to treat respiratory ailments, in exchange for a gift of fruit.[123] Other more oblique references to the exchange of recipes between Caterina and her correspondents further testify to her standing as a provider of secrets. Even letters of strictly political content contain brief mentions of remedies and prescriptions. One such missive, which focuses primarily on the continued political maneuverings under way regarding Caterina's efforts to regain possession of Imola and Forlì, closes with Caterina assuring her unnamed interlocutor that she will give him the recipe he requests upon their next meeting.[124] Recipes were indeed a form of currency, used here to reciprocate important political intelligence and counsel. They thus become an integral and deliberate element of Caterina's carefully crafted, authoritative persona as a kind of prince-practitioner. Early modern women exchanged recipes for a spectrum of purposes, and, as we see in the case of both Caterina Sforza and her contemporary Isabella d'Este, their interest in experiments was not just abstract. Caterina, like Isabella, had a practical understanding of the components of her recipes, and took an active hand in both their production and deployment.

Two final examples demonstrate still more clearly Caterina's activity not just in the collection of secrets, but with respect to alchemy specifically—not only for the production of medicines, cosmetics, or counterfeit coinage, but also for making alchemical gold. We have seen that Caterina's interest in practical alchemy is very much in evidence in her *Experiments;* her correspondence confirms that this was a real pursuit, not just a textual exercise. A letter from an unidentified correspondent who signs himself only "That faithful servant" *(Quel fidel servo)* alludes to an exchange of secrets the two parties have agreed upon, and links together the cosmetic and the alchemical as equally valuable areas of knowledge. For his part, Caterina's correspondent supplies a special cream for the face *(quel lustro del volto)* that he has promised her, obtained through a lady of the Gonzaga family (possibly Isabella d'Este herself). In return, he awaits Caterina's recipe for making nineteen-carat gold: "I beg Your Excellency to attend to that which you promised me; that is to send me the recipe for making nineteen-carat gold, and to send me the complete recipe and as soon as possible."[125] His anxious tone and insistence that Caterina should immediately provide the recipe in its entirety, with no part left out, speak both to the difficulty and the excitement of the experiment he seeks to undertake, and suggest that Caterina was known for performing such experiments herself. Indeed, as noted earlier, her *Experiments* provides several recipes for tinctures of gold.

A second, more obscure letter is addressed to Caterina by Lorenzo de Mantechitis, a correspondent who was imprisoned after playing a trick on a self-professed alchemical "master" *(maestro)* who thinks too highly of his own skills. Lorenzo had assumed (wrongly) that Caterina and her second husband, Giacomo Feo, would appreciate the joke: "[I was] hoping that Your Excellency and messer Jacomo would join me in poking fun at him, since he fancies himself a grand philosopher of this art and very intelligent."[126] The trick itself seems to have centered on proving wrong the maestro, who doubted the veracity of a recipe to make alchemical silver, by performing a kind of alchemical fraud in his presence:

I was careful . . . that no one should know what we were doing, just as Your Ladyship advised me; even at the end I said it was nothing, and that I did not believe it was possible but that I was trying to show the master, who was certain that it is impossible to make silver or to make it by alchemical means, so that in showing him this calcinated silver and reducing it, in his presence, with black soap [and] salt niter or rather borax, which in that form seems like ash, that he would be amazed at seeing that powder reduced to silver.[127]

Lorenzo pleads with Caterina to be freed and returned to her service, insisting he carried out the prank only to please her. Lorenzo suggests that if indeed he were a successful alchemist, he would not find himself in these present straits.[128] At the same time, however, he also seems to refer to some prior success in this area, mentioning a little book that now constitutes the sum total of his activity in this area and promising to use his knowledge of alchemy only in Caterina's service. "If Your Ladyship or indeed messer Jacomo [Feo] should wish for some proof of those things I have which I consider useful, I will provide it; and especially with regard to what is contained in that little book . . . because I have no intention of ever trying anything beyond that again. And if you don't want [proof], I will let it be and will do only that which Your Ladyship and messer Jacomo command."[129] While difficult to interpret without further information about Lorenzo and his relationship to Caterina and Giacomo, the letter suggests that Caterina had a strong interest in experiments but also that she prized secrecy and discretion in their undertaking; and that she disdained the fraudulent use of alchemy—at least when it was not to her benefit. It also reiterates that alchemical experiments brought together men and women of varied position, allowing for a certain fluidity in social relations. Lorenzo clearly identifies himself as Caterina's "servant," of much lower social status than she, but refers to his interactions with Maximilian I, king of the Romans and Caterina's relative by marriage, and notes that Caterina's auditor was surprised to learn of this connection: "I also told him that I was employed with King Maximilian: incredible that someone like me would have reason to work with so exalted a *signore*."[130] Lorenzo's anxious promises never to attempt another alchemical experiment without the permission of Caterina and Giacomo, however, establish a clear hierarchy of authority between prince and subject, master and assistant.

As noted previously, such experiments—especially the transmutation of base metal into gold or silver—would have held particular appeal for one in Caterina's position, as well as considerable risk. Indeed, the dangers of alchemy were not only legal or moral but also physical: a century later, the Jewish writer Leone Modena would describe his son's death from the noxious fumes produced by his efforts in the alchemical laboratory.[131] With these kinds of experiments, Caterina likely sought—despite the risks—tools that could strengthen her financial position, and therefore her prospects for retaining (or regaining) title to her states in the Romagna. Thus she seeks recipes to create precious metals, or at least to simulate them, without succumbing (as specified in the *Experiments*) to any pangs of conscience.

From her military and diplomatic strategies to her strategies of self-fashioning as a female prince, Caterina's history has been examined from many angles. Her alchemical activity, however, has remained largely in the background, a supporting actor in the larger story of her political significance. Both her *Experiments* and her epistolary exchanges over the last decade of her life demonstrate that the pursuit of beauty, health, and wealth through alchemical means was an ongoing and central focus for Caterina even in the midst of political and personal turmoil, one that occurred within the private spaces of the family and in the public venue of the court and that engendered a complex web of exchange and reciprocation. Indeed, even as Caterina was preparing for siege by Cesare Borgia—a pivotal moment in her political life—she was thinking of her next alchemical experiment. A letter addressed to her confessor, Francesco Fortunato (Fortunati), asks him to send her a series of glass vessels made to particular specifications, along with other ingredients known for their medicinal—and potentially poisonous—properties, as soon as possible: "Send us three round glass balls with small openings that can hold two *bucali* of liquid and [send us] twelve sea onion bulbs, called squill."[132] The letter is dated November 2, 1499—only a few weeks before Borgia's conquest of Imola. The glass balls were most likely alchemical vessels (which were generally transparent so the alchemist could observe changes to their contents during experiments); while sea onion bulbs (officinal squill) possess diverse diuretic and emetic properties. In small doses, they were used as a remedy for asthma and dropsy; in large doses, as a poison capable of provoking violent vomiting, seizures, and even death.[133] Pharmaceutical and alchemical methods were sometimes employed in political contexts such as this one: in 1601, Vincenzo Gonzaga, duke of Mantua, launched poison gas into enemy territory in a desperate (and unsuccessful) attempt to alter the course of a battle.[134] Caterina's request to her confessor certainly raises the prospect that she was entertaining some creative solutions to her dire political situation, and putting her knowledge of secrets to use. By restoring Caterina's activity as an alchemical practitioner and a collector of alchemical, medicinal, and cosmetic recipes to the broader picture of her historical importance, we can begin to see her in a more nuanced way: one that integrates her political significance with her involvement in scientific culture.

The recipes collected by Caterina, moreover, offer insight into the role of scientific inquiry in the lives of early modern women and provide important information regarding specific ingredients and processes used in treating illness, creating cosmetics, and performing alchemical experiments. They also shed light on the relationships and networks established while doing so, which in Caterina's case involved both men and women from her immediate circle and farther afield; and from the highest echelons of society as well as from marginalized communities such as the Jews. It is clear that we must expand and reformulate our understanding of early modern laboratory practices, and particularly the notion that this was an exclusively male domain. Not only did female practitioners collaborate and share secrets among themselves; there was also a noteworthy degree of crossover between male and female practitioners. Caterina's alchemical activity involved the participation of at least one of her husbands, Giacomo Feo, and she collaborated with both her apothecary and her confessor in collecting recipes.[135]

Caterina's example reverberated in her Medici descendants. We have seen that she passed her collection of *Experiments* on to her son, Giovanni de' Medici, who prized it. Given the value Giovanni placed on the manuscript, coupled with the fact that its transcriber, the condottiere Lucantonio Cuppano, passed into the service of Cosimo I after Giovanni's death (remaining loyal to the Medici family until his own death in 1557), it is likely that Caterina's manuscript was handed down, in turn, to Giovanni's son—the first Medici Grand Duke—and even to Francesco I and Don Antonio de' Medici after him.[136] Indirect evidence for the influence of Caterina's experiments may also be seen in the circulation of secrets attributed to her descendants in sixteenth-century Florence. A manuscript compilation by the apothecary Stefano Rosselli carefully notes the origins of its most valuable alchemical-medical secrets, singling out some as personal gifts from the Grand Dukes Cosimo and Francesco and attributing others to Cosimo's daughter, Isabella de' Medici.[137] Where Caterina collected recipes for her own use and experimented with them on a small scale in her workshop in Forli, Cosimo and Francesco poured their considerable resources into greatly expanded endeavors. In addition to the private spaces they each constructed in the Palazzo Vecchio, they also created larger laboratory spaces with commercial potential: Cosimo, the foundry in Palazzo Vecchio; Francesco, the Casino adjacent to the Church of San Marco, a hub of technological activity that worked to produce glass, porcelain, and pyrotechnics, among other things. In a letter to the Venetian senate, one ambassador described the Casino as "a little arsenal composed of various rooms, in which different masters work on different things, and this is where [Francesco] keeps his alembics and all his equipment"; the letter goes on to recount that Francesco spends virtually all his time here.[138] The extensive inventories made of the Casino di San Marco after the death of Don Antonio de' Medici, Francesco's son with Bianca Cappello, show that he possessed a number of books of secrets and alchemical works as well as materials for experiments: a "book of chemical recipes," a "sheaf of papers containing many secrets," and "a little book bound in lambskin marked 'elixir of silver and gold' "; along with "four little vials marked 'alchemical powder' . . . a phial marked 'powder of quintessence of coral,'" and quantities of antimony, silver, and other alchemical materials.[139] If Caterina's *Experiments* were indeed preserved, amended, and utilized by her Medici progeny, she can be situated as an influential founding figure—a kind of alchemical mother—at the origins of a well-documented Medici interest in scientific investigation that continued into the seventeenth century.

The bonds of scientific community—not to mention the corresponding financial obligation—could run deep. When Caterina succumbed to quartan fever in 1509, not only did she leave behind a substantial debt to Lodovico Albertini for the materials she had used in her experiments, but she also left her apothecary with a great sense of sadness for the partner he had lost. In a letter dated June 3, 1509, Albertini shared his grief with one of Caterina's Riario relatives, writing, "never did I feel greater sorrow, along with Bastiano and my whole family, and never again will I be content for I have lost my sweet mistress and I grieve all the more because I was not there at the end."[140]

Caterina's example, finally, is also important in contextualizing the wave of printed books of secrets that flourished throughout the sixteenth century. As this analysis of her *Experiments* has shown, the therapeutic and cosmetic solutions offered in manuscript compilations such as hers did in many cases find their models in quotidian practice, and even some of the more strictly alchemical recipes had roots in women's everyday lives, in the household economy, and—in Caterina's case—in political concerns. As we will see in Chapter 2, printed books of secrets, often dedicated to women or filled with prescriptions pertaining to female health and beauty, made a concerted effort to reach—and to reflect—a specifically female audience. The blurring of boundaries between diverse areas of knowledge—alchemical, medical, and cosmetic—is also in evidence in these printed texts. Indeed, the similarities between the kinds of recipes recorded by Caterina and those that appear in print in the later sixteenth century further underscore the continuing links between books of secrets and women practitioners of alchemy and medicine.

2

⌘

The *Secrets* of Isabella Cortese

Practical Alchemy and Women Readers

As the sixteenth century progressed, scientific culture continued to flourish in Italian courts and academies as well as on the printed page. Against the backdrop of the "new" empiricism and the corresponding enthusiasm for investigating (and manipulating) nature, the acquisition of secrets—for creating medical remedies, producing cosmetics, and effecting alchemical transformation—became an ever more inviting (and competitive) pursuit.[1] Secrets were a form of currency, exchanged to pay debts or establish and increase social, intellectual, and political standing. In this early modern context, the term "secret" was virtually synonymous with "experiment," referring not to something unknown but rather to something that was proven.[2] The most prized secrets were not necessarily the most unusual, but rather those said to have been tested (at least ostensibly) and found reliably to produce the desired result. The demand for secrets, whether among court circles or within the wider world of print, crossed boundaries of gender and class, resonating with women as well as men.

In Italy, the fashion for secrets, or "experiments," which already circulated in manuscript form and collections compiled for private use, found its broadest expression in the printed works known as *libri di segreti,* or books of secrets.

These works—eclectic collections of medicinal, cosmetic, and alchemical recipes—functioned as repositories of knowledge that mirrored, on a textual level, the increased accessibility of scientific culture to a broader public and captured the imagination of early modern readers. Editors and publishers actively promoted books of secrets, impelled not least by the early modern debate over the *questione della lingua,* or "language question," and the resulting efforts to bring vernacular works of all genres to press. Many sixteenth-century books of secrets were thus directed less at an elite, highly educated, or specialized public (as the Latin medical manuals on which they were often partially based had been) than at a bourgeois audience that was becoming increasingly literate in Italian. While some works were aimed at a courtly audience (for example, Giambattista della Porta's influential *Magia naturalis* [*Natural Magic,* 1558], composed in Latin but soon translated into Italian), others, such as Giovanni Battista Zapata's *Maravigliosi secreti di medicina e chirurgia* (Marvelous secrets of medicine and surgery, 1586), clearly differentiated their contents from those addressed to wealthy, upper-class readers, stressing instead the accessibility and reasonable cost of the ingredients used in their recipes.[3] Works such as Evangelista Quattrami's *La vera dichiaratione di tutte le metafore, similitudini, & enimmi de gl'antichi filosofi alchimisti* (The true description of all the metaphors, similes, & enigmas of the ancient philosopher-alchemists, 1587), which focused specifically on alchemy and its techniques, aimed to demystify the secret art for the vernacular reader and to combat the threat posed by charlatans or "ignorant alchemists" *(alchimisti ignoranti).*[4]

Books of secrets also addressed an audience of women, who were becoming an increasing part of the market for printed works thanks to the growing diffusion of the *volgare* as the literary language of early modern Italy. While literacy rates for women remained, predictably, far lower than those for men, the dual factors of vernacularization and the rapid spread of print had an enormous impact on women's reading and writing practices throughout Italy.[5] This demographic shift did not escape the attention of authors and publishers. As an examination of sixteenth-century books of secrets reveals, many of these works targeted female readers by paying close attention to subjects including feminine beauty and its enhancement, domestic management, and women's health—particularly with regard to fertility, pregnancy, and childbirth. Although some of this medical material was aimed as much at male practitioners as at women themselves, the centrality of women's concerns in books of secrets is unmistakable.[6] In this respect, books of secrets seconded a much wider array of vernacular literary works that had taken up various facets of women's experience, spurred to describe (and prescribe) women's roles and women's nature by the ongoing *querelle des femmes* ignited two centuries prior by Boccaccio's *De claris mulieribus (Famous Women)*, and Christine de Pizan's *Livre de la cité des dames (Book of the City of Ladies)* in response. As the debate over women continued to rage in early modern Europe, it brought under scrutiny all aspects of women's experience, from their intellectual capabilities to their roles as wives and mothers, and their corresponding duty to look after the well-being of themselves and their families. Humanist treatises embedded meditations on pregnancy, motherhood, and whether or not women should nurse their children into a broader discussion of women's social and civic functions; while works such as Agnolo Firenzuola's *Dialogo delle bellezze delle donne (On the Beauty of Women, 1548)* examined female beauty from a Neoplatonic perspective.[7] The fifteenth-century letters of Laura Cereta integrated a litany of medicinal and cosmetic remedies based on Pliny's *Natural History,* including some aimed specifically at women and children, into a narrative of her own intellectual and literary development.[8] Books of secrets, for their part, offered practical (and sometimes impractical) advice to women on how to take control of health, hygiene, and household through medicine, alchemy, and cosmetics. Such texts, which share a common interest in teaching women how to negotiate their worlds, reflect the reality of an increasing female readership by the middle of the sixteenth century as well as the keenness of writers and editors to benefit from it.

Books of secrets thus sought to capitalize on the allure of remedies and prescriptions relating to health, beauty, alchemy, and household management, while also demonstrating the merits of the *volgare* as a vehicle for disseminating scientific information among a wider audience. These collections circulated with enormous success in early modern Italy (and, indeed, throughout Europe and England); producing, at the apex of the genre's popularity, "bestsellers" such as *De' secreti del reverendo donno Alessio Piemontese (The Secrets of Alexis of Piedmont),* published in 1555 with dozens of editions and translations.[9] So great was the demand for *libri di segreti* that publishers frequently repackaged existing medical or technical manuals for the new market in secrets, as in the case of *I secreti medicinali di Pietro Bairo* (Medicinal secrets of Pietro Bairo, 1561), which had first circulated in Latin; or plundered proven successes for recyclable material.[10] In most cases, books of secrets were produced not only by scientific adepts, but by professional writers and editors accustomed to adapting their skills to the demands of the market; or, in some instances, by "celebrity" doctors seeking to promote themselves and their cures: such was the case of Leonardo Fioravanti, the Bolognese surgeon, alchemist, and author of several books of secrets.[11] In the case of the *Secrets of Alexis,* generally attributed to Girolamo Ruscelli, the author likely fit into both camps: Ruscelli, a *poligrafo* who worked for the publishing house Valgrisi in Venice, claimed to have founded a society for scientific experiment in Naples called the *Accademia segreta,* which he described in a book of secrets later published under his own name (*Segreti nuovi* [New secrets], 1567]).[12] As we will see shortly, Ruscelli may have been behind other successful books of secrets as well, perhaps in an effort to create a textual representation of the scientific collaboration fostered by his *Accademia segreta*—a kind of virtual scientific academy.[13]

Interspersing therapies for medical conditions—from rashes and rotting teeth to impotence and infertility—with recipes for face creams and perfumes, methods for preserving food and wine, and instructions for removing stains or dying fabrics, books of secrets marry utility with entertainment. They aim to entertain and instruct the reader, but make no attempt to explain the scientific processes that underlie the recipes. Fundamentally utilitarian in nature, books of secrets are less interested in the theoretical underpinnings of the "secret" knowledge they purport to reveal than in its practical application to daily life, its tested efficacy, and its clear results. As William Eamon points out, *libri di segreti* were likely read for pleasure as well as for instruction.[14] Linked to one another only by their common anchor in the scientific domain, books of secrets combine the most disparate of subjects. Yet they share a pragmatic mystery characteristic of Renaissance scientific culture: a delight in the possibilities of nature, and in the ways it can be turned to human service.[15] The enthusiasm for scientific and technical marvels showcased in books of secrets—from creating invisible inks to manufacturing precious gems and metals—parallels the curiosity that drove collecting and museum development in the early modern period.[16]

Books of secrets also share a pervasive reliance on procedures and ingredients associated with alchemy, an arena of scientific experimentation that was intricately linked with the practice of medicine in the early modern period and, increasingly, applied to cosmetic secrets for both sexes.[17] In addition to employing alchemy in medicinal and cosmetic secrets, books of secrets also devote specific and extensive attention to true alchemical experiment, as in the *Secrets of Alexis,* which includes a chapter on alchemy and metallurgy that contains detailed technical descriptions of working with alchemical materials (part 1, book 6). The presence of women as practitioners of alchemy to a variety of medicinal, cosmetic, and transmutational ends is readily acknowledged in works such as the *Secrets of Alexis,* which explains that a recipe for mercury sublimate (a staple of alchemical experiment) is commonly used by "goldsmiths, alchemists, *women,* and for many medicinal purposes" (italics mine).[18] Likewise, the proem to Ruscelli's *New Secrets* states that women visited the laboratory and grounds of his scientific society in Naples and elaborates that the society wished its experiments to benefit "every sort of person, rich and poor, learned and unlearned, *male and female*" (italics mine).[19] Many books of secrets explicitly address women in their dedications, their titles, or in other paratextual material. One of the earliest such works, Eustachio Celebrino's 1525 *Opera nova intitolata dificio de ricette* (New work entitled house of recipes), for example, devotes sections to health and hygiene as well as to perfumes and food preparation, and also includes a series of parlor tricks, such as making a candle burn under water, or writing on paper without ink.[20] A professional author of works for the vernacular editorial market, Celebrino cannily highlights the female readership of his text as well as the centrality of alchemy in its contents by interspersing recipes addressing menstruation, fertility, and makeup with illustrations of alchemical vessels used for distillation: the "primary procedure" of alchemy and a key element of these recipes.[21] Likewise, Celebrino's 1551 *Opera nuova . . . per far bella ciaschuna donna* (New work . . . for making every woman beautiful), offers advice on menstruation, pregnancy, and various female complaints, and links alchemical practice to the enhancement of feminine beauty by evoking distillation in the preparation of rouges and face creams: "Here are colors white and red, / diverse concoctions of waters distilled."[22] Celebrino's text addresses female readers directly in an opening sonnet: "Ladies, who wish to make yourselves beautiful" *(Donne che desiate farve belle)*.[23] Timoteo Rossello's *Della summa de' secreti universali in ogni materia* (Summa of universal secrets of every kind, 1561), which also intertwines cosmetic and medicinal secrets, addresses itself to both men and women "of great intelligence" *(huomini & donne di alto ingegno)*.[24] The *Secrets of Alexis* offers recipes for rouges and hair dye (which could be used by both sexes) alongside general remedies for burns, bites, and worms; and treatments for lactation problems or disorders of the uterus next to herbal remedies for toothache and stomachache. Similarly, *I secreti della signora Isabella Cortese* (The secrets of Signora Isabella Cortese, 1561), the only book of secrets currently ascribed to a woman author in this period, dispenses advice deemed useful to "every great lady" *(ogni gran signora),* as stated in the work's subtitle.[25] This includes medical and cosmetic remedies, advice for managing the household, and—reflecting the alchemical element that runs through virtually all books of secrets—turning base metal into gold. Evidently of great interest to readers, Cortese's volume went through seven editions by 1599.[26] Although, as we will see further on, the authorship of the *Secrets of Signora Isabella Cortese* is contested, the work's deliberate invocation of a female audience is not. Of course, issues related to women's health, in particular, were of important interest to male medical practitioners (who had by this point largely displaced their female counterparts) as well as to women readers, and on the whole books of secrets were composed of a varied mixture of material: some of interest to men, some to women, and much that was useful to both sexes. However, the presence of medical material related specifically to women's bodies, cosmetic recipes to enhance female beauty, and recipes for food preservation and domestic management, together with titles and dedications addressed to women, strongly suggests that books of secrets indeed sought out female readers.

The notion of secrecy, with its patina of the occult, the obscure, and the novel, was carefully highlighted in the titles of these works and integral to their marketing and editorial success. The allure of the secret as esoteric, protected knowledge had an established textual tradition by the sixteenth century, thanks to influential medieval works such as the pseudo-Aristotelian *Secretum secretorum,* which had circulated in Latin and in various translations from the twelfth century.[27] Artisans' manuals containing technical information related to craft activities—metallurgy, goldsmithing, perfumery—constituted another kind of privileged or secret knowledge.[28] The *libri di segreti* of the sixteenth century unite these varied traditions, combining technical knowledge with secrets relating to the care of the body, household governance, cosmetics, and practical alchemy. The preface to the *Secrets of Alexis* reflects this conflation of provenance in its claim to have gleaned its contents not just from "men of learning," but from artisans, farmers, and women.[29]

Central to the preface of the *Secrets of Alexis* is the privileged status of knowledge derived not from study, but from those with the most direct connection to nature itself—including women. Much of the information presented in books of secrets is, in fact, constructed as both highly secret and specifically gendered. As Katharine Park argues, the very notion of secrecy assumed increasingly gendered connotations throughout the medieval and early modern periods, becoming progressively linked to the mysterious landscape of the female body and to the kind of knowledge conferred on its possessor.[30] From the influential thirteenth-century treatise, *De secretis mulierum* (*Women's Secrets,* falsely attributed to Albertus Magnus), to a spate of manuscripts such as the fifteenth-century *Le segrete cose delle donne* (The secret conditions of women), to works such as the *Secrets of Signora Isabella Cortese* in the mid-sixteenth century, instructions and advice for tending to women's bodies, women's health, and women's worlds were situated within the context of a gendered form of knowledge that derived from direct experience, rather than study.[31] As Park points out, these secrets had connotations that were at once epistemological and physical.[32] They referred increasingly to the physiology of the female body (specifically, to reproduction and generation), signifying that which was anatomically hidden and even shameful. At the same time, women were presented as being in possession of powerful knowledge with regard to these functions, a knowledge that they shared only with one another—for example, how to provoke miscarriage or determine the sex of a fetus. The linkage between gender, authority, and empiricism was clearly etched in medieval works such as the texts of the *Trotula*.[33] As male medical practitioners took on a greater role in tending to women's reproductive health, women's special position as "knowers" of the female body would be gradually supplanted by male observations and speculation about that body, and these "secrets" disseminated in literary form.[34] In taking up the notion of "privileged" knowledge as deriving from women's more direct relationship to the physical world and the body, books of secrets thus meld common early modern notions about sex and gender with empirical culture's emphasis on praxis. They situate, in effect, gender at the "crossroads" of scientific knowledge.[35] This is most evident in a work such as the *Secrets of Signora Isabella Cortese,* where alchemical knowledge is attributed to a female author in an effort to underscore its value and authenticity.

Marketing themselves as unique and privileged, even as they sought a wide readership, books of secrets highlighted the empirical provenance of their pages as they vied to persuade readers that their own contents were more valuable than those of their competitors, their secrets derived from laymen, travelers, and women. In seeking a broader readership, however, a necessary tension was established between the secret nature of the information contained in the text, on the one hand, and its accessibility—through its publication in vernacular form—on the other. Indeed, as Karma Lochrie notes, "[s]ecrecy is never a solitary activity," that is, it can only exist when there is someone from whom a secret must be kept (or to whom it is revealed).[36] Decades prior to the *Secrets of Alexis* and the *Secrets of Signora Isabella Cortese,* Laura Cereta, describing the variety of medicinal and cosmetic recipes to be derived from the body, blood, milk, and dung of the ass, wrote that nature itself "lets nothing remain a secret."[37] This tension is neatly encapsulated in both the *Secrets of Alexis* and the *Secrets of Signora Isabella Cortese.* The preface to the *Secrets of Alexis,* for example, explains that after seeing a man die because he would not disclose a life-saving secret, the author is now making amends by revealing everything he knows.[38] In Isabella Cortese's case, a traveler dies in her home in Olmütz (Moravia). He leaves behind a letter to a learned friend in Kraków that contains a precious alchemical secret, which she now includes in her book. Both examples reveal the tensions inherent in promoting knowledge as "secret" while at the same time rendering it accessible to a wide public. Indeed, the *Secrets of Signora Isabella Cortese*—the express purpose of which is to divulge and circulate alchemical secrets—cautions that "to reveal secrets causes them to lose their power."[39]

This tension between secrecy and divulgation is reflected in Tomaso Garzoni's description of the "professors of secrets"—the authors of books of secrets—in his encyclopedia of professional categories, the *Piazza universale di tutte le professioni del mondo* (1585).[40] Garzoni's decidedly ambiguous consideration of this category, which references the authors of the most successful works (including Girolamo Ruscelli, Leonardo Fioravanti, Giambattista della Porta, Timoteo Rossello, and Isabella Cortese), reveals that the textual practice of secrecy had in fact become a "profession" like any other, one that revolved around the dissemination and circulation of secrets to an ever larger public.[41] It also suggests that books of secrets could be interpreted according to a generic code, a hierarchy of "secrecy" that could aid the reader in distinguishing valuable information from more ordinary lore. Referencing Girolamo Cardano's *De secretis* (1562), a work solely devoted to the nature of secrets, Garzoni first seeks to define the secret, explaining that it is "an obscure, veiled, and occult thing, whose explanation is not clear enough to be known to all, but by nature is made manifest only to a very few."[42] He then goes on to outline six ways in which such hidden knowledge is uncovered: through learned insight *(speculatione d'un intelletto perito);* through comparison of similar things *(adattando un simile all'altro);* from other teachers; by traveling the world *(cercando, e investigando varie, et diverse cose);* by having enough financial stability to direct such investigations from one's home; and finally, through Fortune, because sometimes "miraculous things happen by chance" *(qualche volta a sorte succedon cose miracolose)*. Attempting to further define the various types and qualities of secrets, Garzoni proposes an elaborate paradigm that reflects quite accurately the terminology used by the *professori di segreti* themselves. For example, he notes that with respect to their goals, secrets can be "great" *(grandi)*, as in cures for plague; "mediocre" *(mediocri)*, as in cures for quartan fever; or "trivial" *(leggieri)*, as in cures for common skin ailments. With respect to their rate of success, they can be "perfect" *(perfetti)*, succeeding every time; *ut in pluribus*—successful most of the time; or "rare" *(raro)*, usually not successful due to various unavoidable complications. Difficulty and cost also come into the equation: is the secret very expensive to carry out and based on hard-to-find ingredients, or does it rely on economical and ubiquitous materials? Garzoni concludes, with Cardano, that good secrets must meet seven necessary conditions: they must be authentic *(che non sian fallaci)*; they must be useful and profitable *(che arrichino utile e guadagno grande)*; they must not impinge on the conscience *(che non nuocano alla coscienza)*; they must be composed of easily acquired elements *(che siano di cose facilmente vendibili)*; they must not take too long *(che non sian di longhissima aspettatione)*; they must not be overly labor-intensive *(che non v'intervenga fatica intolerabile)*; and finally, they must relate to a goal worthy of a noble soul *(che versono attorno a cose degne d'huomo nobile)*.[43]

As Garzoni's hierarchy and definitions suggest, a "secret" is a recipe: one that is easy to comprehend and follow, has already been tried and tested, and will prove successful and profitable for anyone who undertakes it. Indeed, the association between secrets and recipes is so strong that in some cases a "secret" may be completely devoid of the connotation of "secrecy."[44] Allison Kavey notes that recipes, concise and recognizable to early modern readers, "packaged unfamiliar and exotic materials and ideas in a familiar and accessible form."[45] Recipes communicated even the most arcane information in a manner that was approachable, manageable, and predictable. They underscored the new conviction that one might, through logical and reasoned procedure, manipulate one's world to achieve certain desirable and replicable results.[46]

What connotations, then, did such secrets (or recipes) have for women, to whom publishers were increasingly turning to market their wares? If the recipe was the chief vehicle for conveying information in early modern books of secrets, its extensive use as a narrative and epistemological framework suggests that these texts reached out to women not only in their content but also in their structure. Women were familiar with the recipe's straightforward format, the order in which it presented information, and the tools for which it called.[47] They understood how to read recipes that were incomplete at best, leaving out measurements or detailed explanations of cooking times and preparations and thus relying implicitly on a reader able to fill in the missing pieces, bringing to bear her own experience to complete it successfully. Indeed, the cookery books produced for and by women that would proliferate throughout the sixteenth and seventeenth centuries, particularly in England, bear many resemblances to early modern books of secrets.[48]

The powerful "secret knowledge" of women—associated with knowledge about the inner workings of the body (and, by extension, the generative endeavors of alchemical experiment) and with the structure of the recipe—gained an undeniable currency in the literary marketplace by the mid-sixteenth century. So pronounced was the interest in women's secrets among writers and readers of both sexes that they were packaged in a variety of forms, from true *libri di segreti* to dialogues, treatises, and epistolary collections. These representations of women's knowledge were filtered, however, through the male authorial voice, raising questions of literary authority and appropriation that echo those surrounding women's scientific activity itself in early modern Italy.

A "Lady's" Text: The *Secrets of Signora Isabella Cortese*

The *Secrets of Signora Isabella Cortese,* first published in 1561, furnishes a particularly illuminating example not only of the diversity of recipes characteristic of books of secrets but, especially, of the gendering of medical, alchemical, and cosmetic secrets and the influence of female readers in the production and publication of such works. It also embodies many of the complexities surrounding the authorship of these texts. Although the *Secrets of Signora Isabella Cortese* is often accepted uncritically as the only woman-authored book of secrets from this period, no biographical evidence for the existence of a historical Isabella Cortese exists, and clear textual overlaps with other, coeval works—especially Timoteo Rossello's *Summa*—suggest that the question of authorship may be more complicated.[49] As Massimo Rizzardini notes, the name "Cortese" itself, which is not a common one, is an anagram for *"secreto."*[50] It also calls to mind an aristocratic, authoritative woman author—literally, a "courtly" woman, a literary persona evocative of female purveyors of secrets, such as Caterina Sforza, who populated Italian cities and courts. By inviting such an association, Cortese's text establishes a deeper patina of authenticity. Some scholars have raised the possibility that Cortese might in fact be another pseudonym of Girolamo Ruscelli, to whom the popular *Secrets of Alexis* is also generally ascribed.[51] There is some evidence for such a hypothesis. The strongest support comes from Garzoni's 1585 entry on the "professors of secrets," which states that the name "Isabella Cortese" was likely "invented, along with that of Alexis, by Ruscello."[52] In the nineteenth century, Armand Baschet and Felix-Sebastien Feuillet de Conches argued that stylistic similarities between the *Secrets of Alexis,* the *Secrets of Signora Isabella Cortese,* Rossello's *Summa,* and Ruscelli's *Secreti nuovi* suggested that all four works had sprung from the same pen—that of Ruscelli.[53] If this is the case, the publication of Cortese's book would indeed demonstrate the strong interest among sixteenth-century writers and publishers to produce works by women (genuine or not) that appealed to readers interested in women's experience, inventing such an author where female *secretiste* were in short supply. Ruscelli, as we saw earlier, was a professional writer and editor attuned to the tastes of the reading public; his *Secrets of Alexis* were, by any measure, a resounding success that he may have wished to replicate with further publications (including the *New Secrets* published under his own name in 1567).

The *Secrets of Signora Isabella Cortese* also shares similarities with Rossello's *Summa,* printed in Venice in the same year (1561) by the same publisher, Giovanni Bariletto. Both works are dedicated to the archdeacon of Ragusa, Mario Chaboga (Caboga), suggesting enduring intellectual links to this important port city that had once been under Venice's dominion.[54] Caboga (1505–1582) taught at the University of Padua, a hub of international scientific networking and exchange, and was a member of the Accademia dei Confusi in Viterbo.[55] Through any of these scientific and literary activities, Caboga could have come in contact with Ruscelli (himself a native of Viterbo). Alternatively, the archdeacon may have entered the scene through his contact with another figure from the Venetian editorial world, Curtio Troiano di Navò, the brother-in-law of Bariletto. In 1560, Navò applied for publishing privileges for the *Secrets of Signora Isabella Cortese* as well as for parts 2 and 3 of Rossello's *Summa.*[56] Navò also published the third edition of Vanuccio Biringuccio's *Pyrotechnia,* which, significantly, is also dedicated to Caboga; as well as Alessandro Piccolomini's *Raffaella* (1539), discussed shortly, which includes cosmetic and hygienic recipes for women similar to those found in books of secrets.[57] The documented presence of Navò in Ragusa, where he traded books between 1559 and 1560, suggests that a "Ragusan connection" between Navò and Caboga is "highly probable," making Navò a likely candidate for the editorial engine behind the two works (perhaps together with Ruscelli).[58]

In addition to sharing a common dedicatee, the dedicatory letters to the *Secrets of Signora Isabella Cortese* and Rossello's *Summa* present similar perspectives on the nature and importance of secrets. Cortese, for example, asserts that it is human nature to seek to understand, imitate, and even outdo nature: "man is not content with investigation alone, but seeks in all things and through all things to ape nature, even surpass it, as he attempts to accomplish what nature cannot; and the truth of this can be seen in the secrets that one hears of every day and sees put into practice."[59] Likewise, Rossello notes that "investigating secrets" is natural for human beings: "So natural is it to seek out the secrets of nature that ancient histories tell us learned men have always engaged in this practice . . . all the good knowledge we have of natural things can be seen as stemming from this desire to investigate secrets."[60] Their emphasis on human curiosity and the study of nature is echoed in Ruscelli's *New Secrets,* which states that the goal of natural philosophy is "diligently to interrogate and make a kind of true anatomy of the things and actions of Nature."[61]

Although they each emphasize man's desire to understand and dominate nature, Cortese and Rossello differ in their attitude toward the past. Rossello expresses reverence for the ancients. Cortese is instead firmly focused on the [early] moderns, eschewing the study of prior models in favor of direct experience.[62] Indeed, Cortese does not hide her scorn for the alchemical canon: if one wishes to practice alchemy, one must do—not study.[63] Cortese's conception of alchemy resonates with the new culture of testing, assaying, and experiencing: "if you wish to follow the alchemical art, and practice it, you must leave off studying the works of Geber, Ramon [Lull], Arnaldo [of Villanova], or other philosophers, because they said nothing truthful in their works except through figures and puzzles."[64] Cortese herself laments having wasted thirty years of her life in such fruitless study, rather than devoting herself to practice.[65]

Other connections between the works of Cortese and Rossello are suggested by a number of similarly worded recipes that appear in both texts.[66] Overlap between books of secrets was common, however, and it can be very difficult to trace any one recipe back to a sole "original" source. Ultimately, the question of authorship is of less significance here than the content of these texts and their targeted readership. Publishers understood that there was a market for secrets among women readers, and for "women's secrets" among readers of both sexes. In the case of the *Secrets of Signora Isabella Cortese,* they further grasped that a woman-authored book of secrets would fill a crucial lacuna among existing offerings and bestow a particular aura of authority onto the contents of the work through the established associations of gender with secrets, alchemy, and empirical culture.

Although Cortese's work is dedicated to "every great lady," this does not mean that her recipes were of use only to women. On the contrary, a brief overview of the work's contents reveals that the recipes, which are divided into four books, encompass a wide range of subjects, and include numerous recipes useful to either sex, with those specifically directed at women clustered primarily in book 4. Book 1 focuses on medical recipes, and includes treatments for plague, antidotes to poison, and a number of remedies for syphilis *(mal francese)*—a source of great anxiety throughout early modern Europe.[67] Several recipes illustrate how to heal wounds of various kinds, and a single recipe addresses helping women heal after childbirth.[68] These medical recipes, like those found in other books of secrets, incorporate alchemy in various ways—through distillation, multiplication, and other operations central to alchemical transformation—reflecting the pervasiveness of iatrochemistry by the mid-sixteenth century. A recipe to treat sores caused by venereal disease, for instance, calls for distilling oil in an alembic in progressive stages, each one more effective: "Note that the more it is distilled, the stronger it will become."[69] A remedy for mouth sores employs mercury sublimate, an alchemical substance.[70]

In book 2, devoted to recipes intended to produce gold, elixir, or the philosopher's stone, alchemy moves to center stage. In keeping with the text's utilitarian bent, Cortese's instructions to the alchemical adept are notable for their pragmatism. Unlike the highly cryptic or allegorical alchemical texts of the medieval tradition, these precepts are clear and accessible. She advises her reader(s) to follow a few simple rules: work alone; acquire strong terra-cotta or glass vessels; become familiar with the basic materials of alchemical experiment; watch the fire; keep tongs at the ready; do not discuss the Art with anyone or allow anyone into the laboratory; acquire a faithful and trusted helper; and, most important, when you have successfully completed your experiment, give thanks to God by making charitable donations to the poor.[71] Such instructions recall those exchanged between Cortese's "real world" antecedent, Caterina Sforza, and her assistant Luigi Ciochi, who hastened to assure his mistress of his own experience, loyalty, and discretion.[72] Having established these rules, Cortese moves on to the recipes, which range from true alchemical experiments (seeking to create the philosopher's stone) to what Garzoni might term *secreti leggieri,* such as tricks for creating invisible ink. Recipes for creating and manipulating various metals abound: reductions of silver; gilding iron or softening it; turning copper to gold and making potable gold.[73] Practical in nature and short on commentary, these alchemical recipes are not addressed specifically to men or women. Nonetheless—and despite the author's claim in this section to have received her information from a Polish traveler—the knowledge receives a stamp of authenticity from the female author who conveys it, thereby inscribing women within the circle of alchemical authorities.

Book 3 is composed largely of the kinds of "marvelous" secrets that held as much fascination as true alchemical experiments: how to make mirrors, affix gems to wood and crystal surfaces, create gold lettering for books, or dye leather in various colors. This section also includes recipes for hair color (with a preference for a reddish tint), along with numerous recipes for soaps and stain removal: "To remove all oil and grease stains from cloth"; "Soap balls to remove stains"; "To remove stains from any fabric in any color."[74] Such recipes could be used, of course, by men as well as women, but the sustained attention to domestic concerns again suggests an audience that included women. Interestingly, this section also includes a single remedy for impotence, a mixture of quail testicles, ants, musk, and amber: again a recipe that could be of interest to either sex.[75]

The final section, book 4, is devoted to cosmetics and beauty enhancers. Lip and cheek color, face creams, skin lighteners and teeth whitening powders, hair color, and perfumes all figure prominently. Although this is the section of the book most clearly addressed to the needs and concerns of women, men are not excluded: recipes for dying the hair could also be used for the beard, and men could make use of the perfumed oils and moisturizing lotions just as women did. However, many of the recipes specifically state that they are meant for women ("lip color for women"; "sublimate for women"; "depilatory for women") and the general emphasis on beauty in this section certainly reflects an audience of women.[76] This section is also the lengthiest of the four parts, containing 221 recipes (as compared to 30 in book 1, 70 in book 2, and 99 in book 3). These cosmetic recipes reflect standard early modern notions of female beauty echoed in lyric poetry and treatises on the female form:

white teeth, smooth skin, a fair visage devoid of freckles. Particularly numerous are recipes for beauty waters—often prepared and sold by apothecaries, as we will see shortly—and oils, suggesting that these products were among the most sought-after and commonly used. They rely on common organic ingredients such as lemon, egg white, and goat's milk, as well as on inorganic, alchemical ingredients like litharge, alum, silver, and gold. A recipe to "give color to one who is pale" calls for boiled pigeon distilled with sugar and gold ducats in an alembic:

> Take two pigeons with white feathers and feed them on pinenuts for eight, or rather fifteen, days; then butcher them and throw away the head, feet, and guts; put [the rest] in an alembic and distill with half a loaf of sweetened bread and four ounces of true silver, three gold ducats, four heels of white bread that has been left to soften in goat's milk for six days . . . distill all of this over low heat, and it will produce a most perfect water to give color to a pale complexion.[77]

A rather complicated recipe, this beauty water would also have been costly to produce: a "most perfect" recipe like those categorized by Garzoni in the *Piazza,* but certainly in violation of his warnings there regarding labor and expense.

As in the previous three sections, the cosmetic recipes in book 4 employ a variety of alchemical procedures and vessels. A recipe for another "perfect aqua vita" is distilled from wine in a carefully specified series of recipients:

> distill it in a container with a neck the length of one *bracio* and a half,[78] in a bain-marie with its lid, well-sealed at every juncture, and when you see that nothing more can be distilled it means that the essence [*spirito*] has been coaxed forth; when you see that sign, immediately take the container [from the heat] and you will empty the water into a small glass container, adding a little bit of the essence at a time. . . . Then do this again with more wine to produce the same as the first time around, and continue in this fashion until you have as much of it as you desire.[79]

The integration of alchemy into cosmetic recipes, with the resulting association created between women's concerns and alchemical practice, is also characteristic of other books of secrets. A recipe for women's rouge in a work spuriously attributed to the physician Gabriele Falloppio (*Secreti diversi et miracolosi* [Diverse and marvelous secrets], 1563), for example, calls for mixing ingredients in a clean glass alembic and allowing them to rest two days before boiling them until all the ingredients have dissolved, a kind of cosmetic transmuta-

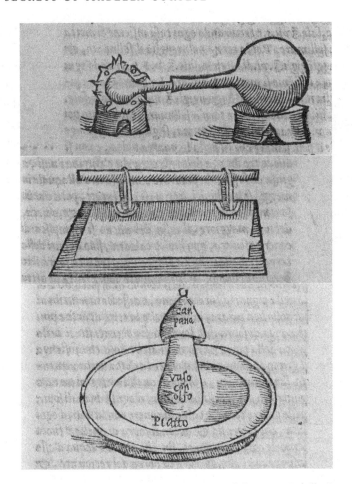

Illustrations depicting alchemical instruments, *I secreti de la signora Isabella Cortese, ne' quali si contengono cose minerali, medicinali, arteficiose, & Alchimiche, & molte de l'arte profumatoria, appartenenti a ogni gran Signora. Con altri bellissimi Secreti aggiunti* (Venice: Appresso Giovanni Bariletto, 1574), pp. 50, 57, 187

E. F. Smith Collection, Kislak Center for Special Collections, Rare Books and Manuscripts, University of Pennsylvania

tion. A recipe to enhance the beauty of women's hair likewise calls for the prolonged distillation of ingredients in an alembic.[80] Such instructions abound in sixteenth-century books of secrets. To underscore the importance of such alchemical procedures, many are accompanied by illustrations. Celebrino's *House of Recipes,* for example, includes images of vessels being heated at the hearth, explicitly evoking the alchemical laboratory; while the *Secrets of*

Signora Isabella Cortese depicts alembics, plates, lids, and other requisite receptacles.

As the *Secrets of Signora Isabella Cortese* makes evident, alchemy functioned as an accepted and familiar aspect of everyday science, becoming part of experimental culture as it was applied to quotidian concerns such as the management of personal health, the household, and physical appearance. It is also apparent that women, like men, had become well-versed in the language and tools of such experiments, possessed of the necessary skills to navigate a recipe and produce the desired result. Works such as the *Secrets of Signora Isabella Cortese* demonstrate that books of secrets recognized a gendered element to the authority and appeal of secrets as well as the presence of a female audience eager to read about them. So great was the popularity of books of secrets, and of "women's secrets" in particular, that their influence spilled over into other literary genres throughout the sixteenth century.

Women's Secrets across Literary Genres

Even prior to the success of the *Secrets of Alexis* and the *Secrets of Signora Isabella Cortese,* as vernacular books of secrets were still finding their footing, the contents of such texts had begun to infiltrate vernacular dialogues, treatises, epistolary collections, and satires, particularly those directed at women and their social roles and comportment. In some cases, authors sought to warn women away from medical and arcane knowledge, in others to claim ownership of it. Some works, such as Alessandro Piccolomini's *La Raffaella, dialogo della bella creanza delle donne (Raffaella . . . A Dialogue of the Fair Perfectioning of Ladies,* 1539), focus primarily on women's production and use of cosmetic recipes, using artificial embellishment as a metaphor for moral corruption and, more generally, political decay. Others, such as Ortensio Lando's *Lettere di molte valorose donne* (Letters of many valorous women, 1548), weave alchemical secrets into a larger polemic about learning, education, and religion. If *libri di segreti* such as the *Secrets of Signora Isabella Cortese* sought to appropriate and market the "secrets" of women to a newly expanded reading public eager for such texts, these literary texts attempted a similar endeavor, using medical, cosmetic, and alchemical secrets to comment on gender and the new science, as well as on women's roles more generally. The remainder of this chapter will consider how the model of the book of secrets is reinterpreted and manipulated in two very different literary frameworks.[81]

With *Raffaella,* Piccolomini—in other cases an ardent defender of women and a supporter of the vernacularization of scientific texts to their benefit—satirizes feminine morality and makes a parody of Renaissance comportment manuals through the advice of the elder Raffaella to the newlywed Margarita.[82] Echoing Aretino's Nanna of *I ragionamenti* (1534–1536), Raffaella encourages Margarita to take a lover while she is still young enough to enjoy one, and instructs her on how to maximize her beauty to achieve this goal.[83] As a cautionary tale about the moral perils of cosmetics, *Raffaella* echoes classical works such as Xenophon's *Oeconomicus,* which condemns the use of makeup as deception; Xenophon's text had served as a model for Leon Battista Alberti's fifteenth-century dialogue, *I Libri della Famiglia (On the Family,* 1433–1434), which includes the tale of a woman who is literally disfigured by the arsenic she uses to lighten her face. It was Piccolomini who translated Xenophon's *Oeconomicus* into Italian as *La economica* (1540), just months after the publication of *Raffaella;* some scholars have suggested that these two works—each dedicated to the Sienese intellectual Frasia de Venturi—must have been composed as a pair, with Raffaella as a counterpoint to the good wife in *La economica.*[84] Other works of the period also take a negative view of cosmetics, such as Francisco Delicado's *Ragionamento del Zoppino* (Dialogue of Zoppino, 1539), which decries the foul-smelling makeup used by prostitutes along with the lotions and oils they use to soften their wrinkled skin.[85] As part of a wider satire of the institution of marriage, Ariosto's *Satira V,* too, warns against women who employ cosmetics. Drawing on a widespread cultural association that linked Jews to the production of cosmetics, Ariosto's condemnation of makeup is imbued with anti-Semitic claims regarding such products ("Don't you know that makeup is made from the spit of the Jewesses who sell it; even the addition of musk can't mask its stench . . . how many other foul things, alas, are contained in what they apply to their face?").[86] Later in the century, Annibal Guasco's *Discourse to Lady Lavinia* (1586) reminds the author's daughter that cosmetics are "an offense against God" and urges her to use only clean, fresh water on her face.[87] Piccolomini's dialogue is likewise intended as a rebuke to women's sexual appetites and their reliance on artificial enhancers of beauty to ensnare lovers (or at least as a parody of conduct books that presented such images of women), while also working on a broader scale to craft an analogy between the domestic household and the state, in which the use of cosmetics is symptomatic of the same idleness that works against good governance.[88] Nonetheless, the inclusion in *Raffaella* of detailed recipes for beauty waters, oils, lotions,

and teeth whiteners clearly links this work to books of secrets and the enthusiasm for experiment behind them. Where Alberti's dialogue mentions cosmetics only to condemn them, and Ariosto's *Satira* refers disparagingly to the
unpleasant odor they confer on the women who use them, Piccolomini's
work includes lengthy, detailed, and potentially useful recipes and ingredients. On one level *Raffaella* can be read as a satire of feminine morality and
vanity; on another level, it functions as a reflection of the same daily concerns
and practices of early modern women that found expression in the *libri di
segreti*. Raffaella does not discourage Margarita from using cosmetics: on the
contrary, she insists that Margarita use the best and most effective ones, and
at her urging, provides the recipes ("ricette perfettissime e rare"). Raffaella's
language echoes that found in books of secrets, which routinely classify recipes as "perfect" or "unique." Piccolomini's use of such descriptors underscores his efforts to link this satiric dialogue to that other, highly popular genre.
The conversation between Raffaella and Margarita, moreover, offers insight
into how women actually used and produced the sorts of beauty waters and
other cosmetic recipes described in books of secrets. From Margarita's explanation that she obtains her current beauty water from a preferred apothecary
("uno speziale, che sta alle Costarelle"), we learn, for example, that women
did not necessarily concoct these treatments themselves, but often purchased
them outside the home.[89] Raffaella's acknowledgment that the beauty water
in question is a big seller for the apothecary affirms that particularly prized
and effective remedies were often associated with specific pharmacies or
doctors.

Although Raffaella concedes the popularity of Margarita's beauty water,
she nonetheless maintains she knows a better—albeit more expensive—recipe.
Echoing the tension between secrecy and accessibility characteristic of the *libri
di segreti,* as well as the hierarchy of secrets detailed by Garzoni in the *Piazza
universale,* Raffaella agrees to describe this more costly recipe to Margarita,
but withholds certain details, insisting that only she can prepare it for her.[90]
Raffaella's version requires a range of animal, vegetable, and mineral ingredients, including pigeon parts and precious gems, similar to those found in the
Secrets of Signora Isabella Cortese cited earlier. Raffaella instructs Margarita:

> I first take a pair of pigeons plucked asunder, then Venice turpentine, lily
> flowers, fresh eggs, honey, seaperiwinkles, pearls ground and camphire.[91]
> All these I embody together and put them within the pigeons in a round

glass vial on a slow fire. Afterwards I take musk and ambergrease[92] and yet more pearls and shreds of silver and grind these last together fine upon the porphyry slab; then I put them in a bag of linen cloth, and tie them to the beak of the limbeck with the receiver below. I then clarify the lotion and it becomes a most rare thing.[93]

The recipe, like several more that immediately follow, is characterized by a mixture of precision and vagueness. Raffaella describes the ingredients required but not the necessary quantities; the procedures to undertake, but not the duration of the process. Having access to a recipe does not ensure that one will be able to successfully replicate it without prior knowledge or particular expertise.

Like the cosmetic preparations offered by Isabella Cortese and other "professors of secrets," Raffaella's recipes rely on alchemical materials and processes. The actions required for production (mixing, heating, pounding, distilling) belong to the realm of the laboratory, as do many of the ingredients, such as the camphor, pearls, and silver in the previously mentioned water. Another recipe provided by Raffaella requires the distillation of mercury, a staple of alchemical experiment, in a bain-marie:

> One takes the finest true silver and quicksilver passed through buffin cloth, and when blended together they are ground for a day in the same direction with a little fine sugar. Then I take it from the mortar and grind it on a painter's porphyry slab, and I embody therein shreds of silver and pearls. Then anew I grind all the things together upon the porphyry and set them back in the mortar, and next morning early I slake them with foam of mastic with a little oil of sweet almonds; so when the liquid has stood for a day I slake it all again with water of dittany and put it in a flask and bring it to the boil in a Mary's Bath lymbeck. Then having done this four times, ever casting out the water, the fifth time I conserve it . . . and at the bottom the sublimate remains.[94]

In addition to beauty waters, Raffaella offers other recipes for personal hygiene, including lotions to keep the complexion supple or lighten the skin, teeth powders, and hand lotions. Although Piccolomini's dialogue implicitly takes a stance against the use of cosmetics by women by associating makeup with interlocutors of questionable character such as Raffaella and Margarita, the detail of the recipes provided complicates this satiric aim, rendering the dialogue a book of secrets in itself. If cosmetics are explicitly associated with physical (as well as moral) disfigurement in works such as Alberti's *On the Family,* here the reader might easily ignore this cautionary subtext and, on the contrary, follow Raffaella's instructions for maintaining beautiful skin and teeth. In this respect, *Raffaella* aligns itself with works such as Giovanni Marinello's *Ornamenti delle donne* (Ornaments of women, 1562), which (in more straightforward fashion) enthusiastically supports women's right to enhance their physical appearance.[95] It is worth noting that Piccolomini went on to pen an *Orazione in lode delle donne* (Oration in praise of women, 1549) in a likely correction to this early satiric work and in keeping with his more general and characteristic defense of women.[96]

Like Piccolomini's *Raffaella,* Ortensio Lando's *Letters of Many Valorous Women*—the only sixteenth-century anthology devoted exclusively to women's letters—liberally incorporates material from books of secrets into its epistolary structure.[97] Published in Venice by Giolito, a major promoter of women's texts, the *Valorous Women* capitalized simultaneously on the Renaissance vogue for epistolary works, the growing interest in secrets literature, and a general fascination with works addressing all aspects of women's experience. It presents advice and opinions on subjects ranging from the utility of education for women (a fundamental issue of the early modern *querelle des femmes*) to the preparation of cosmetics, medicinal remedies, and alchemical experiments. Although the recipes and advice offered in the anthology are attributed to women, the *Valorous Women* was authored in its entirety by Lando, as evidenced by numerous internal clues, including a Latin postscript that names him as their compiler.[98] By writing in the female voice, Lando endeavored to authorize both his epistolary style—letter writing was typically construed as a "feminine genre"—and the authenticity of the secret information he presented, which he attributes firmly to women. In this respect, Lando's anthology proves a suggestive locus for observing the kind of epistemological shift described by Park in her study of gender, sexuality, and the history of human dissection. There, as noted earlier, Park notes an increasing appropriation of the "secrets of women"—that is, information about the female body and its reproductive operations that women are thought innately to possess—by male medical practitioners and authors.[99] Transferring the notion of "women's secrets" from a strictly medical context to the literary realm of the *querelle des femmes,* Lando's collection similarly presents female knowledge as private and privileged, passed along from mother to sister to friend. However, these networks of information are recreated and revealed by a male author who, through the impersonation, demonstrates his own mastery over these arenas of discourse. The female impersonation of the *Valorous Women* thus constitutes a multilayered appropriation of specifically feminine domains, from the domestic space of the household to the physical realm of the body.

The association between gender, secrecy, and the exchange of information among women is rendered explicit in Lando's anthology, continuously invoked by a community of epistolarians drawn from a range of female patrons throughout Italy as well as invented interlocutrices. A letter attributed to "Argentina, countess of Rangona" (in Verona), divulging a recipe for an elixir that can cure many kinds of illness, explains that only the bond of female friendship motivates her to share her secret: indeed, this recipe is so powerful that she would not reveal it to her own son.[100] Although Lando's epistolarians have special access to secrets, they are not always able to protect this valuable knowledge. A letter from Livia Beltrama reprimands Adria della Rovere for her inability to safeguard a secret, stating that "all the misfortune that has recently befallen you derives from your inability to keep secret that which was told to you in secret."[101] Livia uses a gendered metaphor to drive home her point, likening Adria with her incontrollable loquacity to a woman in the advanced stages of pregnancy: "Truly you would have exploded had you not given birth to this paltry little secret."[102] Despite constant invocations of and admonitions to secrecy, the female community recreated by Lando proves unable to keep them. Underlying Lando's reiteration here of the trope of the female gossip is another commentary, directed not at feminine loquacity but at the popularity of books of secrets themselves and at the inherent and problematic tension between openness and secrecy. Lando's *Valorous Women* offers a critique of the utility of these works as well as a warning about their certain capacity for misuse.

On the one hand, Lando's letter-writers exchange useful medical recipes for facilitating conception and promoting healthy pregnancy, much like those found in the pages of books of secrets. A cure for infertility echoes the widespread opinion that spa waters could sometimes remedy sterility and offers a prescription based on the dried uterus of the hare similar to those provided in works such as Bairo's *Medicinal Secrets*. The wife of a physician, the writer of the letter confesses to "having done my share of reading," proffering advice she could indeed have found by consulting any number of *libri di segreti:* "if you will come with your husband to the Villa baths, I will make sure you know from which of you the problem stems; should I find there is no defect in either of you, and [knowing that] you desire a son, I will take the uterus and female parts of the hare and dry them, and you will drink the powder with a little wine, and without a doubt you will become pregnant."[103] Further echoing the claims of many books of secrets, the letter notes that this concoction will ensure the conception of a son.[104] A letter on a related theme from "Mamma Riminalda" explains how to prepare an antiabortifacient from celery seed, cardamom, mint, and other herbs. A valuable recipe for women readers, it reflects the heavy reliance on botanical ingredients common to early modern books of secrets, as well as the enduring influence of another set of texts with a constructed female authorship, the medieval *Trotula:* "because you tend to miscarry, have your apothecary prepare this powder: three drams each of celery seed, amomum, mint; three drams each of mastic, garofilum, cardomon, blackberry root, plus three drams each of zedoary, castoreum, iris . . . five drams of sugar, and take this powder with honey and mix up three scruples of it in wine every time, and it is certain that [you will be] protected from miscarriage."[105] As in the case of Piccolomini's Margarita, the recipient of this letter is instructed to have the remedy compounded by her pharmacist. Advice on avoiding miscarriage abounds in books of secrets, from Bairo's *Medicinal Secrets* to Giovanni Marinello's *Delle medicine partenenti alle infermità delle donne* (On medicines pertaining to the illnesses of women, 1574), which includes a recipe composed of many of the same ingredients as those cited by Riminalda.[106]

Finally, a letter attributed to "Madama la Grande" details the causes and remedies for maladies of the matrix, or uterus, considered by early modern medical theory to be a principal component of female illness.[107] Madama la Grande's letter diagnoses a friend with a dislocation of the womb stemming from an imbalance of bodily humors. She offers a number of typical remedies, including uterine fumigation conducted with a mixture of amber, balsam, and musk (another prescription outlined in the *Trotula*).[108] Significant in the letter is Madama La Grande's assertion that her friend should seek advice not from a male doctor, but rather from other women, any of whom can offer medical guidance on a par with Galen himself: even a "country girl" *(contadinell[a])* is as "learned as any physician found today in Padua or in learned Bologna."[109] Just as the broader scientific climate of the mid-sixteenth century sought to elevate empiricism and experiment along with, if not above, theoretical speculation, so too direct experience takes precedence over study for Lando, a critic of humanism and indeed all the learned professions.[110] His appropriation of the "secrets of women" underscores the importance of praxis over theory. It designates women's knowledge as derived from direct experience, in contrast to that of men, which comes from study.

While Lando may have been making a point about the efficacy of practical knowledge over formal learning, he did not offer an uncomplicated endorsement of this kind of "secret" knowledge, no matter how direct its source. Greatly influenced by the *spirituali* movement of the 1540s, Lando held strong views regarding faith and the centrality of scripture over all other forms of study.[111] Accordingly, even the "direct" and practical knowledge of women was a target for Lando: although it could be useful, when exercised judiciously by virtuous women, it could just as easily devolve into superstition and magic. If some recipes in the *Valorous Women* detail useful remedies for common physical ailments, others for artificially enhancing the complexion or permanently reversing the aging process are given a satirical spin. A letter addressed to Cassandra Lanfreducci takes aim at her excessive preoccupation with smooth skin: Cassandra uses more egg yolks on her face than the entire "Certosa of Pavia" (the famed monastery in northern Italy) could consume, not to mention whole barrels of horse's urine and dried hare's blood. The use of egg yolk for the skin was common in books of secrets—it appears in Celebrino's *New Work,* among others; hare's blood is recommended to remove blemishes from the face in works such as Marinello's *Ornaments of Women.*[112] Whereas books of secrets provide such cosmetic recipes without comment, Lando makes his opinion clear: Cassandra's search for physical beauty has led her to neglect her wifely duties. In many cases, Lando's writers even acknowledge that the knowledge in which they traffic is improper for good Christian women, but proceed to offer it nonetheless. Lucretia Cuoca condemns an antiwrinkle ointment as an unnatural attempt to halt the natural progression through life toward death and salvation, but is persuaded to reveal the recipe regardless. Even Sister Lucretia, a nun, possesses and reveals such recipes, stating that after loyalty to Christ, her duty is to serve "honorable women").[113] These women trade their secrets back and forth, the knowledge a form of power and, indeed, currency, exchanged for status. The more secret the recipe, the more prestige it confers, and one secret must be repaid with another, as explained by the Countess of Rangona, who notes, "I want to repay you for the secret you recently sent me, with another secret of no less value."[114] Such exchanges indeed took place with regularity among Italy's aristocratic women, for example at the courts of Caterina Sforza or Isabella d'Este.

Lando's anthology, finally, creates further intertextual links to books of secrets in a series of recipes explicitly based on alchemical practice. The notion
of secrecy was central to alchemical discourse, and alchemy was, as we have
seen, a principal element of Renaissance books of secrets. Newly accessible
thanks to the development of printing technologies and the diffusion of books
of secrets, alchemy was embraced by many; yet others—including Lando—had
only disdain for what they considered fraud. Alchemy—and especially its prime
goal, the transmutation of base metal to gold—met with deep suspicion, due
in no small part to fears about charlatans and, especially, counterfeit currency. The production of alchemical gold was increasingly subsumed in the
early modern period by other therapeutic or spiritual transmutatory goals
(for example, the creation of distilled elixirs, the fabled "quintessence" that
would prolong life; or, for Ficino and the Florentine Neoplatonists, the transformation of the soul in harmony with the divine universe).[115] A skeptic,
Lando targeted alchemy in his anthology as both quackery and a distraction
from spiritual growth. In one letter, Giulia Gonzaga—a major figure in the
spirituali movement in Italy—paints alchemists as charlatans and their fol-
lowers as greedy fools, here targeting the quest for alchemical gold: "I was
deeply disappointed to hear that you were visited by a wicked alchemist, who
confused you with false promises and made you believe that the substance of
elements can be transmuted, and silver made from copper or converted into
gold. It's really something, how these lying, flea-ridden quacks want to make
everyone rich."[116] In other letters, the satire is subtler and applied to the therapeutic and practical aspects of alchemy, as in a missive attributed to Isabella
Sforza describing a distilled elixir that can simultaneously cure leprosy, remove
stains from linens, clear the vision, and restore eternal youth: "the most miraculous water ever made by man or woman."[117] Central to Isabella's recipe,
which employs alchemical ingredients like silver and gold foam, is the practice of progressive distillation, by which a substance was thought to acquire
different and increasingly powerful properties over a period of days, capable
at each stage of curing various complaints: "on the first day soak these things
in the urine of a virgin youth, on the second in warm white wine, on the third
in fennel juice, on the fourth in egg white, and on the fifth in the milk of a
woman who is nursing a son."[118] After the final stage, which should be preserved in a golden vase, the results will be incontrovertible: "and when you
have clearly seen the result, you will learn to trust those who know, by virtue
of age and experience, more than you, and to revere the burners and alembics

[of the alchemical laboratory]."[119] Books of secrets routinely offered such mul-tipurpose recipes, but Lando problematizes remedies that could run to sev-eral pages, as well as their alchemical underpinnings and the equipment needed to undertake them.[120] Isabella—who serves elsewhere as an example of female virtue and spiritual rectitude—is chided here for wasting her time distilling such potions under the tutelage of a certain "mastro Christophoro," an alchem-ical expert who has revealed to her "the great secrets of the most secret part of Philosophy."[121] No longer devoting herself to the study of scripture, she has come under the spell of alchemy's false promise. Alchemical recipes are empty currency—ineffective and even dangerous when exchanged among women, who compete to outdo one another while deviating from the path of virtue and salvation.

In presenting Isabella Sforza as a practitioner of "domestic" or practical alchemy—alchemy applied to the demands of everyday life—Lando was drawing on well-established gendered foundations of alchemical discourse. He thus favors women as possessors of this secret knowledge, even as he under-scores the inherent fraudulence of alchemical practice. Lando uses alchemy to target not only the vanity and credulity of women but also to critique what he perceives as popular superstition and magic. Books of secrets, like other genres by which knowledge is textually transmitted, constitute a distraction from what should be the true focus of virtuous people: that is to say, scrip-ture. A book of secrets in new guise, the *Valorous Women* is rooted in the con-text of cultural and religious renewal that would form and inform an ever greater part of Lando's work. That Lando should have chosen the secrets of women to help convey his broader cultural and religious critique only serves to underscore the pervasiveness of medical-scientific discourse and the popu-larity of the sixteenth-century *libri di segreti* among readers of both sexes.

Despite the claims to privileged knowledge put forth by their titles, books of secrets—unlike their more arcane Latin predecessors—were accessible and pragmatic manuals, spilling "secrets" to a wider, nonspecialist audience of men and women even as they promoted the exclusivity of their contents. Whether offering recipes for conducting true alchemical experiments (such as producing alchemical gold) or proposing formulas for skin creams and medi-cines, vernacular books of secrets emphasized practice over theory, extolled the virtues of trial and error, and prized efficacy most of all. With their expansive catalogues of remedies for ailments from rashes, worms, and rotting teeth to sciatica, epilepsy, and cancer, books of secrets offer rich territory for exploring a variety of early modern concerns about the body. Their attention to recipes for manufacturing gems, transmuting metals, and producing youth-giving "quintessences" or elixirs, moreover, reflects the increasing stature of alchemy within early modern scientific culture.

The inclusion of women in the audience for books of secrets—whether in titles, authorship, introductions, or recipe headings—situates them squarely within the vibrant intellectual context of early modern science. Practical alchemy: experimentation applied to all facets of daily life, from health to appearance to perfumery and metallurgy, involved and intrigued women as well as men. So prevalent was the association of women with secrets that natural philosophy would become an integral part of the Renaissance debate over women, addressed in myriad literary works that explicitly take up the defense of women's intellect, ability, and education, as we will see in Chapter 3.

3

⌘

Scientific Culture and the Renaissance *Querelle des Femmes*

Moderata Fonte and Lucrezia Marinella

In the second part of Moderata Fonte's dialogue on female superiority, *Il merito delle donne* (*The Worth of Women*, 1600), the character Leonora interrupts the conversation of her female interlocutors to question why they have digressed from their principal subject and instead have begun debating the causes of various natural phenomena (earthquakes, thunder, currents), along with the medicinal properties of plants, herbs, and spices (fennel for the eyes, rhubarb for fever, cardamom for pregnancy). "What have the kinds of things we've been talking about got to do with us?" she asks, with respect to the discussion of medicinals. She continues, "Are we doctors, by any chance? Leave it up to them to talk about syrups and poultices . . . it's absurd for us to be discussing them."[1]

To this point, the conversation of the seven women featured in Fonte's polemical work, which she divides into two parts or "days," has focused not on women's medical or scientific knowledge, but rather on exploring the origins of gender inequality in sixteenth-century society.[2] As the women gather together at Leonora's Venetian palazzo on the first day, they question how men have come to dominate women in virtually every capacity, despite their essential inferiority to the female sex. As the character Corinna (a textual stand-in for Fonte herself) argues, this unequal state of affairs results from custom, not natural law. Implicitly calling into question an ingrained Aristotelian tradition of argument against women, Corinna states "[men's] pre-eminence is something they have unjustly arrogated to themselves."[3]

Stretching back to Boccaccio's *De claris mulieribus* (*Famous Women,* ca. 1360), the "querelle des femmes," or debate over women, was a constantly evolving cultural skirmish that was waged throughout Europe.[4] Humanist writers and, eventually, a much broader array of vernacular authors, sought to describe and define the qualities and roles of women, and argued variously for and against their intellectual, moral, and civic potential, with works such as Galeazzo Flavio Capra's *Della eccellenza e dignità delle donne* (*On the Excellence and Dignity of Women,* 1525) and, especially, Henricus Cornelius Agrippa's *Declamatio de nobilitate et praeccellentia foeminei sexus* (*On the Nobility and Preeminence of the Female Sex,* ca. 1529) serving as influential models on the pro-woman side.[5] The debate could be sharp, as evidenced by the early seventeenth-century context in which Fonte's *Worth of Women* circulated.[6] Like another central text of the "woman question," Lucrezia Marinella's *La nobiltà et eccellenza delle donne co' difetti et mancamenti de gli huomini* (*The Nobility and Excellence of Women and the Defects and Vices of Men,* 1600[7]), Fonte's dialogue was brought to press in response to a particularly virulent antifeminist treatise by Giuseppe Passi, *I donneschi difetti* (*The Defects of Women,* 1599). A misogynist diatribe that condemns female vanity, greed, sexual excess, and other alleged flaws, Passi's work went on to additional and expanded editions in 1601, 1605, and 1618. However, it also inspired some writers to rebut it.[8]

Among Passi's many accusations about women is a lengthy chapter devoted to their activity within the spheres of science and medicine, which he sees as nothing more than various forms of witchcraft. What for men might be characterized as valid scientific knowledge—for example, about astrology, an accepted branch of Renaissance natural philosophy; or about the use of plants and herbs in medicine—becomes dangerous superstition or even magic viewed from Passi's misogynistic lens.[9] Passi was not alone in his dismissive characterization. As we saw in Chapter 2, other writers also sought to warn women away from such knowledge or to deride certain kinds of scientific learning as superstitious, frivolous, or even dangerous. Yet not everyone agreed with this view. Consequently the topic of women's participation in scientific study and practice, from natural philosophy to pharmacy, was increasingly deployed over the course of the sixteenth century in Italy in the context of the debate over women. As the best-known contributions to the debate by women writers in the late Renaissance, both Fonte's *Worth of Women* and Marinella's *Nobility and Excellence of Women* have attracted much critical interest for their proto-feminist arguments.[10] Considerably less attention has been paid to the role of scientific discourse as a rhetorical tool in crafting and expanding those arguments.[11] In these and other works, however, Fonte and Marinella take ownership of scientific discourse, claiming it as an area in which women can demonstrate their merit and as a fundamental aspect of the defense of women.

The use of science as a tool in the *querelle des femmes* and as a category for highlighting women's intellectual capabilities is evident in the most direct contributions to the debate by Fonte and Marinella. Other works by these two writers—Fonte's unfinished chivalric romance *Tredici canti del Floridoro* (*Floridoro*, 1581); Marinella's pastoral *Arcadia felice* (Happy Arcadia, 1605) and her epic poem *L'Enrico; overo Bisantio acquistato* (*Enrico; or, Byzantium Conquered*, 1635)—can be said to reflect a similarly pro-woman sensibility, and these texts also demonstrate a deep engagement with science.[12] An exploration of scientific discourse in these pro-woman, if not strictly speaking protofeminist, narratives illuminates how the literary culture that accompanied the new science as well as the established traditions of natural philosophy converged with the attention to sex and gender that drove the early modern *querelle des femmes*.[13] The work of Fonte and Marinella broadens the scientific conversation to include female interlocutors, protagonists, and observers, revealing how scientific culture reverberated at the very core of the Renaissance debate over women.

Moderata Fonte

While Fonte is best known for her *Worth of Women*, she was a well-rounded writer with an established reputation long before completing her final work.[14] Cristofano Bronzini praises her sharp mind and prodigious memory in his defense of women, *Della dignità e nobiltà delle donne* (*On the Dignity and Nobility of Women*, 1625), in which Fonte's compatriot Lucrezia Marinella also figures as an example of female erudition.[15] Fonte's mentor and biographer, Giovanni Niccolò Doglioni, recalls that she "excelled in everything she tried," from poetry to arithmetic.[16] Although the primary focus of this chapter will be the *Worth of Women*, which most clearly reflects Fonte's interest in natural philosophy as well as the influence of the books of secrets tradition, other works by Fonte also demonstrate an engagement with scientific discourse. In particular,

VERA MODERATÆ FONTIS EFFIGIES,
ÆTATIS SVÆ ANNO XXXIIII.

Portrait of Moderata Fonte, *Il merito delle donne. Scritto da Moderata Fonte In due giornate. Ove chiaramente si scuopre quanto siano elle degne e più perfette de gli huomini* (Venice: Presso Domenico Imberti, 1600), unnumbered but a4-v

Courtesy of the University of Pennsylvania Kislak Center for Special Collections, Rare Books and Manuscripts

Floridoro—among Fonte's earliest and most ambitious works—weaves together the discourses of science and protofeminism in a manner that prefigures their more overt twinning in the later *Worth of Women.*

Much of Fonte's biography comes to us through the account of Doglioni, whose *Vita della Sig.ra Modesta Pozzo de' Zorzi . . .* (Life of Signora Modesta Pozzo de' Zorzi . . . , 1593) circulated as a preface to the *Worth of Women.*[17] It recounts the story of the young Modesta Pozzo (later to assume her more elegant pen name), orphaned as a child. Placed in the convent of Santa Marta as a boarder, Modesta soon returned to the home of her maternal grandmother, Cecilia di Mazzi, and her second husband, Prospero Saraceni. When Prospero's daughter Saracena married Doglioni, a writer and member of the Accademia Veneziana, Modesta followed, remaining in her friend's new household until marrying Filippo de' Zorzi (1558–1598), a Venetian lawyer, at twenty-seven years of age.[18] Literary historians have often characterized Fonte's marriage as curtailing her literary output, noting that most of her major works were published prior to 1583, and the *Worth of Women* only posthumously. However, Doglioni's biography asserts that Fonte composed "countless sonnets, canzoni and madrigals" as well as various *rappresentazioni,* suggesting that she continued to write throughout her marriage.[19]

Not only did Fonte's association with Doglioni foster her participation in the Venetian literary world (facilitated though her mentor's connections to the Accademia Veneziana), it undoubtedly encouraged her immersion in a context of natural inquiry. A respected *letterato* who wrote works in praise of various Venetian luminaries, Doglioni was the author of wide-ranging and ambitious scientific works including the *Anno riformato* (Reformed year, 1599), a cosmological treatise; and the *Compendio historico universale* (Universal historical compendium, 1594), a sort of scientific *selva* or encyclopedia. In the second day of the *Worth of Women,* Fonte specifically references her mentor's scientific writings, singling out Doglioni by name both for his loyalty (uncommon virtue in a man) and his knowledge of natural philosophy. Here, Corinna, the most learned of the seven ladies, cites a sonnet originally contributed by Fonte herself to the *Reformed Year* in praise of its author (attributing it to a "gentleman who is a great friend of his"). Employing the language of natural philosophy, the sonnet praises Doglioni's thirst to understand the mysteries and movements of nature: "What wisdom! That a mortal can decipher and distinguish each power of the heavens, each power below: limit, state, motion, site, and source; time, elements, heavens, nature, and art."[20] This auto-citation serves to create a direct link between Fonte and the scientific culture with which she engages here, implicitly establishing her as a protagonist in the conversation that follows. Corinna then goes on to praise other men of science, including Giovanni Antonio Magini, a prominent astronomer and professor of mathematics at Bologna; as well as Claudio Cornelio Frangipani, Giovanni Padovani, and Annibale Raimondo, all of whom published works on astronomy and meteorology.[21]

Although Fonte does not mention by title Doglioni's other well-known work, the *Compendium*—a hefty encyclopedia of historical events and a catalog of princes, popes, and other historical figures—she does echo Doglioni's linking there of scientific knowledge to patriotic interests in a specifically Venetian context. In his letter to the reader, Doglioni states that Venice should embrace the learning ("scientia") of its people as part of its larger civic identity and reputation.[22] His compendium, Doglioni explains, is intended to further this project by compressing a vast breadth of information into one easy-to-use reference resource. Such a goal is typical of the *selve* and books of secrets of the sixteenth century, which sought to make historical and/or arcane knowledge accessible to a wide audience. Scholars have increasingly highlighted the centrality of such efforts, amid a wide array of reference tools, to manage an unprecedented influx of information in early modern scientific culture.[23] In many senses, as we will see further on, Fonte undertakes a similar project in her *Worth of Women*. There, access to knowledge (scientific, literary, political) becomes a key element for achieving equality for women; and arguments for women's intellectual merit, in all disciplines, are tied into the rhetoric of Venice's civic mythology as a sovereign Republic and site of justice and equality.[24]

Early glimmerings of Fonte's interest in harnessing scientific discourse to pro-woman commentary can be found in *Floridoro*, an unfinished chivalric romance in thirteen cantos. Published in 1581 and modeled on Ariosto's *Orlando Furioso, Floridoro* reworks canonical representations of female characters in the warrior Risamante and—most significantly for our purposes—the *maga*, or enchantress, Circetta. Fonte's "feminist" epic would serve as a model for other women writers, notably Marinella and the poet Margherita Sarrocchi, who also undertook new approaches to this traditionally "male" genre.[25] Despite the work's title, derived from a male character who does not appear until canto V, the book revolves around the heroine Risamante's quest to wrest her rightful inheritance back from her sister Biandaura. Unlike Ariosto's Bradamante (Fonte's obvious model), Risamante is not motivated

by love: despite the requisite stanza in canto III predicting her dynastic function (her eventual marriage and the birth of a daughter, Salarisa [not a son]),
she retains her autonomy throughout the poem.[26] Similarly, Circetta—an
amalgam of Ariosto's female characters Alcina, Logistilla, and Melissa—
subverts tradition (not to mention misogynist characterizations of woman
as witch) as a "maga" who is virginal, wise, and devoid of the destabilizing
carnality typical of the enchantress figure.[27] Fonte uses her female characters
to make the point that, if afforded the same opportunities as men, women are
capable of the same achievements in both arms and letters: one of the most famous stanzas in *Floridoro* borrows the imagery of metallurgy to describe wom-
en's potential as "buried gold," needing only to be mined and refined.[28]

With its reinterpretation of archetypical chivalric female characters and
themes, *Floridoro* ranks alongside the *Worth of Women* as an exercise in protofeminism, albeit in a different generic framework.[29] In this respect, Fonte's
decision to dedicate *Floridoro* to Bianca Cappello (1548–1587), along with Francesco de' Medici (1541–1587), acquires added resonance. Having fled Venice at
fifteen to marry a Florentine banker—earning a death sentence from the Venetian senate in the process—Bianca Cappello transformed her initial noto-
riety into a powerful position as consort to the Medici grand duke.[30] Although
Fonte may not have known of Bianca's early scandal—the story was conve-
niently erased from public consciousness as Bianca's stature improved and her
nuptials approached—she likely appreciated her compatriot's achievements.
A 1580 letter by Fonte to Francesco de' Medici regarding *Floridoro,* which takes
pains to reassure Francesco that he, not Bianca, is the poem's primary dedicatee, only serves to underscore the prominence Fonte awards to Bianca
throughout the poem.[31] While Fonte's primary motivation for the dedication
may have been political, timed to celebrate the new links between Venice and
Florence, it is tempting to postulate that she may also have had in mind the
well-known scientific interests of Bianca and Francesco, which were pursued
in dedicated spaces such as Francesco's studiolo in Palazzo Vecchio and the
laboratories of the Casino di San Marco.[32]

If Risamante, woman warrior and Medici progenitrix, symbolizes female
strength and perseverance, Circetta—like her mother Circe before her—
embodies female erudition. Fonte's imagined product of the union between
Circe and Ulysses, Circetta appears in five cantos, occupying a central position in the poem. Circetta serves to illustrate the capacity of women to be
learned in natural philosophy (as well as other disciplines, particularly political

history) without impinging on the "natural" feminine modesty and chastity of their sex. In contrast to Circe, her more powerful—and threatening—mother, Circetta's example reassures the reader that women can be measured in their deployment of such knowledge, using it only for the good of society and never for its destabilization—themes to which Fonte will return in the *Worth of Women*. Circetta also serves an ideological function as the prophet of the Medici dynasty (up to the reign of Francesco) and of Venice itself (represented by Bianca Cappello), offering a lengthy panegyric culminating in the union of Francesco and Bianca in 1579.

Circetta is introduced in canto V when the knights Clarido and Silano, having landed upon a mysterious island, witness her transforming other men into trees, in an echo of Ariosto's Alcina.[33] Unlike the episode of Alcina, however, these men are not cast-off lovers, but simply unworthy to enter Circetta's palace or break the enchantment placed on her by Circe, which has frozen her in time and space. As Circetta explains, she must exist suspended in youth until a knight who is the equal of Ulysses in "cleverness and valor" can liberate her (VII, 25); the knights transformed into trees are those who attempted—and failed at—this task. Despite her unsettling powers, then, Circetta is not a completely autonomous character, for she must await a male rescuer. Honest and good, Circetta is described as "grave and severe," "courteous and most kind," and later, "honest and marvelously wise"—not to mention sexually pure.[34] Although she commands all the beasts of the forest with her words alone, the knights are reassured when Circetta assumes the more traditional and less threatening role of gracious hostess, taking them under her protection and serving them a lavish meal in her palace, assisted by three handmaidens.

Circetta is an enchantress, but her skills pale in comparison to those of her legendary mother, from whom she gleaned her wisdom and whose story she relates in canto VIII. Circetta's account of her mother's command of herbs and spells and her power over nature itself paints Circe as a more formidable and destabilizing locus of knowledge than her daughter, to whom she has imparted only the "good which can come from that art."[35] The rest of Circe's wisdom regarding the domination of nature—the more dangerous part—is recorded in writing *(possenti carti)*, suggesting the power, and perils, of the written word as well as the widespread early modern paradigm that figured nature as a book to be read and deciphered.[36] Circe's knowledge derives from both study and practice; the source of her wisdom is ancient: "Circe in olden times by the virtue of herbs and words, / with lofty lore not revealed to anyone today, / could obscure the illustrious face of the sun."[37]

The melding of magic and science in Circe's supernatural insight into (and command over) nature's elements recalls in many respects the description of the philosopher-*magus* described by Giambattista della Porta in *Magia naturalis* (*Natural Magic,* 1558), a work Fonte surely knew. Della Porta's influential treatise, reprinted numerous times throughout the sixteenth and seventeenth centuries, offers a generalized philosophical framework for "natural magic" (in essence, science) along with a pragmatic series of applied examples.[38] Careful to distinguish his paradigm from superstition and black magic, della Porta clarifies that "natural magic" is nothing more or less than the observation of nature and its workings, allowing the philosopher to discern its hidden secrets: "Magick is nothing else but the survey of the whole course of Nature. For, while we consider heavens, the stars, the elements, how they are moved, and how they are changed, by this means we find out the hidden secrets of living creatures, of plants, of metals, and of their generation and corruption."[39]

Like della Porta's *magus,* Circe studies the forces of nature to discern their secrets and, by extension, exert mastery over them. Abandoned by Ulysses, she initially gives vent to her hurt and fury by harnessing her command of the elements and covering every city with dense fog (VIII, 19). Later, after Ulysses's death, she decides to honor him by creating a mausoleum for his ashes on the island now inhabited by Circetta. Circe deploys her magic to force the wind to "penetrate into the center of the earth," causing the earth to swell and give way to a mountain, upon which she creates a perfect temple, "without an architect, / with the virtue of her magic incantation."[40] Circe's creation of the mountain is echoed in Fonte's analysis of earthquakes in the second day of the *Worth of Women,* in which Corinna explains that "wind, when . . . it gets trapped underground . . . cannot find a way out. And since by its nature, it cannot stay enclosed, it puts all its energy into trying to escape, and it is this force that agitates the earth so violently."[41] Fonte's source for this explanation was Aristotle's *Meteorology,* but the theory circulated, with variations, among many sixteenth- and seventeenth-century authors.[42]

Like her daughter, Circe, too, possesses the power to transform men into other forms: not trees, but more fearsome beasts such as bears, wolves, and bulls. Circe's transformative actions—and those of Circetta after her—are, on the surface, fantastic in nature, a remnant of the archetypal enchantress figure of chivalric romance. Yet they, too, reflect something of the new scientific climate, here linked to Fonte's condemnation of male perfidiousness. They are, in effect, alchemical in nature, resting on the notion that all substances share a common element and may be manipulated in form—a central precept of alchemical thought and experiment. Fonte, however, draws on the imagery of alchemy to make a bigger point, using it to comment upon the unreliable and dishonest nature of men. Employing a rhetorical structure she will repeat in the *Worth of Women,* Fonte ties Circe's magical transformation of men into beasts to men's real deceitfulness. Interrogating humanist tropes of self-determination as a path to positive self-transformation and (semi) divinity, Circetta argues that "in our age men in erring / are *transformers of themselves*" (italics mine).[43] With no need of spells or potions, men ably and constantly shift forms, to the dismay and detriment of women:

> I see them go around changing themselves
> With such facility, without employing verses or potions,
> That I esteem that art little,
> Since our century so frequently takes part in it.
> Each man is so good a magician with his own form
> That [Circe] could not match their skill at that time,
> When she employed to change our image
> So many herbs, so much study, and so much time.
> Each man is so eager to stray from himself
> That he does not thereafter find the time to return.
> Not of all, but of most men I discourse,
> Who like to seem what they are not.[44] (VIII, 3–4)

What takes alchemists years of study and experiment to achieve is thus innate to men with their facile capacity for (negative) self-transformation. In this formulation, alchemy serves as a vehicle to illustrate a broader *querelle* point. Circetta's analysis of men's dissimulation links the discourse of alchemy and natural philosophy to the polemics of the *querelle des femmes,* prefiguring an even more extended use of this rhetorical tactic in the *Worth of Women.* The digression also sets up the conclusion of Circe's story, in which she herself falls victim to male falsity when Ulysses abandons her—after first duplicitously persuading her to reveal all her knowledge to him, thereby depriving her of power and rendering her unable to prevent his departure: "she, unable to withstand his graceful manners, / with various magical verses and various ampullae, / made every science plain to her lover."[45]

Having recounted Circe's story—and clarified that Circetta, unlike her mother, possesses only the "good" part of Circe's scientific knowledge, Fonte turns her attention to the lifelike sculptures adorning the walls of Circetta's palace. The first set of sculptures, all astrological in nature, are still linked to the scientific discourse of the earlier cantos. They depict the various seasons along with the activities appropriate to them, and display the corresponding zodiac sign above each one: "Nor above them did they lack their own signs: / Aries, Taurus, Gemini, and the rest. / It seemed then that each planet dwelled and ruled / Over the houses appropriate to it."[46] An accepted arm of early modern science, astrology included not just the study of the stars, but cosmology and meteorology more generally.[47] It was an area of great interest to Fonte, who returns to it in the second day of the *Worth of Women,* as we will see further on. In *Floridoro,* it serves to yoke Fonte's attention to science to her panegyric intent: first with respect to Venice's literary panorama and next to its political history and ties to Florence's Medici dynasty. Indeed, the next series of sculptures described by Circetta depict illustrious Venetian poets and writers, among them Fonte herself, modestly depicted as a "solitary young woman."[48] Also named here are some who contributed prefatory material to *Floridoro,* such as Bartolomeo Malombra and Cesare Simonetti.[49]

Having fulfilled her obligation to her supporters and fellow poets, Fonte now turns her attention to an extensive overview of Venice's history and political structure. Fonte draws on the civic mythology of Venice as a site of republican liberty and sovereignty, themes she takes up later in the *Worth of Women* where, inevitably, they must exist in tension with the actual status of women in Venetian society. Stretching over two cantos and encompassing centuries of events, Fonte's historical reconstruction is unique among early modern Italian women writers, and prefigures in "ambitiousness, uniqueness, and breadth" her appropriation of scientific tradition in Day Two of the *Worth of Women.*[50] Both areas—science and political history—constitute arenas in which women can demonstrate their intellectual prowess and civic identity.

In *Floridoro,* therefore, we find early evidence of Fonte's interest in natural philosophy, as well as a clearly articulated argument regarding the importance of education for women in a variety of disciplines. Even more significantly, in Circe and Circetta we see the roots of Fonte's use of scientific discourse to advance and enhance her protofeminist agenda. If *Floridoro* constitutes an initial experiment with this rhetorical strategy, it is in her final work, the *Worth of Women,* that we see the culmination of Fonte's use of scientific learning within the context of the *querelle des femmes.* Moreover, where *Floridoro* depicts Circe and Circetta as natural philosophers in the della Portian sense of the *magus,* the *Worth of Women* more clearly exhibits the combined influence of the authoritative texts of natural philosophy together with the pragmatic, popular vernacular "books of secrets" that circulated in sixteenth-century Italy, making a whole spectrum of medical and scientific experimental knowledge accessible to early modern readers. The outlines of the Scientific Revolution were not drawn as sharply as is sometimes thought: rather, the boundaries between the old, Aristotelian, text-based science and the "new" culture of empiricism were blurred. As Paula Findlen writes, "the Scientific Revolution grew from the constant mediation between the old form of knowledge and the new."[51] Fonte's discussion of various therapeutic remedies and natural phenomena in the *Worth of Women* reveals the depths of her learning and her familiarity with canonical philosophers such as Aristotle and Pliny, as well as the influence of the *libri di segreti*.

In the second day of Fonte's dialogue, her interlocutors address many of the topics present in vernacular books of secrets: balneology, pharmacology, nutrition, and the healing properties of stones and gems, to name just a few. In Fonte's hands this information is given new form and focus, harnessed now in the service of her social commentary. If most practical "handbooks," such as those of Timoteo Rossello, Pietro Bairo, Alessio Piemontese, and Isabella Cortese (discussed in Chapter 2), offered myriad recipes but little in the way of narrative framework, in Fonte's case, the "secrets of nature" are woven into the narrative established in the first day of the dialogue, and constitute a key element of her feminist project. Remedies and recipes serve both to educate the reader and demonstrate the medical expertise of Fonte's speakers—now active divulgers, not passive recipients of knowledge—and also, repeatedly, to underscore the shortcomings of men, as when Corinna—echoing Circetta's dim view of male honor—concludes that all the medicinal prescriptions in the world "wouldn't be enough to protect us from men's malice."[52] Each area of knowledge touched on by Fonte's speakers comes to signify something larger, functioning on at least three levels: first, as medically or scientifically useful information made available to the reader; second, as evidence of women's intellect and capacity for scientific reasoning; and finally, to further emphasize the moral failings of men. Although critics often describe the second day as an "encyclopedia" of sorts, stressing the diversity of its topics and its dissimilarity to the more linear first day, it is perhaps more accurate to term it *encyclopedic* in nature, as it lacks a clearly defined organizational structure beyond the repeated linking of scientific discourse to protofeminist argument and hovers between the pragmatic and the marvelous. What is clear, however, is that scientific knowledge—here encompassing both medicine and natural philosophy—falls under the umbrella of Fonte's more general call for women's equal access to education in all arenas.[53]

Despite its obvious structural differences, the second day opens where the first leaves off, with the question: if men treat women so badly, why do women continue to love them? The question establishes the parameters of Fonte's pro-woman argument. Notwithstanding the separate female space she has created with this all-woman garden gathering, Fonte recognizes that women and men coexist—and manage to love one another—despite enduring and institutionalized gender inequity. She does not seek to alter this fundamental aspect of human relations. She does, however, link it immediately to science and the natural world. Responding to the query that has been posed, Corinna—who continues in the second day to function as the work's intellectual engine—immediately turns the discussion to astrology, signaling a shift in direction. Proceeding through all facets of the question (What do we mean by love? What differences are there in the ability or predisposition of men and women to love?), Corinna, echoing other early modern opinions on the question, posits that it is the influence of the stars, which imprint on women's receptive hearts, that causes them to love: "Many women are quite conscious of men's unworthiness and ingratitude . . . and yet they still find themselves disposed to love, through the force of astral influence, which contributes more than any other factor toward both the inception and the continuance of amorous passion."[54] Careful not to stray from the Catholic doctrine of free will, Corinna clarifies that "the heavens can influence but not compel" people to love. Pointing to the particular receptiveness of women to love, Corinna continues to underscore her argument for women's superiority to men, linking it to their "innate" goodness: "Women—or women's hearts—because of the innate goodness of their nature, are perfectly disposed to receive the imprint of true love. But men are both by nature and by will little inclined to love, and so . . . they can be influenced only to a limited extent by the stars."[55]

Corinna's explanation of the astrological aspects of love lead the women into a more general discussion of friendship, central to the fabric of both the social and natural worlds.[56] Citing a series of examples drawn from Antiquity, the women agree that true friendship "is the cause of all good" and "keeps the world alive."[57] Again, this assessment of friendship is tied to Fonte's overarching pro-woman argument as well as to the pervasive scientific discourse that characterizes the second day. Not only is friendship the basis of happy and stable marriages (a recurring topic of concern for these female interlocutors), but it is also the principle that governs the health of the body and the environment at large: "the friendship and bonding of the elements maintains health in our bodies, and brings fine weather to the air, calm to the sea, and peace to the earth, so that cities can be built, kingdoms grow to greatness, and all creatures live in comfort."[58] The perversion of *amicitia,* by contrast—occasioned by men's propensity for discord—leads to war, which in turn produces disharmony among the elements: "in the air, disharmonies of the elements produce thunder and lightning; at sea, they provoke storms; and on earth, earthquakes."[59] Natural disasters, and earthquakes in particular, were the subject of much attention in early modern Italy, particularly in the wake of those that devastated Ferrara between 1570 and 1574. A plethora of vernacular dialogues and treatises, produced in both university and courtly settings, explored this aspect of meteorology and its implications for political and religious culture (Are earthquakes the result of God's will, or accidents caused by the confluence of certain natural actions?). Works by Lucio Maggio (*Del terremoto dialogo,* 1571), Giacomo Buoni (*Terremoto dialogo,* ca. 1571), and Gregorio Zuccolo (*Del terremoto, trattato,* 1590) offer various explanations for the causes of earthquakes. As Craig Martin has argued, accessible and informative texts of this kind combined "practical utility" with the "entertainment of the wondrous," seeking a wide audience that often included women.[60] Girolamo Borro's dialogue on tides (*Dialogo sul flusso e reflusso del mare,* 1561), for example, is dedicated to Elisabetta della Rovere, marchesana of Massa; a 1577 version depicts Borro himself using the tides to explain Platonic philosophy to Giovanna, the Grand Duchess of Tuscany. Niccolò Vito de Gozze's *Discourses on Aristotle's Meteora* (Discorsi, sopra le Metheore d'Aristotele, 1584) contains a pro-woman dedication attributed to the author's wife, Maria Gondola, and directed to the Ragusan noblewoman Fiore Zuzori.[61]

Fonte's all-female cast of characters adds a new twist to such considerations: linking natural phenomena such as lightning and earthquakes to the destructive behavior of men, they give voice to a protofeminist argument through the language of natural philosophy. This line of reasoning, initially inspired by the question of love raised on the first day, leads the women to turn their conversation more attentively to the discourse of natural philosophy and, in Corinna's terms, the "great secret[s] of nature."[62] When Leonora, playing her usual role, objects to the distinctly scientific direction of the discussion, her outburst only serves to underscore Fonte's linkage of these two central threads: "If you want an example of something unstable, why not talk about men? If you're interested in natural disharmonies, men will do fine . . . what's the point of all this astrology? This kind of talk and this kind of study have nothing to do with us."[63] Leonora's interruption neatly sets up Corinna's rejoinder that natural philosophy—and astrology in particular—is "a very distinguished discipline, worthy of the loftiest intellects" (for example, Fonte's mentor Doglioni, cited by Corinna here).[64] Corinna's modest disclaimer that she herself has "not studied any discipline . . . and this least of any" is belied by her deft and pointed manipulation of the discussion throughout the second day and her informed contributions to it.[65]

The women's discussion of natural phenomena such as earthquakes and the cycles of the moon soon gives way to a wide-ranging consideration of living creatures and their properties: birds, fish, shellfish, and other sea creatures.[66] In these pages we clearly see the integration of authoritative texts on natural history and philosophy (in particular Pliny's *Natural History*) with the *libri di segreti* tradition that had by now taken deep root.[67] An examination of eagles, for example, encompasses typology (specimens may be gray-brown or white); the noble nature traditionally attributed to these birds; a popular myth regarding how eagles recognize the legitimacy of their offspring; and finally, whether they make a nutritious, or even tasty, meal.[68] Similarly, Corinna's detailed description of fish, including bass, sturgeon, eel, flounder, tuna, mullet, and many more varieties—surely inspired in part by Paolo Giovio's *De piscibus* (*Of Roman Fish*, 1524)—serves not only as a springboard for speculating as to whether or not fish outnumber other species, but more practically, for the consideration of the nutritional aspects of fish and (in accordance with most early modern Galenic medical theory), its suitability for people of various complexions, or humors.[69] In this way, Fonte intertwines the wisdom of influential sources such as Aristotle and Pliny with the practical approach to nature's bounty found in books of secrets and, especially, handbooks of alimentary regimes (for example, Castor Durante's *Il tesoro della sanità*, 1588]).[70] In addition to demonstrating the speakers' knowledge, such an exchange of information has a practical purpose. By sharing these facts—about food, fish, medicines—among themselves, Fonte's women help one another to become educated consumers in the Renaissance marketplace, an increasingly important skill.[71] In this respect, science has a pragmatic function that will serve them well in their daily lives. Once again, however, the discussion eventually loops back to the central question of men's unreliable nature and unfair treatment of women. After Helena laments that women need men's help to hunt and catch many of the species they have discussed, Cornelia quips that men excel in "trapping, deceiving, ensnaring: that's their specialty, in fact."[72]

As the women's conversation moves into the curative waters of spas, popular destinations in the early modern period for those seeking to cure a variety of ailments, their discussion not only demonstrates their command of the relative properties of different types of spas and their broader knowledge about the movements of springs and rivers, but also is again explicitly tied to their pro-woman argument.[73] Many sixteenth-century books of secrets praised the healing virtues of waters and in particular their capacity to treat women's physical complaints (such as infertility) and enhance their beauty.[74] Fonte's own discussion of spas, however, is a seamless extension of her broader narrative and polemical purpose. Despite the various types of spas (cold, tepid, hot, and boiling) and their healthful properties, Leonora notes, they cannot cure those most "pernicious" of male defects—namely, deception, ingratitude, and infidelity—indispositions which, as Corinna adds, could not be cured by all the waters of the ocean.[75] Marveling that water, despite being a heavy element, manages to rise high enough to feed mountain springs, Leonora further comments that in this respect it is similar to men, who although inferior and "heavy" in nature, yet find the means to rise and rule—an example of Nature breaking her own laws:

> Just think of the way in which men, who are inferior to us and so should by rights stay below us, in a lowly and humble position, manage to rise above us and dominate us, against all reason and against all justice. So you shouldn't be surprised if water, too, though such a base element, is presumptuous enough to ascend to the level of mountains. At least water flows back to its natural level again, whereas men remain obstinately fixed in their stern position of eminence.[76]

The discussion of the medicinal properties of the spa thus serves simultaneously to impart information to the reader, to demonstrate the women's grasp of the subject, and to fortify their argument about male inferiority.

If the women's discussion of the natural world—celestial, animal, and oceanic—demonstrates a clear familiarity on Fonte's part with the some of the authoritative works devoted to natural philosophy, the remainder of their conversation engages topics likely most familiar to readers primarily through the newer tradition of vernacular books of secrets as well as pharmabotanical references such as Pierandrea Mattioli's 1544 translation of and commentary on Dioscorides's *Materia medica*. These include an examination of balsam—the miraculous resin panacea that was a staple of early modern *libri di segreti*, the use of foodstuffs to treat physical ailments, and cosmetic recipes. For Fonte, however, balsam is not only a precious medicinal tool, but—more important— an illuminating example of male intractability. Corinna describes how this "divine liquor" revives the spirits, restores health and youth, and acts as a "preservative against putrefaction and decay."[77] Driving home her point about men's deeply rooted prejudices, however, Corinna adds that balsam is the "ideal remedy for everything," except the disease the women are talking about (namely, men).[78] Likewise, even the most precious of therapeutic stones—coral, sapphire, pearl—capable of staunching blood, restoring vision, and nourishing the sick, respectively, prove unable to effect a change in men's behavior.[79] As Corinna, echoing a widespread conviction in the early modern period, points out: "There are many kinds of stones that have curative powers," but as to the problem under discussion, "I'm not sure a stone exists that could perform that particular magic."[80]

In the same vein, various foodstuffs are prized for their efficacy in treating common ailments. Fonte's interlocutrices spend several pages discussing many such examples: rhubarb for fever and choler; licorice for chest ailments; and sweet wine for the stomach. Their discussion touches on a variety of healthful fruits, as well as meats, bread, and herbs. Rosemary, the women note, is "amazingly effective against all kinds of ailments."[81] Remedies such as as colocynth (also known as bitter apple) are prized for their efficacy in numerous conditions at once: "colocynth is also good against hardening of the spleen (the marrow of the plant, that is, mixed with a solution of hart's tongue fern) or you can make it up into a decoction with vinegar as a cure for toothache, or a poultice to be used against worms. It's not to be used alone, though, as it's poisonous."[82] Their discussion returns, however—yet again—to their primary topic. As Leonora notes, colocynth is poisonous on its own—just as men are poisonous without women to serve as an antidote.[83] Like the other women, Leonora does not miss the opportunity to tie together her own knowledge about medicine with her efforts to demonstrate female superiority—despite her insistence early on that matters of "syrups and poultices" are best left to male doctors. In these pages, therefore, Fonte turns to the recipe—that staple of scores of books of secrets—to underscore her arguments about men.

Some of the substances under discussion serve dual purpose as both a medicinal and cosmetic product, as in the case of cows and oxen, which Cornelia regards as "the most useful animal nature has provided us," furnishing "all kinds of milk products, necessary to human life. . . . Everyone knows how useful their skin is for various purposes: we eat their flesh, and we even use their horns and hooves for various things."[84] In addition to these healthy uses, Cornelia explains that calf's foot, when boiled down over a period of forty days, creates a gelatin that eliminates wrinkles and restores the user's skin to that of a "girl of fifteen."[85] Multifaceted recipes such as this one, addressing medical and cosmetic concerns simultaneously, are common in vernacular books of secrets. Handbooks devoted solely to cosmetic recipes also abounded in sixteenth-century Italy (for example, Giovanni Marinello's *Gli ornamenti delle donne* [*On the Ornaments of Women*, 1562]), together with Neoplatonic treatises on female beauty (such as Agnolo Firenzuola's *Dialogo delle bellezze delle donne* [*On the Beauty of Women*, 1548]). Nonetheless, detractors of women often seized upon the use of cosmetics as evidence of female vanity and moral inferiority to men. Fonte, however, uses this medical-cosmetological recipe to make a pro-woman point, for it occasions Corinna's comment that it is far easier to understand the curious, varied, and sometimes occult properties of beasts, such as oxen, than to make sense of the falsity of men: "Women aren't as aware of men's failings as they should be. . . . In fact, it's easier for us to understand the properties of irrational animals (even though they should be more mysterious to us, since their nature is alien from ours and they can't speak) than to understand these false creatures who are close to us in nature, but quite different in their character and desires, and who never speak the truth."[86] The secrets of nature, from the hidden medicinal properties of herbs and stones to the medical power and cosmetic potential of foods, are ultimately easier to understand and unravel than the process by which men, who are inferior to women, have managed to establish dominion over them—the central question of Fonte's dialogue.

Although the status of women is Fonte's primary focus in the *Worth of Women,* her discussion of the various properties of food and herbs is not limited to their usefulness to women. Apart from a handful of references to remedies for the discomforts of pregnancy (nutmeg for the stomach, eaglestone against miscarriage), most of the medicinal therapies offered by Fonte's interlocutors are general in nature—efficient simples capable of serving many purposes, and both sexes. Aloe, for example, treats choler, phlegm, melancholy, nerves, dropsy, constipation, and blurred vision—all problems that affect both men and women. In keeping with her inclusive view of the inevitability (if not desirability) of male–female social and sexual relations, Fonte provides information that is of value to both sexes. Her characters' medical expertise does not avail only women, but rather functions as evidence of the rational and intellectual capacities of women that make them superior to men and worthy to counsel both sexes equally wisely in this arena as in others. Women's knowledge about the body, here, transcends their own bodies. In this respect, I would suggest, Fonte not only responds to the association of women with the "secrets of nature" but also seeks to reestablish the seriousness and utility of women's medical knowledge in the face of efforts to satirize or diminish it. Furthermore, Fonte's women are quite aware that they are delving into an arena of medicine for which there are no models among male medical authorities. Leonora, for example, complains that although there are remedies aplenty against "bad blood and choler," there is no such recipe to cure the dangerous naiveté of women, who, blinded by natural compassion, continue to care for their "sick [that is, male] companions." She is answered by Corinna, who quips, "There's a remedy you won't find in Galen. And none of the other authorities on the subject seems to have discovered it either—or if they did, they certainly didn't record it." As Corinna explains, it would hardly be in men's interest to teach women to destabilize the normal social structure in which women care for men, as daughters, sisters, wives, and mothers; and men abuse women in all their corresponding roles.[87] Not only does Fonte demonstrate in the second day that she has learned and absorbed the lessons of the great authorities on medicine and science—along with their vernacular permutations in the books of secrets—but she also recognizes the need for a new, feminist practice of medicine for which no models exist. To take control of their future, women themselves will need to invent new forms of "medicine."

If Fonte's treatment of medicine serves to underscore and elaborate her argument about the superiority of women, finally, the same can be said of her treatment of alchemy. As we have seen, alchemy was a subject of great interest throughout the early modern period and constituted a major element of many books of secrets; it entered into the territory of the *querelle des femmes* through its critique in works such as Ortensio Lando's *Lettere di molte valorose donne* (Letters of many valorous women, 1548), where it is meant to reflect allegedly "feminine" flaws such as greed and gullibility.[88] In contrast to such works, where alchemy functions as an example of female folly, Fonte divorces alchemical practice—and the pursuit of alchemical gold in particular—from the world of women, associating it instead most clearly—and negatively—with men.

While acknowledging that certain materials central to the alchemical process have important and useful medicinal functions—for example, mercury, which has "the power to dissolve substances," or litharge (froth from gold and silver), which "is good for cleaning, closing, and healing wounds"—Fonte draws a distinction between the therapeutic use of alchemy and the pursuit of riches. Gold is a "life-giving substance" but also a "death-dealing one" as a result of men's greed.[89] Alchemical fraud—a punishable offense at courts throughout Europe—is therefore linked in the *Worth of Women* to men, not women, and described as a further manifestation of their moral defects.[90] Alchemy, Corinna states, is an obsession, one that reduces men to nothing. There is no better alchemy, she concludes, than for "a man to study hard and develop skills and then to work hard to make money through the sweat of his brow. That, for me, is an alchemy that never fails."[91]

Although Corinna dismisses alchemy's pretensions to the production of gold, Fonte does not dismiss alchemy out of hand as a structuring scientific principle. Instead, her characterization of the workings of the natural world owes much to the language of alchemy, and her discussion of medicinal preparations reflects the deep integration of alchemy and medicine among Aristotelians and Paracelsians alike. For example, Corinna's description of the notion of the perpetual conversion of one element into another is, in essence, a presentation of the concept of transmutation, alchemy's core operation: "one element converts into another over time, so we find the earth gradually turning into water, water into air, and air into fire. And we also find the opposite: fire becomes air; air, water; and water turns back into earth."[92] Discussions of different preparations of balsam or colocynth necessitate references to distillation, another procedure inherent to alchemical experiment: balsam can be transformed from liquid to ointment, to varied effect; colocynth may be decocted with vinegar.[93] Fonte's objection to alchemy, therefore, has to do specifically with alchemical fraud, which provides yet another opportunity to underscore the shortcomings of the men who fall prey to it. Alchemy in a broader epistemological sense, as a way of structuring understanding of the natural world, is not at issue. Indeed, the sonnet recited by Corinna in praise of Doglioni, discussed earlier, concludes with the telling verse, "[God] inspires your intellect, transforming you, microcosm that you are, into himself."[94] As Cox points out, Fonte's characterization here of her mentor as a "microcosm" *(picciol mondo)* reflects the generalized Neoplatonic idea, common to Renaissance humanism, that man carries within himself the same physical and spiritual forces of the wider universe and has the capacity to become godlike through his intellect.[95] Fonte's use of the expression "picciol mondo," however, also calls to mind the Paracelsian conceptualization of man and nature as "small" and "great" worlds, or microcosm and macrocosm, separated only by the barrier of skin.[96] For the physician and alchemist Paracelsus (1493–1541), who pioneered the use of minerals and chemicals in medicine, health and illness are but reflections of the state of harmony between man (microcosm) and Nature (Macrocosm). The foundation of science and medicine lies in the innate connectedness of human beings with the cosmos, and all of Nature's machinations are a form of "natural alchemy."[97] Fonte's own views, expressed here, of alchemy as useful primarily within a medical context are evocative of those of Paracelsus, a prolific author whose theories were widely circulated by the time Fonte penned the *Worth of Women*. Although Fonte associates the misuse of alchemy with men's weak natures and thus their mistreatment of women, her discussion of alchemy, like that of medicine, furnishes an opportunity for her female interlocutors to demonstrate their knowledge, good judgment, and moral and intellectual superiority, while also showing that they do not seek to use this knowledge to any kind of revolutionary purpose or to upend the social status quo.

Lucrezia Marinella

To an even greater degree than her compatriot Moderata Fonte, Lucrezia Marinella was immersed in the scientific culture of Seicento Venice. Her father, Giovanni Marinello, was a doctor and the author of works on the medical and cosmetic concerns of women: *On the Ornaments of Women* (1562) and *Le medicini partinenti alle infermità delle donne* (On the diseases of women, 1563).[98] Her brother, Curzio Marinello, was a physician and the author of a work on pharmacology: *Pharmacopaea, sive de vera pharmaca conficiendi, & praeparandi methodo* (1617). Relatively late in life Marinella married Girolamo Vacca, also a physician.[99] Marinella benefited from her father's support of her learning and likely from her brother's as well; traces of their influence on her intellectual formation can be found, perhaps, in those works that most clearly display Marinella's engagement with science. Like Fonte, Marinella's erudition in the scientific as well as literary disciplines was remarked upon by seventeenth-century admirers: Bronzini, for example, describes her as "exceptional" in prose, poetry, and spiritual compositions and as a "supreme expert in moral and natural philosophy."[100] The following pages will examine the role of scientific discourse in three of Marinella's works: *The Nobility and Excellence of Women* (1600); *Happy Arcadia* (1604); and *Enrico; or, Byzantium Conquered* (1635). In each of these works, Marinella makes use of the language of natural philosophy, linking it to the cultural controversy over the status of women in early modern society through a variety of rhetorical and narrative approaches.

Marinella's most well-known work is the *Nobility and Excellence of Women*, a polemical pro-woman treatise that, together with Fonte's *Worth of Women*, constitutes one of the major textual contributions to the Renaissance debate about women.[101] Where Fonte's work was published posthumously as a contribution to the *querelle* reignited by Passi's inflammatory *Defects of Women*, Marinella's *Nobility and Excellence of Women*—probably commissioned either by Lucio Scarano, to whom it is dedicated, or by its publisher, Ciotti—was a deliberate, point-by-point response to Passi's treatise. If Fonte's *Worth of Women* cloaks its arguments for women's intellectual equality and moral superiority in the lively form of the dialogue, the *Nobility and Excellence of Women* uses philosophical proofs and examples drawn from ancient and contemporary history to present a precise rebuttal of Passi's misogynist accusations, followed by a presentation of the shortcomings of men. Despite the clear differences in rhetorical organization, the two works share similarities in their agendas: in their defense of women's equality and even superiority to men, most obviously; but more specifically in their calls to grant women the same access to education as men (indeed, Marinella cites Fonte's *Floridoro* on this point) as well as their rejection of the "natural law" argument for male dominance over women.[102] Like Fonte, finally, Marinella introduces scientific learning as a category in the debate over women. She does so most formally in the *Nobility and Excellence of Women*, where scientific learning *(scienza)* constitutes its own category as a criterion of analysis, but scientific discourse pervades many of her other works as well, particularly her pastoral romance *Arcadia* and her epic poem *Enrico*.[103] Marinella's use of scientific discourse in these texts demonstrates her command of this arena of knowledge while also illuminating, like Fonte's *Worth of Women*, a gradual epistemological shift in approaches to scientific culture as the "new" empiricism gained traction alongside older Aristotelian paradigms.

The *Nobility and Excellence of Women* is dedicated to Scarano, a *letterato* identified by Marinello as a "physician and philosopher" *(medico & filosofo)* and the bridge between Marinella's own world of study and the public world of Venetian literary culture.[104] The work is organized into two parts: the first, in six chapters, focuses on women's excellence and dismantles points made by Passi; the second, in thirty-five chapters, examines, in a reversal of Passi's own title, the "defects and vices of men." Scholars have afforded much attention to the protofeminist arguments of the *Nobility and Excellence of Women*. Here, we will instead highlight some areas that reflect with particular clarity Marinella's conviction that learning in the sciences is as central to the debate over women as discussions of their moral character or any other category or criterion.

Marinella first turns her attention to the question of education and learning in a chapter titled, "Of women's noble actions and virtues, which greatly surpass men's, as will be proved by reasoning and example."[105] In these pages she echoes—and explicitly cites—Fonte's sonnet in canto IV of *Floridoro* regarding the equal capacity of women to excel in both arms and letters, given equal opportunity (it is interesting that Marinella references *Floridoro* here—a model that would influence her own later *Enrico*—but does not mention Fonte's most overtly protofeminist work, *The Worth of Women,* of which she likely knew at least by 1601, when the expanded edition of the *Nobility of Women* was printed).[106] In underlining the problem of access to learning, Marinella seconds Fonte in arguing that men fear reprisals, were women to be granted the same tools of education ("men, fearing to lose their authority and become women's servants, often forbid them even to learn to read or write").[107]

In the next section, Marinella clarifies the kinds of learning to which she is referring: education in both literature and the sciences. Using the catalog structure characteristic of many early modern defenses of women, Marinella devotes this section to "learned women and those who are illustrious in many arts."[108] Decrying the commonly held view that there are no learned women in the arts and sciences, Marinella counters such claims with dozens of examples drawn from both ancient and recent history to support her argument for women's intellectual achievements—a "vast tapestry" of learned women spanning centuries.[109] Her examples are carefully chosen specifically to highlight natural philosophy as a category of intellectual assessment. From Hypatia of Alexandria, skilled in astronomy, to Sosipatra of Ephesus, possessed of a quasi-divine command of all disciplines, Marinella underscores the breadth of knowledge mastered by these women. Her examples also underscore the diverse and wide-ranging forms and functions of natural philosophy, which, as we have seen, encompassed all disciplines focused on the study of the natural world. It had a higher function as well, tied to theology, as the example of Catherine of Alexandria serves to illustrate. Said to have converted the pagan emissaries of Emperor Maxentius with her superior philosophical arguments in favor of Christianity, Saint Catherine was known for her intellect and command of all the arts and sciences. With this example, Marinella illustrates the capacity of natural philosophy—understanding of the natural world—to bring us closer to God, allowing us to transcend our "bodily veil" to touch the wonders of earth and heavens. Underlining this fundamental function of learning, Marinella cites these lines from a panegyric of Saint Catherine: "But the learned disciplines that reach so high, and take with them our bitter, earthly senses, so that enclosed in our bodily veil we might then know about the earth and sky."[110]

Marinella also draws on more recent examples from the Italian context to showcase women's learning in the sciences. Costanza Varano (1426–1447), wife of Alessandro Sforza, was studious in both philosophy and poetry; the renowned Cassandra Fedele (1465–1558) took part in public debates and wrote a work on "the order of the sciences."[111] Isotta Nogarola (1418–1566), like Fedele, a humanist author, is also described as "learned in philosophy."[112] Of particular interest is Marinella's mention of Lucrezia d'Este (1535–1598), Duchess of Urbino, said to excel in both philosophy and poetry. To commemorate Lucrezia, Marinella turns to a verse in her honor by Giulio Camillo (1480–1544) that makes deliberate and evocative use of the language of natural philosophy, praising Lucrezia's gift for investigating the secrets of nature: "It is you, only you with your lofty mind surpassing reason in everything, who extracts nature's secrets from her. And as Apollo and his muses stand attentive to your learning and style, already glorious you surpass the philosophers and poets."[113] Marinella thus situates women within a lineage of scientific learning stretching back to Antiquity, while positioning scientific culture within a specifically literary framework through the invocation of Apollo, the patron god of poetry. It is a heritage that encompasses mastery of the authoritative foundations of the discipline while also embracing the new directions of early modern science in its practical investigation of "nature's secrets."

Indeed, scholars have pointed to the work of both Marinella and Fonte as helping to establish such a lineage of female intellectuals and inaugurating a new era of imitatio "al femminile."[114] The references in Marinella's *Nobility and Excellence of Women* to women from ancient and modern history certainly serve this purpose. Marinella's mention of *Floridoro* creates an even more immediate and more explicit intertextual resonance; her *Arcadia* and *Enrico* are also in clear dialogue with Fonte. Marinella's catalog of famous women in the *Nobility and Excellence of Women* is not entirely original. It draws, as scholars have noted, on obvious *querelle* models such as Boccaccio's *Famous Women*.[115] In her chapter on learned women, Marinella also turns to another source: a Renaissance repertory text by Ravisius Textor, *Theatrum poeticum et historicum sive Officina* (1520), which contains an encyclopedic array of examples in various categories ranging from types of animals, fish, and plants to illustrious men, faithful women, and, in chapter 12, "Learned Women" *(De mulieribus doctis)*.[116] It is this chapter that serves as a reference for Marinella.

As I have discussed elsewhere, Marinella's is not the first *querelle* text to rely on the *Officina* for material.[117] As Paolo Cherchi has shown, it was common practice for early modern writers to recycle repertory sources in this manner.[118] The number of corresponding examples that appear in both Textor and Marinella, the temporal range they cover, and the precise resonances in wording confirm that we may add the *Officina* to the list of sources with which Marinella worked closely. Indeed, nearly half of Marinella's more than sixty entries in this section can be found among Textor's "Learned Women," ranging from classical examples (including Sosipatra, Praxilla, Leontia, Hypatia, and Heptachia) to modern (such as Costanza Sforza, Cassandra Fedele, and Isotta Nogarola).[119] A brief comparison of Marinella's text with that of Textor confirms the parallels between these examples of female erudition:

Textor, *Officina*, 558	Marinella, *Nobility of Women*, 84
Leontium Graeca adolescentula, philosophicas disciplinas adeò sectata est, ut etiam non dubitaverit cum magna sui nominis laude in Theophrastum philosophum alioqui laudatissimum, scribere.	Leontium, a young Greek girl, was very distinguished at philosophy and did not hesitate, much to her credit, from writing against the highly praised philosopher Theophrastus. (Leontia giovinetta greca fu molto chiara nelle filosofiche discipline, & non dubitò con sua gran laude di scrivere contro Theophrasto filosofo lodatissimo [38]).

Textor, *Officina,* 563

Constantia Alexandri Sfortiae uxor
numerator inter claras mulieres, &
eccellentes doctrina. Ab infantia
bonis disciplinis operam dedit, adeò
ut imparata quoque disertissimè
loqueretur.

Marinella, *Nobility of Women,* 90

Costanza, the wife of Alessandro
Sforza, is celebrated among
illustrious women. As a young girl
she was most studious at philosophy
and poetry.
(Costanza moglie di Alessandro
Sforza è celebrata fra le chiarissime
donne, & essendo fanciuletta diede
opera a' buoni studi, come alla
filosofia, & alla Poesia, 40]).

Textor, *Officina,* 563

Isota Novarrola, Veronensis,
philosophiam verbo & opera
professa est. Scripsit quandoque ad
Nicolaum Quintum, & Pium
secundum pontifices. . . . Quibus
literarum laudibus adiecit
perpetuum virginitatis votum.

Marinella, *Nobility of Women,* 89

Let us now mention Isotta Nogarola
Veronese, who was learned at
philosophy and led a philosophical
life. . . . She corresponded with
Popes Nicholas and Pius, and
remained a virgin.
(Hor diciamo di Isota Novarrolla
Veronese, la quale di filosofiche
dottrine era adorna, faceva vita
filosofica . . . Scrisse a Nicolao
Pontefice, & a Pio, et sempre si
conservò vergine [41]).

As Paola Malprezzi Price and Christine Ristaino point out, in the *Nobility
and Excellence of Women,* Marinella uses past models to reconstruct "a new
definition of woman to serve as a model for future paradigms."[120] Her effort
places her in dialogue with an array of male authors who weigh in on women,
including Ercole Tasso, Torquato Tasso, Sperone Speroni, and Boccaccio (not
to mention encyclopedists such as Textor); but her boldest philosophical defense of women's superiority challenges the authority of Aristotle himself, whose
views fueled much of the debate. Rejecting the Aristotelian notion of woman
as an imperfect creation, Marinella argues against the widely accepted notion
that women's inherently colder and moister humoral complexion renders them
inferior to men. She does not, however, dismiss Aristotle out of hand. At some
points, she uses Aristotle to support her argument; for example, refuting the

notion that women are the cause of sin and vice by quoting Aristotle's own statement that women are more compassionate than men.[121] Aristotle still occupied a position of authority for early modern thinkers, and even his greatest critics agreed with some aspects of Aristotelian thought. Marinella undoubtedly admired Aristotle—her scientific discourse often appears rooted in the Paduan Aristotelianism that developed the empiricism of Aristotle's doctrine—but she does not hesitate to challenge him on this terrain.[122] Marinella's deft use of Aristotelian arguments about sex and gender demonstrates her mastery of the philosophical underpinnings of the controversy over women as well as her gift for philosophical argument. Marinella was well aware, as Letizia Panizza notes, that "offensive attitudes of Aristotelian origin about women" were experiencing a resurgence.[123] Her decision to take on such arguments about female inferiority is a courageous one that serves not only to demonstrates the author's own literary and philosophical bravura but also to lay a new foundation on which future writers might build pro-woman defenses.

The *Nobility and Excellence of Women* offers Marinella's most direct philosophical argument in favor not only of women, but of women's engagement with the sciences as evidence of their intellectual merit and potential. However, it is in the pastoral romance *Happy Arcadia* and, later, the epic poem *Enrico,* that Marinella begins to experiment with scientific discourse as a creative tool and a more fully integrated narrative facet of her work. Although *Arcadia* and *Enrico* do not position themselves explicitly as defenses of women, both works offer positive characterizations of women and revisionings of traditional female roles that align them with the *Nobility and Excellence of Women* as prowoman texts and echo Fonte's depiction of women's engagement with science in *Floridoro* and the *Worth of Women*.[124] Key female characters such as the nymph Erato in *Arcadia* and the *maga* Erina in *Enrico* are positioned as fonts of knowledge about Nature and about past and future events: their wisdom derives from their connection to the natural world and its workings.[125] Nor is Marinella's recourse to scientific culture limited to her female characters. In *Arcadia,* the shepherd Erimeno, with his grotto full of marvelous inventions, is both *magus* and rationalist, philosopher and inventor, reflecting a continuing early modern anxiety about the blurred lines between magic and science. In addition to functioning as an important facet of Marinella's defense of women's moral and intellectual nature, scientific discourse serves more generally in these narratives to position them within an evolving cultural transition from

Neoplatonic approaches to natural philosophy and "natural magic," to the rationalism that characterized Galileo's "new" science. Marinella's characters illuminate this path, but stop short of following it through to its conclusion. Observers and guides, Marinella's women present a world of scientific discovery and wonder that is opening up to them, but in which they are still finding their place.

Published in 1605, just a few years after the *Nobility and Excellence of Women, Arcadia* is dedicated to Eleonora Gonzaga—daughter of the same Francesco de' Medici to whom Fonte dedicates *Floridoro* and wife of duke Vincenzo Gonzaga of Mantua, known for his own wide-ranging interests in natural philosophy (especially with respect to alchemy).[126]

Although the pastoral genre may have been an unusual choice for Marinella, given that it had largely fallen out of vogue by this time, it presented an opportunity to reinterpret familiar elements from a pro-woman perspective while at the same time incorporating the new language of scientific invention.[127] Recasting pastoral tropes, *Arcadia* recounts the story of the princess Ersilia, forced through a series of events to cross-dress as a shepherd and flee to the safety of Arcadia. Here she acquits herself, in male guise, at all manner of games and activities, only to be unmasked as a woman before Arcadia's ruler, Diocletian. As a condition for remaining in Arcadia as Diocletian's surrogate daughter, Ersilia is made to resume dressing as a woman and to accept his patriarchal authority. Despite Ersilia's intellectual merit and valuable skills, she is in the end forced to accept an inescapably subordinate status—begging comparison to Marinella's final work, *Essortationi alle donne e agli altri* (*Exhortations to Women and to Others*, 1645), which has vexed scholars with its seeming acceptance of women's traditional roles.[128] Even in Arcadia, Marinella seems to say, women are not truly equal or free.

Despite the pragmatic realism of *Arcadia*'s conclusion, the idealized natural setting of pastoral and its remove from the "real" world made it inherently receptive to the influence of scientific culture, as well as to Marinella's still pro-woman agenda. By tradition, pastoral literature pays close attention to the natural beauty of the environment. Jacopo Sanazzaro's *Arcadia* (1502), for example, details the many natural attributes of its setting, including grassy plain, brook, and more than a dozen different kinds of trees.[129] Similarly, in *Arcadia felice* we find vivid descriptions of the many species of trees and flowers that grow in Diocletian's garden: olive, pear, peach, fig, almond, cherry, laurel, prune, and more. Marinella's delineation also incorporates recent advances in

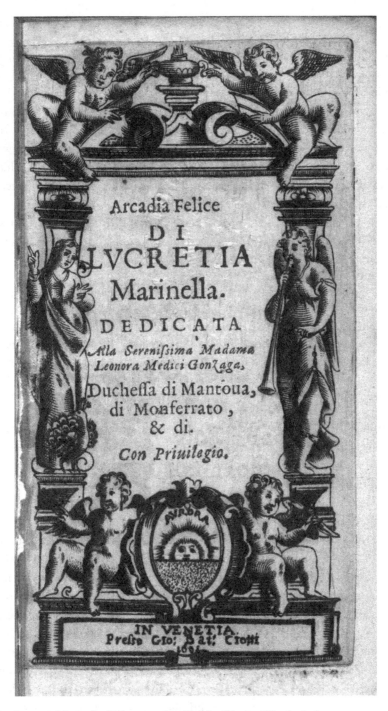

Arcadia Felice
DI
LVCRETIA
Marinella.
DEDICATA
Alla Sereniſsima Madama
Leonora Medici Gonzaga,
Ducheſſa di Mantoua,
di Monferrato,
& di.
Con Priuilegio.

IN VENETIA.
Preſſo Gio: Bat: Ciotti

Lucrezia Marinella, Title page, *Arcadia felice* (Venice: Ciotti, 1605)

agricultural science and exhibits the mark of the new philosophy. We learn, for example, that some of these marvelous trees—which bear a diverse array of fruits on a single branch, or stone fruit without pits—are the product of meticulous experiments in grafting: "the variety of fruits upon a single branch, the diversity of flowers in a single plant, derives, simply put, from grafting."[130] The subject of several early modern treatises on agriculture, grafting made its way into treatments of nature's marvels such as della Porta's *Magia naturalis,* which devotes a full chapter to the procedure.[131] Nature is indeed marvelous; careful study of its workings and organized experimentation can improve upon it.

This attention to the natural world extends to the heavens. Shepherds read the sky to make meteorological predictions, noting that its color predicts the advent of rain or hail that can be damaging to crops.

> Sometimes we see different colors flash across the sky and (if my observations have not been in vain), I know that azure patches announce abundant rains; and that if it is lit up by fiery red you will undoubtedly see the forceful winds hurl unripe apples to the ground. Should you see, in its splendid appearance, certain azure notes begin to mix together with a darker color, all will soon erupt in noisy hail, heavy rains, and fearsome winds; and let the poor sailor beware who places his trust in the untrustworthy sea.[132]

The shepherds stress the importance of observation and analysis: they are not passive observers of nature's signs, but empiricists who gather data and apply it to improve their daily lives. Observing the natural world around them and making predictions about its behavior based on those observations is central to "pastoral wisdom," and parallels the work of the natural philosopher as well as that of the scientist who conducts empirical experiments.[133]

If these shepherds exhibit an interest in nature that is to some extent typical of pastoral in general, it is the figure of the wise Erimeno who embodies most clearly the sometimes competing facets of scientific knowledge that had begun to converge by the early seventeenth century. Erimeno occupies a transitional space between his own teacher, Ciberione—a true *magus* described in almost supernatural terms—and Erato, Ciberione's astronomer-daughter. Explaining to the amazed shepherds the science behind his unique trees, for example, Erimeno's comments bring wonder and reason into harmony, reminding them that there are often rational explanations for marvelous things: "I don't deny . . . that there are many—even infinite—things that cause us to marvel, but which are actually very ordinary and unworthy of special consideration."[134]

Erimeno, who can understand the voices of all living creatures, is also a collector, a figure who assumed increasing prominence in early modern scientific culture. As Paula Findlen has shown, the practice of collecting was an "activity of choice" among the educated and elite in sixteenth- and seventeenth-century Italy, a period in which the first science museums also appeared.[135] Collectors prized objects of "scientific worth," broadly conceived and encompassing the ordinary as well as the exotic.[136] Among his own treasures, Erimeno counts new technologies such as a camera obscura, a hydraulic organ, and a magic lantern, all inventions of the early modern period. His most incredible prize, which frightens the shepherds, is an automaton: a representation of a shepherd so realistic that it "lacked only breath."[137] Hero of Alexandria's *Pneumatics,* an account of ancient automata, was translated into Latin and Italian in 1579, and debates on the nature and status of "self-movers" continued throughout the sixteenth and seventeenth centuries. Again recalling the anxieties that surrounded scientific innovation, the shepherds fret that this marvelous object ("not a real man, but a false image and representation of a man") must be the work of demons.[138] It impinges too closely, that is, on the line separating art from nature, magic from science.

As in an early modern science museum, in which fantastic or exotic objects might be interspersed alongside other less remarkable specimens, Erimeno's collection contains a spectrum of items.[139] To reassure his visitors, Erimeno draws their attention to a collection of objects more closely associated with the familiar natural world, although possessed of unfamiliar powers: gems, stones, roots, and herbs. His explanation of the virtues of the specimens in his collection draws upon a range of likely sources, from Pliny's *Natural History* to Dioscorides's *Materia medica* as well as contemporary sources such as della Porta's *Magia naturalis,* Pietro Bonardo's *La minera del mondo* (The mine of the world, 1569), and any number of sixteenth-century books of secrets. Erimeno's description underscores the magical properties of these objects and substances, but his explanation is presented in an expository structure more suggestive of the medical-pharmacological culture with which Marinella was familiar, not least via her father and brother. Translucent "radiano" (probably radium, an alkaline earth metal), for example, as Erimeno explains, worn close to the heart, ensures that the wearer will be respected and honored by all, no matter his true character. Chelidonium, placed on the tongue in a honey mixture, as commonly prescribed in remedy books, allows one to predict the future.[140] Artemisia is useful for both fatigue and protection from enchantment (although artemisia was commonly prescribed to for various "female" ailments and said to bring on menstruation, Marinella does not include these nonmagical, gendered medicinal uses here).[141]

Erimeno explains to the shepherds that his knowledge about the properties of these plants and stones derives from his teacher Ciberione, who imparted to him, "upon the highest mountaintops," their "hidden virtues."[142] A magical, quasi-divine figure, Ciberione's learning was so immense that "only the immortal gods exceeded him"; he traveled in a "chariot pulled by four black coursers, which could traverse the world in just a few hours."[143] Ciberione's daughter, Erato, who devotes her time to the study and observation of nature and, especially, astronomy, is possessed of an intellect as sharp as that of her father. If Ciberione connotes a magical connection to the natural world, and Erimeno a middle ground between magic and science, the figure of Erato points the way toward the emerging new direction in science, with its emphasis on the direct experience of nature.

Significantly, we learn of Erato's great intellect and deep scientific curiosity primarily through the accounts of Erimeno and, further on, from her companion Armilla. Erato herself makes a brief, albeit striking, appearance in part 4, appearing as if conjured from nowhere to intone a prophecy before Diocletian and his men: "the glorious nymph Erato appeared in the midst of the beautiful room, with unbound hair and fiery gaze and visage . . . with a resonant and terrible voice she said, 'Diocletian, and you his worthy companions, ask what you wish to know.' "[144] Erato's ambiguous textual presence/absence suggests that she presents a complex problem for Marinella, who herself had a complicated relationship to her own erudition and literary production, if we consider the seeming shift in her thought between the *Nobility and Excellence of Women* and the later *Exhortations to Women and to Others*. Certainly, Erato embodies Marinella's model for the learned woman. Chaste and virginal, her feminine reputation is irreproachable (indeed she goes largely "unseen" in the text); her mind is exceptional. A Diana-like figure, she is described as both hunter and "virgin sovereign" or "virgin oracle"; she is also a new Hypatia who contemplates and interprets the secrets of the heavens from her garden, located above the clouds.[145] As Armilla explains, together the two women observe how the clouds produce rain and lightning and how hail is formed from condensation, recalling the earlier meteorological conversation of the shepherds. Marinella's depictions of Erato's cosmological interests is likely motivated by the observations of new astronomical phenomena recorded throughout early modern Europe, such as Tycho Brahe's "new star" of 1571 and the comet of 1577, described in dozens of Italian texts.[146] Erato observes the behavior of comets and differentiates between various kinds (some have a "golden mane," others a "shining tail")[147] and hypothesizes—in opposition to Aristotle—that the Milky Way is composed of a multitude of stars and does not result from the Earth's "dry and constant exhalation."[148] Erato's conviction that the stars predict and influence human behavior and events reflects the prevailing acceptance of astrology as a valid scientific discipline, a view held by many prominent scientists including Galileo.

Erato's observations about the behavior of clouds, vapors, comets, and stars are revisited in canto V of Marinella's epic poem, *Enrico: or, Byzantium Conquered.* As the work's title suggests, Marinella's *Enrico* is conceived on the Tassian model, treating events in the distant—but not too distant—past (the Fourth Crusade, freely reinterpreted from Marinella's Venetian perspective), incorporating a Christian viewpoint, and respecting an Aristotelian unity of action.[149] The only real exception to Marinella's primary focus on the Crusade comes in those cantos devoted to Erina, a *maga*-like figure who recalls both Erato and Fonte's Circetta.[150] As wise and chaste as these literary predecessors, Erina, too, recasts typical chivalric models to function as a learned woman accomplished in the ways of nature, but devoid of any erotic function. The episode of Erina, which—like that of Circetta—unfolds over five cantos (5–7, 21–22), serves as an important interlude in which the Venetian crusader knight Venier, finding himself shipwrecked on her enchanted island, learns of the origins and glory of Venice before returning to battle and his destiny. It also showcases, as in *Arcadia,* the range of Marinella's scientific knowledge, now relayed in *Enrico* principally through female characters.

Our first introduction to Erina and her domain, a peaceful, Eden-like, and all-female *locus amoenus,* comes as she greets Venier on the shore: a beautiful, proud, unadorned figure "saintly" in manner and "firmly set against love."[151] No seductress, Erina seeks only to protect Venier and convince him that there is a purpose to all things—even his shipwreck.[152] Indeed, we become divine only by studying the "main principles of hidden causes and wonderful deeds."[153] Erina invites her honored guest to an elaborate dinner prepared by her handmaidens—one hundred beautiful virgins. Among them is the bard Altea, a secondary character, but one to whom Marinella first entrusts the task—or privilege—of showcasing her knowledge of natural philosophy. Like Armilla in *Arcadia,* whose role is to highlight the intelligence of Erato, Altea is an

ancillary figure through whom the wisdom of the primary female character
is channeled. Recalling Armilla's monologue in *Arcadia,* Altea likewise details the secrets of nature, this time in song. Instrument in hand, she sings
about how cold temperatures condense air inside stone, causing rivers to be
born; and the dangers of certain vapors that resemble comets.[154] Echoing
Corinna's words about the perpetual transformation of natural elements in
the second day of Fonte's *Worth of Women,* Altea explains the cycle of production and disintegration that is essentially alchemical in nature: "She sang
about how earth turns into water, how water turns into air, and how air turns
into fire and ascends lightly. She sang about how fire turns into air, how air
turns into water, and how water takes the shape of lowly earth."[155] She sings
of the moon's effect on the tides and (again recalling Fonte's *Worth of Women*),
the characteristics of rivers, and the causes of earthquakes ("a cruel wind shut in
the bowels of the earth that shakes high buildings").[156] In a direct citation of
Arcadia, Altea also sings of the Milky Way and its composition: "She sang
about the unified light of many stars that is reflected and constitutes a white
way, similar to spilt milk, that directs the gods to Jupiter's palace."[157] As in
Arcadia, Marinella's comments here part ways with the Aristotelian theory
that the Milky Way results from an exhalation of vapor (although she continues to reject the Copernican view of the earth's movement around the sun).
Altea's song, like Erato's attention to the heavens in *Arcadia,* further demonstrates Marinella's interest in natural philosophy, and especially meteorology
and astronomy, and suggests her conversance with the new scientific discourse
and discoveries occurring in this period. As we saw earlier, treatises on natural phenomena such as tides, earthquakes, comets, and the like abounded.
Her comments on the Milky Way are suggestive of Galileo's findings about it
and other celestial bodies that circulated in *Sidereus nuncius* (1610) and were
received with excitement even among orthodox circles (Margherita Sarrocchi,
for example, author of the epic poem, *La Scanderbeide* [1623], was an active
supporter of Galileo's discovery of the satellites of Jupiter). As can be seen in
her comments on the organization of the heavens, still Ptolemaic in nature,
Marinella walks a line in her work between a firm adherence to the established
convictions about the universe that were integral to her Counter-Reformation
context, and an openness to the exciting transformations of scientific culture
that were emerging. That openness, however is tinged with disquiet, or perhaps
nostalgia: Erimeno possesses new technological instruments like the prized
camera obscura, but Erato and Erina still rely on their own senses—not a lens

or telescope—to observe and comprehend natural phenomena.[158] In this sense, Marinella's epic shares with Tasso an ambivalence toward the discoveries of the new science.[159]

Learned though she is, Altea is but a handmaiden to the enchantress Erina, who rules this paradisiacal island, and her song sets the stage for Marinella's true example of exemplary erudite femininity. Like the figure of Corinna for Fonte, Erina—devoted daughter, chaste virgin, and dedicated scholar—is perhaps a stand-in for Marinella herself. As in the case of Erato and Ciberione in *Arcadia*, there are parallels here to Marinella's own circumstances, especially when it is revealed that Fileno, Erina's deceased father, was a natural philosopher. Like Erato, Erina credits her father for her disciplined intellect ("You created at the same time my simple spirit and my wise, learned, prudent mind").[160] Although we hear nothing of Erina's mother, she still mourns the loss of her father, and recounts to Venier how Fileno learned "the famous sciences" *(chiare scienze)* from his own father—who learned from the magician Armano—and passed them on to her, "his only daughter and heir of his valor."[161] Devoted to the study of astronomy, Fileno (like Erato) would observe from a mountaintop "the appearance, motions, courses, and rotations of the stars in the starry heavens."[162] Fileno's sharp mind allowed him to proceed "not from effect to hidden reasons, but from causes to manifest signs"— distinguishing him from others and rendering him quasi-divine in his erudition.[163] Like Circe in Fonte's *Floridoro*, Fileno shared with his daughter "the good and beautiful of what he knew."[164] Erina is thus a direct descendant of a long line of learned natural philosophers, transforming what was a male intellectual lineage into one inclusive of women (or, at least, of this woman).

Erina's erudition is put to work here to fulfill the ideological function of showing Venier Venice's origins and future and, ultimately, returning him to his destiny. In this she follows the counsel of Fileno, who appears to her in a dream and explains that she and Venier share the same lineage (VI, 3). Although Fileno counsels Erina to keep Venier on the island, where a "divine power" has led him that he might escape "death and anguish" in battle, he nonetheless advises her to allow the knight to return to his army if he desperately wishes it.[165] Erina's knowledge still contains elements of the fantastic, as in *Arcadia*. To show Venier his destiny she takes him on a marvelous grand tour of the world in a flying chariot reminiscent of that used by Ciberione (although, as befitting a Venetian epic, hers is driven by winged lions); their voyage reflects both a nostalgia for the great age of exploration and Venice's commercial glory, as well as a lingering reluctance to completely embrace the new scientific rationalism.[166] On the one hand, they map continents including Asia and Africa as they fly above them, resorting to the most recent carto-graphic discoveries; on the other, Erina's clarification that "'[t]hough our chariot travels in pure air, we're not in heaven or in the burning sphere" establishes Marinella firmly within the Ptolemaic cosmos.[167] Nonetheless, Erina offers Ve-nier an overview of the Arabian Sea, the Nile, the Egyptian cities of Alexan-dria, Damietta (Dumyat), and Memphis, and parts of Ethiopia and Nubia (Sudan). They fly over Carthage and Algiers, Mauritania and Morocco. They observe parts of Asia, Arabia, and India; Erina always pointing out impor-tant cities, rivers, and, on occasion, a precious spice or substance for which a region is famous and casting "the imperial glance of Venice" over it all.[168] In a telling moment, a marveling Venier turns to Erina and exclaims, "This is wonderful! My celestial guide, who gave names to all these places and areas? Who explored them so that we know about them? Who went to discover all these unknown and treacherous lands?"[169] Venier's inquiry conveys the excite-ment of the era of geographic exploration and also offers Erina the opportu-nity to "foretell" the discovery of the New World (XXI, 62), as in Ariosto's *Orlando furioso* (XV, 21–26) or in Marinella's model, Tasso's *Gerusalemme liberata* (XV, 31–32).

As in Tasso, however, geographical boundaries are not pressed here, and the journey of Erina and Venier soon turns westward once again: toward Eu-rope, that "most beautiful and . . . merriest part that the sky looks down upon."[170] They linger in particular over Venice, which surpasses even Rome in the majesty of its architecture, the expanse of its empire, and the knowl-edge and prudence of its heroes. Erina's grasp of geography and history now serves to show Venier what he is fighting for and to return him to battle, de-spite the certainty that his death at the hands of the Byzantine warrior Emilia is preordained. When their chariot passes above the warring armies, Erina can hardly restrain Venier from jumping out to join his friends, his "soul ablaze in his martial countenance." Erina must follow her promise to Fileno, allowing Venier to return to battle now that she sees how fierce his wish is to return.[171]

Although Erina is possessed of broad knowledge about the natural world—the equal of her father in her command of astronomy and geography—she is none-theless still subject to Fileno's authority, and therefore not a completely autonomous character. Erina is the only female protagonist in a patriarchal intellectual lineage composed of her father, grandfather, and great-grandfather

before him. Fileno instructs her in natural philosophy and advises her in political matters: she values his counsel deeply. For Marinella, this is not necessarily a negative expression of female subordination within patriarchal culture,
but rather the reflection of the deepest bonds of filial respect and affection.[172]
Marinella chooses epic as a vehicle capable of conveying such themes of intellectual kinship and the transmission of knowledge, while also encompassing
the new scientific discourse and discoveries of the seventeenth century. If epic
celebrates the "memorable enterprises of a collective, of which the hero is the
symbolic actor," it is particularly suited to singing the praises of an empire—
Venice—or even an entire age.[173] Indeed, epic poems linked to colonial expansion and the discovery of the New World proliferated in the Seicento, reflecting imperial aspirations and anxieties.[174] Galileo himself was deeply
interested in and influenced by epic; Marinella's near-contemporary Sarrocchi,
who corresponded with Galileo, also penned an epic that served as a model
for *Enrico*, as we see in the next chapter.

The work of both Fonte and Marinella, as we have seen, can be broadly characterized as pro-woman even when the *querelle des femmes* is not explicitly
invoked. Both writers revitalized and reinvented archetypical female characters, such as women warriors and enchantresses, and tried their hands at "male"
genres in which women had only just begun to experiment, such as epic. Each
sought to introduce scientific discourse as a facet of their pro-woman works,
integrating it into broad narrative structure while also linking it more specifically to the display of female intellectual power. With Erimeno's grotto of
marvels and Erato's cosmological contemplation in Marinella's *Happy Arcadia* we see the traditional epic transported into the age of science. In *Enrico*,
Erina gives voice to the wonder of a new era of discovery and exploration,
even while remaining within the bounds of orthodoxy. Although Erina's role
is that of a guide, her knowledge of the natural world makes her indispensable to the resolution of the poem. Finally, in the *Nobility and Excellence of
Women*—so influential it caused Passi to reevaluate and retract many of his
statements—women's excellence in natural philosophy is affirmed through
historical examples.[175]

Likewise, in Fonte's *The Worth of Women*, penned at the height of the Renaissance debate over women, medical and scientific knowledge becomes a
weapon to be wielded in defense of women. This knowledge does not serve a

merely rhetorical purpose. Rather, it plays a fundamental role in women's quest to reclaim their rights and their position in society. When Leonora dismisses medicine as an area best reserved for male doctors, her rebuttal comes from Lucretia, who insists she is wrong. "On the contrary, it's good for us to learn about these things without needing help from men," Lucretia argues. "In fact, it would be a good thing if there were women who knew about medicine as well as men, so men couldn't boast about their superiority in this field and we didn't have to be dependent on them."[176] Women may not be able to change men's hearts. Indeed, Corinna admits at last that, "We've talked about all the stars, the air, birds, rivers, fish, and all kinds of animals, plants, and herbs. And we still haven't found anything with the power to work a change in men's minds."[177] But through their own education, sociopolitical awareness, and the medical knowledge that allows them to care for and counsel themselves and others, women can begin to gain their footing in early modern society. Fonte's seven female characters are fictions, meant to reflect the seven social states of women, but real women were deeply implicated in the new science at a variety of levels: as readers and writers of the kinds of medical and alchemical texts upon which Fonte and Marinella undoubtedly drew to compose their pro-woman works, and as everyday practitioners of the medical and scientific activity they described in such texts.[178]

4

⤙⤚

Scientific Circles in Italy and Abroad

Camilla Erculiani and Margherita Sarrocchi

Scientific discourse—especially with regard to natural philosophy—became an increasingly important element of the *querelle des femmes,* or debate over women, during the course of the later sixteenth and early seventeenth centuries in Italy. The presence of women in the area of early modern scientific discourse was not limited, however, to the printed pages of *querelle* discussions. By the 1580s, women were participating in scientific debate in deliberate, formal and public ways: publishing works on natural philosophy, joining scientific academies, and entering into correspondence with major scientific figures. If, at the turn of the sixteenth century, Caterina Sforza practiced science as a kind of private art, refining and collecting her alchemical experiments in a volume never meant for publication, two women active toward the end of the sixteenth century instead sought to establish themselves as learned contributors to public scientific discourse. The examples of Camilla Erculiani (d. post-1584), an apothecary from Padua, and Margherita Sarrocchi (1560–1617), a writer and *salonnière* in Rome, raise important questions about the nature of scientific communities in early modern Europe and the fluidity with which knowledge circulated across boundaries of both gender and geography. Erculiani had ties to an international scientific community in Padua and circulated her unorthodox views on the Biblical flood, astrology, and meteorology in Poland, where her only extant work was published in 1584. The erudite Sarrocchi, best known for her epic poem *La Scanderbeide* (published posthumously in 1623), forged relationships with members of scientific academies such as the Lincei in Rome, hosted an intellectual *ridotto* in her home, and corresponded with Galileo Galilei about his discovery of Jupiter's satellites and about her own literary work. Although many of Sarrocchi's scientific writings have been lost, her contemporaries unanimously point to her unparalleled excellence in the arena of natural philosophy.

While Sarrocchi, like many of her female contemporaries, encountered some hostility in reaction to her activities in the literary and scientific spheres, and Erculiani was subject to an Inquisition trial for her unorthodox views, there is evidence of increasing support for women's scientific education and activities in this period despite the repressive climate created by the Counter-Reformation.[1] The subject of women's education in the sciences is addressed not only in works specifically devoted to the *querelle des femmes* but also in more strictly scientific texts. Many sixteenth-century treatises on natural philosophy are dedicated to women, and some contain prefatory and paratextual material linking women to scientific knowledge and praising them for excellence in this area.[2] Alessandro Piccolomini (1508–1579), who promoted making Latin works available to a wide audience through translation, addressed the 1552 edition of his *Della sfera del mondo* (The sphere of the world) to Laudomia Forteguerra. In his dedicatory letter he decries women's lack of access to Latin and thus to scientific studies: "being that, in my opinion, the only reason your ladyship was not able to learn certain things is that you were not taught Latin, the result of our unfortunate customs these days, which, since scientific works are not written in Italian, prohibit women from learning the language in which they are written, and keep many women from studying the most important and rare works."[3] Similarly, in the dedicatory letter to his *Instrumento della filosofia* (Instrument of philosophy, 1551), Piccolomini insists that the book is meant to benefit female readers as well as male, and to "be of aid to many whom I know to be of excellent intellect, and suited to philosophy; but [who do not] know any language but their native Italian."[4] Girolamo Borro's dialogue on the movements of the tides, *Del flusso e riflusso del mare* (On the tides of the ocean, 1561) bears a dedication to Elisabetta della Rovere, marchesana of Massa; while commentaries on Aristotle's *Meteorology* by Nicolò Vito di Gozze (Nikola Vitov Gučetić, 1549–1610) and Michiele Monaldi (1550–1590), both Ragusan nobles and academicians, were likewise dedicated to women.[5] Vito di Gozze's *Discorsi sopra le metheore d'Aristotele* (Discourses on Aristotle's meteorology, 1584) is of particular interest in this respect, for not only is it dedicated to Fiore Zuzori, or Flora Zuzzeri (Cvijeta Zuzorić, 1552–1648), a Ragusan woman celebrated for her intellect in her native country and in Italy; but the dedicatory letter written by the author's wife, Maria Gondola (Mara Gundulić), incorporates common threads of *querelle des femmes* argumentation into a defense of women's capacities in the sciences.[6] A Neoplatonic justification of women's superior intelligence that also incorporates typical etymological argument (the significance of the word "donna") and humoral theory (women's humoral composition is more perfect than that of men), Gondola's letter praises Zuzori's beauty and learning and castigates the "fierce blows of the envious" *(fieri colpi de gl'invidiosi)* that led to her friend's departure from Ragusa. It should not surprise anyone, Gondola insists, that she should now seek a woman's protection for her husband's work, because—she argues—women are as capable in the sciences (not to mention in battle) as men:

> [I am] sending my husband's work to you, so that you may act as an unwavering shield against those who, out of innate malignity, are ready to gnaw and tear at the most beautiful and precious things, since you are most virtuous among the beautiful, and most beautiful among the virtuous; although many may marvel at my motive for publishing these *Discourses* under the protection of, or to be defended by, the female sex, perhaps believing that, since we are not by nature suited to arms, we must likewise lack the capacity for knowledge [*scienze*], or for the understanding of things.[7]

With a lengthy catalog, Gondola continues to single out women's accomplishments in the sciences: Arethea, daughter of Aristippus, "held forth publicly on natural philosophy"; Pythagoras learned about natural philosophy from his sister Theoclea, and his daughter garnered more attention in her domestic salon than Pythagoras did when speaking in the public academies."[8] Gondola's foreword—a rare protofeminist work by an early modern Croatian writer—elicited controversy and was removed from subsequent editions of her husband's work.[9] Both Fiore Zuzori and Maria Gondola appear as interlocutors in two other works by di Gozze, *Dialogo della bellezza* (Dialogue on beauty, 1581) and *Dialogo dell'Amore* (Dialogue on love, 1581). The first of these Platonic dialogues, dedicated to Fiore's sister, Nika, puts forth a similarly positive opinion regarding women's intellect, describing women as more innately capable than men in any discipline at all.[10]

The linking together of scientific subject matter with the traditional arguments and topoi of the *querelle des femmes* signaled a new dimension in the evolution of ideas about women's learning. Women's engagement with science had begun to acquire a public facet, not just in dedicatory letters and prefaces to scientific works composed by men, but in the scientific publications and activities of women themselves, as in the cases of Erculiani and Sarrocchi. Whether engaging in hands-on scientific practice themselves, or exchanging opinions with others on matters of natural philosophy, astronomy, or medicine through intellectual networks in Italy and abroad, women continued to add their voices to the early modern exploration of the world.

Camilla Erculiani's *Lettere di philosophia naturale* (1584)

Camilla Erculiani's *Letters on Natural Philosophy*, a treatise packaged in epistolary form, was published in Kraków, Poland, in 1584, the same year as Gondola's polemic in the *Discourses on Aristotle's Meteorology*. Erculiani's *Letters on Natural Philosophy* employs many of the same rhetorical strategies and even some of the same language in making science a new weapon to be wielded in the *querelle des femmes*. Given that Erculiani cites Aristotle's *Meteorology* in her *Letters*—and that di Gozze, like Erculiani, treats the universal flood in his *Discourses*—it is possible that Erculiani may even have known the Gondola/di Gozze text.[11] A slim collection of four letters touching on various problems of natural philosophy—the causes of the universal deluge, the composition of rainbows, the influence of the stars and planets on human temperament and action—Erculiani's *Letters* overtly entwine contemporary scientific debate with the early modern debate over women. In Erculiani's prefatory material, as well as the letters themselves, women's facility for scientific reasoning takes center stage as proof of their equality to men. Dedicated to a powerful female patron, Anna Jagiellon of Poland (1523–1596), and framed as an epistolary exchange with Giorgio Garnero (Georges Garnier, 1550–1614), a Burgundian medical writer, Erculiani's work, with its explicit advocacy of women's participation in scientific culture, also raises questions about the nature of scientific debate and scientific communities in early modern Europe.[12] Where did Erculiani, a self-described *speciala,* acquire her knowledge? How did she come to correspond with Garnier, and why did she dedicate a work of natural philosophy to the Polish queen?

Many questions remain about the identity of Camilla Erculiani and the circumstances in which she produced her book, which eventually drew the attention of the Inquisition for its blurring of boundaries between natural philosophy and theology.[13] The first clue is offered by the author herself on the title page of her *Letters on Natural Philosophy*, which describes her as a *speciala* [*speziala*], or female apothecary, at the Tre Stelle, an apothecary's shop in Padua. While the term "speziala" literally denotes a spicer, it often refers more broadly to spicer-apothecaries, indicating those who made and sold botanical and pharmaceutical remedies (which often included spices). Indeed, early modern *speziali* sold everything from candies and cosmetics to pigments, paper, and medicinals. Erculiani, for example, makes reference to a recipe for theriac, a prized antidote produced from vipers, which she was accustomed to compounding, as I will discuss shortly. As Eleonora Carinci has shown, archival documents attest to the presence of a Tre Stelle *spezieria* in the parish of Sant'Andrea in Padua, and to Camilla's first husband, Alovisio Stella, as its proprietor. After Alovisio's death, Camilla married another *speziale,* Giacomo Erculiani, who then took over the Tre Stelle. Camilla's father, Andrea Greghetti, moreover, was a merchant who dealt in spices, further strengthening her links to the apothecary's world. Camilla's two marriages resulted in six children. Although we do not know the date of Erculiani's death, it was in any case post-1584, after she was questioned by the Inquisition with regard to her book.[14]

The knowledge of natural philosophy Erculiani displays in her *Letters* suggests that she had some education, despite her comments to the contrary: in her letter to readers, for example, she states, "I know people might marvel greatly that, without having had access to [scientific texts], I set myself to publish these paltry, poorly composed lines."[15] Later in the text, she asserts that her knowledge is innate, not learned: "I know this instinctively, without looking at Galen or Aristotle."[16] Although she clearly did know and study both Aristotle and Galen, each cited on numerous occasions throughout the *Letters,* Erculiani's positioning of herself as schooled in experience, rather than having had formal instruction, echoes the empiricism characteristic of sixteenth-century scientific culture. Her experience as an apothecary underscores her claims to direct expertise, and also offers some explanation for the transnational relationships that made the publication of Erculiani's book possible.

A center of scientific and medical culture, sixteenth-century Padua harbored a vibrant community of students and scholars, scientists and philosophers. Home to one of the earliest botanical gardens, established by decree of the Venetian Republic in 1545, Padua attracted a wide and eminent array of visitors.

LETTERE
DI PHILOSOPHIA
NATVRALE, DI CAMILLA
HERCVLIANA, SPECIALA
alle tre stelle in Padoua,

*Indrizzate alla Serenißima Regina di Polonia:
nella quale si tratta la natural causa delli
Diluuij, & il natural temperimen-
to dell' huomo, & la natural
formatione dell' Arco
celeste.*

IN CRACOVIA
Nella stamperia di Lazaro, nel' Anno
1 5 8 4.

Title page, Camilla Erculiani, *Lettere di philosophia naturale* (Kraków: Lazari, 1584)
Courtesy of Biblioteca Universitaria Alessandrina, Rome

Andreas Vesalius taught anatomy at the University of Padua, challenging long-held beliefs regarding the workings of the human body; Cesare Cremonini and Galileo Galilei also held chairs at the university. It is entirely plausible that Erculiani, given her links to the Tre Stelle, would have had both cause and opportunity to interact with the many "doctors and philosophers" passing through the city; indeed the early modern pharmacy, as Filippo de Vivo has shown, functioned as a "shelter for sociability," a place for exchanging news, and—particularly in the Veneto—a hotbed for religious dissent.[17] Moreover, Padua's reputation in the scientific disciplines made it an attractive center for intellectuals from abroad, including Poland: most notably Copernicus, who studied medicine and probably astrology at Padua in the early 1500s.[18] The Polish presence in Padua helps contextualize the publication of Erculiani's work, not in Padua, but in Kraków, under the auspices of the Polish publishing house Officina Lazari; as well as its dedication to the Polish queen, and its internal references to the Polish court. That the Officina Lazari was overseen at the time of this work's publication by Jan Januszowski, who had studied at the University of Padua and published works of natural philosophy and alchemy in Poland, may also be significant.[19] Such details suggest that Erculiani, probably through her work at the Tre Stelle, came into contact with some of Padua's Polish visitors and, seeking the patronage of the queen, Anna Jagiellon, through them arranged for publication of her work abroad.[20] Such a scenario is further supported by the inclusion in the *Letters* of an encomiastic poem by the Silesian academician Andreas Schonaeus (1522–1615), who studied in Padua and would later publish an edition of Albertus Magnus's *Summa philosophiae naturalis* in Kraków. Schonaeus also had editorial ties to the Officina Lazari and likely to the royal court as well.[21] As Erculiani praises Anna by placing her in the company of an ancient tradition of famous women, so Andreas praises Erculiani, comparing her to Hippolyta and Semiramis, bringing her to a level comparable to the queen's own status in intellectual terms.[22] Despite its publication in Poland, the *Letters* did not escape the attention of religious authorities, as we will see shortly.

Anna Jagiellon, to whom Erculiani turned to protect her work, was an important female figure. Coruler of the Polish-Lithuanian Commonwealth, she presided, with her husband Stephen Báthory (1523–1596), over an enormous state (the largest in Europe after Muscovy and Turkey) at the apex of its power.[23] In addition to its strong intellectual links to Italy and, indeed, to Padua,

Poland was experiencing a burgeoning humanist renaissance of its own. The ideas and ideals of the Renaissance in Italy made their way east through expatriates such as Filippo Buonaccorsi (1437–1496, called "Callimachus"); while the University of Kraków, revived in 1400, offered a humanist curriculum of classical languages, history, theology, astronomy, and mathematics.[24] Humanist academies flourished in Polish towns, and a vibrant scientific culture fostered, in addition to the astronomical observations of Copernicus, the production of works of geography, ethnography, and political theory.[25] In Anna Jagiellon, Erculiani sought the support and protection of a powerful female patron who had particularly strong Italian ties: Anna was the daughter of Bona Sforza of Milan (1494–1557), who married Zygmunt I (Sigismund the Old) in 1518. Although Bona Sforza was the subject of hostility in Poland (where she was slandered as a poisoner and practitioner of love magic), she was admired in Italy for her patronage of the arts and her role in spreading Renaissance culture to Eastern Europe.[26] Indeed, Bona brought with her to Poland a sizeable retinue of Italians, including the poet Celio Calcagnini and the astronomer Luca Guarico, demonstrating her strong interest in promoting both literature and science.[27] Bona's stature in Italy is attested to by a number of literary works dedicated to her and, when she visited Venice in 1556, by Cassandra Fedele's public oration in her honor.[28] This lineage, perhaps, led Erculiani to address Bona's daughter—herself the dedicatee of works of poetry and saints' lives[29]—with admiration. In this, she was encouraged by the praise offered by Anna's Polish subjects in Padua, who, as reported by Erculiani in her dedicatory letter, "never tire of proclaiming [the Queen's] virtue, piety, and justice." They promise Erculiani that Anna's interest in science ensures a warm welcome for the *Letters:* "I have been assured by many of your subjects that [the *Letters*] will be well received, since you are known to be most virtuous and a lover of the sciences."[30] The women of the Polish aristocracy did engage with scientific culture, financing the production of scientific works and, in some cases, studying science and medicine themselves. Anna Jagiellon's niece, Anna Vasa of Sweden (1568–1625), for example, was called the "Polish botanist" for her expertise with medicinal herbs.[31] For her part, Queen Anna was known for promoting the education of girls at her court in Kraków.[32] As Erculiani writes in this dedicatory letter, this is all reason to situate Anna as a female example of virtue in every area, from learning and wisdom to justice and piety, and to place her hope in Anna's generosity as a patron.

As the dedicatory letter makes clear, Erculiani intends her work to demonstrate the full range of women's intellectual potential in all disciplines, including natural philosophy. "My goal in my studies," she writes, "is to show the world that we women are as capable as men in every area of knowledge [*scientie*]."[33] An unsigned poetic composition in praise of the author, which precedes the dedicatory letter, further underscores the centrality of gender as a motivating force in the composition of Erculiani's text. Addressed to "students of Philosophy," the composition highlights the sex of the author in the first line ("As a woman, with the greatest ingenuity and skill") and urges readers to spread her praise far and wide.[34] In positioning her work in this way, Erculiani steps onto the battleground of the *querelle des femmes,* that ongoing cultural skirmish over women's intellectual, moral, and civic qualities and duties. Overtly pro-woman works such as Moderata Fonte's *Il merito delle donne* (*The Worth of Women,* 1600), explicitly integrated women's knowledge of medicine and natural philosophy into an overt and wide-ranging defense of women. Erculiani could have correctly assumed her Polish audience to have a similar familiarity with the terms of the debate over women: Agrippa's influential pro-woman volley, *Declamatio de nobilitate & praecellentia foeminei sexus* (*On the Nobility and Excellence of Women,* 1529), was translated into Polish in 1575, and *querelle* texts on both sides of the question were published in Poland from the 1530s on.[35] Some, such as Andrzej Glaber's *Tales about the Harmony of Human Limbs,* criticized men for denying women access to education out of fear and jealousy.[36] Others, such as Andrzej Frycz Modrzweski's influential *De republica emendanda* (*On the Reform of the Republic,* 1551–1554), primarily a political treatise, strongly maintained women's subordination to men.[37] Lukasz Górnicki's *Polish Courtier,* an adaptation of Castiglione's *Il libro del Cortegiano,* or *Book of the Courtier* (1528), while excluding women from textual prominence as interlocutors, nonetheless held them to be the intellectual equals, if not superiors, of men.[38]

Erculiani makes use of many topoi of pro-woman *querelle* texts in her letter to Anna Jagiellon. Shifting her focus from Anna's qualities to her own intellectual merits and those of women in general, Erculiani structures her letter as a veritable catalog of famous women—a commonplace of *querelle des femmes* literature that firmly orients her scientific treatise within that debate. Offering a series of *exempla* for consideration, Erculiani cites the accomplishments of Mirthis, Lidia, and Nicostrata—all staples of the tradition: "Mirthis was called a giantess because of the magnitude of her accomplishments, and Lidia was included among the seven famous kings because of her wisdom and eloquence; Nicostrata, wife of Evandro, would not have been held in such esteem were it not for her exceptional erudition."[39] Erculiani further recalls the example of Cornelia, recognized in philosophy by Cicero himself, who "did not ignore her erudition when he wrote . . . that, were she not a woman, she would have been counted as outstanding among all the philosophers."[40] Erculiani laments that women's intellectual contributions are now ignored or excised from history, stating: "I don't know what malignant star causes [men] to refuse to recognize greatness, except in the things they accomplish themselves."[41] The dedicatory letter closes with a reference to Stephen Báthory, who died in 1586, two years after the *Letters'* publication, and with the curious clarification—given Erculiani's pro-woman discourse—that she might have dedicated the work to the king instead, were his attentions not taken up by matters of war (in fact, he was engaged in the war with Muscovy until 1582).[42]

A second prefatory letter, offered by Erculiani directly to her readers, establishes the *Letters* even more clearly within the context of the debate over women. In this letter, we again find topoi of the genre—for example, a feigned reluctance to publish, coupled with a plea to vindicate women's intellectual potential—as well as innovation, especially with regard to Erculiani's particular emphasis on science and her call to adjust educational practices so that women might have more access to this kind of learning. Her comments here recall those made by Alessandro Piccolomini, whom she had clearly read, on the need to make scientific learning accessible to women.[43] Echoing Piccolomini's conviction that women are as capable of scientific study as men if given the proper tools, Erculiani writes, "It will undoubtedly amaze some that I, a woman, have set myself to write and publish on a subject that does not belong to women (according to the custom of our age); but . . . you will find that women are not without the same abilities and virtues as men."[44] Likely attempting to deflect criticism, Erculiani claims she was persuaded to publish only because someone (she does not say who) threatened to publish her work elsewhere under another name.[45] Nor would she have gone to the trouble, she continues, were it not for her fervent desire to demonstrate, again, "the noble spirit of women today."[46] Keeping her identity—and her responsibilities—as a wife and mother at the forefront, she notes that she has written other works as well, including a (likely unorthodox) composition on the generation of the soul,[47] but fears the reception her writings will find for having come from the pen of a woman: "The work of caring for my children, the burden of running my household, my obedience to my husband, and my fragile health—none of these weighs on my decision to publish so much as the knowledge that many malicious minds will condemn my efforts, and writings, and consider them frivolous and worthless just as they consider women of our age to be such."[48]

The negative connotations cast on the duties of the wife and mother here are striking: these quintessentially feminine tasks are described as "work" *(trava-glio)* and a "burden" *(peso)*. Yet it is not the demands dictated by traditional gender roles that give the author pause in her quest to participate in the public, literary sphere, but rather the malicious and dismissive criticism of the unenlightened and suspicious reader. Nonetheless, Erculiani insists such hostile reactions will not keep her from publishing her work in an attempt to "awaken women" to their own intellectual potential: "Despite all this, I will not cease working to recover the honor [even] of those women who have forgotten it, and perhaps I will be the catalyst for the reawakening of their intellect."[49] Issuing a challenge to potential detractors of women, Erculiani scoffs that they would not dare draw their swords against the women of Padua; she concludes, finally, with an acknowledgment of those readers who will instead admire and enjoy the innovation and intelligence of her work: "And I am certain that if they knew this, no gentlemen would dare come to the brave city of Padua to accuse us women, with sword and lance, of imperfection; and I am further certain that many wise and intelligent readers will not mock the innovation of this work, and will admire my determination."[50]

Such confident references to her own intellect and to that of all women reappear throughout Erculiani's text, which, in addition to the prefatory material, consists of four missives: two letters addressed to Giorgio Garnero; a response to Erculiani's arguments by Garnero; and a fourth letter from Erculiani to Martin Berzeviczy (1538–1596), Báthory's chancellor for Transylvania, who had spent time in Padua in 1568. The *Letters* demonstrate Erculiani's background in natural philosophy—her familiarity with Aristotle and Galen, as well as her willingness to depart from these authorities to advance new theories—and reflect her interest in alchemy, astrology, and Paracelsian medicine. They also repeatedly threaten to veer into theological terrain, despite Erculiani's claims to separate natural philosophy from theology; indeed this distinction would form the crux of her defense in her Inquisition trial. The theme of women's equal capacity for learned and indeed innovative scientific disputation, finally, is evident throughout Erculiani's discussion as well as in the response of her interlocutor Garnero.

Given Erculiani's subject matter and her clear polemical agenda, it is worth pausing for a moment to consider the way in which the *speciala* from Padua structures her text. Although her argumentation is essentially monologic, it is packaged as an epistolary exchange with Garnero, immediately establishing the appearance of a real correspondence—whether or not one actually existed. The tension between these two rhetorical facets is underscored when Erculiani, referring to another work they have discussed and explaining that she has neglected to keep a copy of their correspondence for herself, asks Garnero to return her letter the next time he comes to Padua, "so that I may keep a record of what I have written," correcting it as necessary.[51] Within each individual letter, Erculiani further indicates that she is relaying the content of an ongoing scientific debate between her and a physician identified only as "Montagnana, medico eccelente."[52] Erculiani thus turns to the closely related forms of both dialogue and letter as rhetorical tools (indeed, letters were commonly conceptualized as "one half of a spoken dialogue").[53] As Virginia Cox observes, literary dialogue engaged readers in a unique and open-ended way, provoking them with "argumentation which deliberately eschewed the self-sufficiency of the treatise form."[54] The structure of dialogue allowed writers to strengthen their arguments by anticipating and responding to possible criticisms on the part of other interlocutors, by manipulating the reader's perception of the plausibility of particular viewpoints, and by involving the reader in the production of meaning. In the case of works that advance controversial opinions, dialogue also functioned as a shield or distancing mechanism for authors, who could claim to be merely reporting a conversation, not supporting a specific position. Dialogue was popular among early modern Italian writers for all these reasons, from Castiglione's *Book of the Courtier* to Galileo's *Dialogo sopra i due massimi sistemi del mondo* (*Dialogue Concerning the Two Chief World Systems*, 1632).[55] It was particularly apt for explorations of meteorological and other natural phenomena, given the inconclusive nature of the subject matter: works on earthquakes and flooding such as Lucio Maggio's *Del terremoto dialogo* (*Dialogue on Earthquakes*, 1571) exploit the genre's capacity for ambiguity.[56] Erculiani's explanation at the opening of her first letter to Garnero that she is merely describing a previous discussion with "a most excellent man" *(un eccel[entissimo] huomo),* serves similar purposes, inviting the reader to evaluate and affirm Erculiani's reasoning, while at the same time masking the problematic implications of her statements. Her interlocutor's claim that, were it not for original sin, people would never die, leads Erculiani to respond with the scientific rebuttal that inspires her *Letters* and that she now relays to Garnero. Because she could be perceived (as indeed, she was) as trespassing onto theological territory by arguing the biological inevitability of death, the displacement of the argument onto an innocuous, informal past exchange cushions and nuances the potential consequence of her words.[57]

Letter writing, which provides Erculiani's overt generic structure, shares many characteristics with dialogue. Like dialogue, it allows for the strategic presentation of an argument and consideration of a topic's multiple facets. Unlike dialogue, however, the interpretation of letters is necessarily incomplete, relying on the response of the letter's recipient, which can only be inferred. In the case of published letters, which enjoyed enormous popularity as a literary genre in the sixteenth century, the problem of interpretation takes on additional valence. A published letter addresses an audience of at least two potential readers: what Janet Altman terms "internal" (text-specific) and "external" (general) readers. Each of these readers will bring their own interpretation to bear on the letter, filling in the blank spaces it necessarily presents as "one half" of a conversation.[58] As a stylized literary genre, early modern letter collections were not necessarily—or even usually—reflective of actual correspondence, but instead often composed specifically for publication, or heavily revised before going to press. So great was the demand for epistolary collections that letters were often published without the consent of their authors, and letter writers were often entirely invented.[59] In approaching Erculiani's *Letters,* therefore, it is important to consider her motivations in choosing to adopt this particular generic form. Although Georges Garnier existed as a historical person with clear scientific interests, and Erculiani may indeed have met him in Padua and corresponded with him,[60] it is equally possible that the letter attributed to him in the text was authored by Erculiani herself, and that Garnero (Garnier) serves primarily as a foil against which to shape and polish her argument. Closer examination of Erculiani's letters shows a clear and deliberate division in the function of each interlocutor. When writing in her own voice, Erculiani consistently voices the empirical viewpoint, maintaining the central role of direct experience in the formulation of her arguments. Garnero's letter, by contrast, relies heavily on canonical authorities of natural philosophy and medicine—Aristotle, Plato, Zeno, Democritus, Galen, and others—to support Erculiani's argument, and, at some points, to nuance it. By dividing her argument between these two voices, Erculiani is able to highlight her natural intellect (her knowledge is innate, and born of experience, not study), while at the same time demonstrating that she does, in fact, have a solid grasp of the authoritative texts relating to her argument. Garnero's letter praises and authorizes Erculiani as a devoted and credible natural philosopher, "because [you] take knowledge and virtue to heart, considering all things other than nature [cose naturali] vain and worthless."[61] Perhaps most important, it also serves to deflect potential criticism of Erculiani's argument as theologically suspect, seeking to anticipate and defang such accusations by vouching for Erculiani's spiritual rectitude. As we will see shortly, Garnero characterizes Erculiani's philosophical arguments as "gifts from the Holy Spirit," while at the same time firmly reiterating that they are not theological in nature ("let us leave Theology aside as we return to [your] argument"). Garnero is clear that he engages in discussion with Erculiani—and, by extension, with the Montagnana doctor who ostensibly spurred this debate—as a philosopher, not a theologian: "As to what that excellent doctor said: that is, if man had not sinned, he would have had eternal life—I do not know how to interpret this comment, since it is completely beyond the bounds of philosophy. And since I am not a theologian, you must forgive me if I am uncertain and doubtful in this matter."[62] This clarification notwithstanding, Garnero offers an extensive Biblical defense of Erculiani's position, citing Solomon, Paul, John, and Luke on the soul's desire to be closer, in death, to God.[63] Garnero's epistolary presence— whether real or fictionalized—serves to affirm Erculiani's position as a natural philosopher and to present both a theological and philosophical defense of her claims, freeing her to focus only on argumentation in the letters under her own name.

The major portion of Erculiani's argument is reserved for the explanation of the causes of the great flood: "la natural causa del Diluvio."[64] Whereas from a theological perspective, the universal flood represents the punishment visited upon man as a result of original sin, Erculiani seeks to demonstrate that according to the laws of nature (as separate from those of theology), the deluge can be explained by behaviors known to exist for all substances and by a disturbance in the natural balance of the elements. Floods—both universal and particular—were the subject of vigorous debate in sixteenth-century Italy, particularly surrounding the publication in 1522 of Tommaso Rangoni's *De vera diluvii pronosticatione anni 1524* (Prognostication of the true flood of 1524), which predicted that a great conjunction in the house of Pisces would cause a universal or near universal flood on February 1, 1524.[65] By contrast, Agostino

Nifo used Aristotelian arguments to demonstrate that such a universal flood was physically impossible because of the natural stasis of the elements; therefore, the Biblical flood must have been caused by divine intervention and could not recur naturally. Such debates were driven by anxieties both local and general. On the one hand, the effects of deforestation were beginning to have real consequences in central and northern Italy: between 1492 and 1550 the area around Modena, for example, suffered numerous floods and washouts.[66] Vernacular works on the topic proliferated, many addressed to the general public and published in low-cost pamphlets and broadsheets. Religious tensions also played a role in such discussions, against the backdrop of the Protestant Reformation. A 1524 treatise by Johannes Ploniscus, for example, condemns Martin Luther for spreading panic about the flood in Germany.[67] Questions and answers about the causes of natural phenomena such as the universal flood may also be considered in relation to the tradition of the *problemata,* an Aristotelian genre rivaled only by books of secrets in the "bulk and range of their success."[68] As Ann Blair notes, such exercises do not question the existence of the "facts" they investigate—presuming phenomena such as the universal flood to be so well known as to need no explanation—but focus only on explaining their causes.[69] It is against this spectrum of natural philosophical debate that we may situate Erculiani's *Letters.*

The substance of Erculiani's argument is that all compound elements are subject to corruption, because they vie with one another for superiority. Only simple elements do not degrade. Given that man is composed of more than one element—as evidenced by variations in individual temperaments, by man's capacity for adaptation over time or in reaction to circumstance, and by his susceptibility to astrological influence—all men must necessarily die: it is a natural consequence of the matter from which they are formed. Drawing on Aristotle and Galen, but formulating her own explanation of the interdependent relationship between elements, humors, and the influence of the heavens, Erculiani further argues that all creatures are informed by a primary, dominating element that determines their place and function in the natural world: in people, the dominant element is earth, in birds, it is air, in fish, it is water, and so on. Upon the death of any living creature, the elements of which it is composed are returned to the earth, which thus replenishes itself constantly. At the time of the flood, Erculiani hypothesizes, men were living too long, and as a result the earth itself had become starved, losing actual physical mass as a result of an imbalance in the earth element that was no longer being regularly replenished. This allowed the water element to take over in the flood: "[the deluge] occurred because men's presence on earth had increased so much in number, size, and lifespan, that around the time of the sin, the earth element, which dominates in men, was much diminished; nor had it been replenished for centuries, with the result that [the earth] was so greatly diminished that it was necessarily swallowed up by water, which had contributed little of itself to [the composition of] men."[70] The flood, she concludes, was thus a consequence of an imbalance in the elements, with water eventually overpowering and overtaking earth. Erculiani diverges from the explanations offered by other diluvian works in suggesting that such a catastrophic imbalance of elements could occur as a natural phenomenon.[71] Now linking the elements to the humors, Erculiani explains that the flood waters necessarily obliterated everything in their wake: "water, which putrefies, annihilates, and consumes all those things that are made with it or placed in it: and it does this because of its cold and moist nature, since its coldness removes the vegetative virtue, and its moisture putrefies and consumes, reducing almost everything it encounters to its own nature; and this happens quickly, and more quickly with some materials than with others."[72]

Erculiani's arguments are marked not only by the learning she displays—she is well versed in humoral theory and willing to advance her own theories and hypotheses—but also by her experience in pharmacy. Her discussion of simple and compound elements, as well as her observations regarding the effects of water upon the substances with which it comes in contact, reflect the familiarity with medical-alchemical processes necessary to an apothecary, from distillation to the compounding of remedies. She herself points out, in response to correction by Garnero, that her knowledge is based on her own observations of the natural world, without reference to any scientific authority.[73] Her comments on the composition of human beings and the elemental reciprocity and balance between man and nature also have echoes of Paracelsianism. Her observation to Garnero that the body is a "little world" *(mondo picciolo)* traces a direct link to Paracelsus's conception of the intimate and eternal connection between microcosm and macrocosm, the body and the universe: the body is not "made simply of earth . . . but all the other elements are also present; as a result, given their opposite qualities, one must always eventually overcome the other, and this body, which is called a little world, must decay; and so each of the elements must return to its proper place. And if this did not occur, this machine we call the world would be annihilated."[74] Like Paracelsus, Erculiani believes that man and nature—microcosm and macrocosm—are deeply entwined and must exist in a state of harmony and balance. The four major pillars of Paracelsus's theory of medicine—philosophy (knowledge of nature, specifically earth and water), astronomy (knowledge of the cosmos, specifically air and fire), alchemy (knowledge of all four elements), and the virtue of the physician (indispensable for fulfilling the other three pillars) all assume positions of importance in Erculiani's own explanation for the causes of the flood and her observations regarding the balance of the natural world.[75]

Paracelsus was known for his melding of alchemical and medical practices, particularly in the compounding of medicines, something with which Erculiani, as a self-described apothecary, would have been experienced.[76] His views on toxicology and what we would today term homeopathy—using like to cure like—also led him to seek out balsams and antidotes, as in the case of the substance known as theriac. Theriac was produced throughout Italy and Europe through complex recipes and still more complicated rituals, which were increasingly regulated by states in order to prevent fraud. In the Veneto, its preparation included some sixty-four different elements, the principal component deriving from vipers collected from the hills surrounding Erculiani's native city of Padua.[77] Erculiani had direct experience with the preparation of this valuable compound, as evidenced by her comments to Garnero at the close of her first letter. She is sending him a vial of theriac, she writes, prepared with her husband that very year. Not only does Erculiani produce theriac along with her husband, but, as she also confides to Garnero, she plans to compose a treatise on the precious antidote, in which she will describe the "nature, properties, and qualities of the ingredients used [in it], and how they act favorably against poison."[78]

Amid Erculiani's discussion of the flood and its causes, along with later discussions of the origins of rainbows (where she again diverges from Aristotle's *Meteorology*), and her hints about future works she is planning, the points made in the work's prefatory materials about women and the sciences remain central in the body of the text, articulated most explicitly in Garnero's response to Erculiani. Garnero admires Erculiani's erudition and wishes more women—and men—would devote themselves to natural philosophy. If the Stoics encouraged women to study it, he laments, why has the practice fallen so out of favor?

Not only did these Stoics want . . . men to devote themselves to philosophy and the contemplation of the natural world, but they also wanted all women to do the same: in truth, an endeavor highly praised by all our ancestors, and one that is now—I don't know for what most frivolous reason—despised, and practically exiled from the world, scorned not just by women, but by men themselves, as if studying the natural world were something very base.[79]

Erculiani, by contrast, represents a positive example of those who devote themselves to the study of the natural world at the expense of wealth, recognition, or other worldly reward. Addressing Erculiani in filial fashion (she is "as dear and honored as a mother"), Garnero praises his interlocutor and all natural philosophers in a single breath:

Even more worthy of praise are those few men, and very few women, who with great ingenuity dedicate themselves completely to learning and the sciences, laboring . . . with all their power, and scorning every other public and private pursuit just to undertake the study of the truth of things, believing that it is far more worthy to investigate and understand the reasons behind natural and divine things than to acquire wealth, honor, and other impermanent and vain things. Among whose number you must be counted, [since] you care so deeply for the sciences and for virtue, deeming all other things other than the natural world vain and worthless.[80]

The defense of women and its coupling to scientific learning is made more personal still in Erculiani's letter to Martin Berzeviczy, Báthory's chancellor for Transylvania, dated April 9, 1581.[81] Although no letter from Berzeviczy is included in the *Letters*, Erculiani's missive makes clear that she responds to his suggestion on a prior occasion that her opinions on the flood are drawn from other, more authoritative minds. She finds the barb intolerable, retorting: "I will respond to your question, and tell you plainly: I did not read about this in the work of any other author, nor do I consider it praiseworthy to write the opinions of others.[82] Quick now to clarify that she is nonetheless well-read in the authorities, Erculiani emphasizes the value she places on independent thought: "I don't deny that I read various authors, considering their explanations . . . whence, marveling at their ingenuity and the range of their opinions, I determined that I, too, should write down my own."[83] In the remainder of the letter Erculiani reiterates her theory regarding the origins of the flood, adding an astrological digression attributing the eventual recession of the flood waters to the pull of Venus. (Significantly, elsewhere in the *Let-*

ters, Erculiani corrects the Ptolemaic placement of Venus in the cosmos, favoring an essentially Copernican position by placing it third, after Mercury.[84]) Regretting that illness prevents her from offering this response to Berzeviczy in person, Erculiani concludes her letter with a bold statement: her theory is the only possible explanation for the flood; no philosopher or astrologer could disagree. Although her explanation differs from that offered by theologians, Erculiani trusts in the workings of both God and nature: "it is true that the doctors of the church and the divine theologians have put forth different causes and reasons; but for me, it is enough to say that God and Nature herself do not contradict those causes, but He makes use of her in His works."[85]

Erculiani's efforts to argue science without discounting religion are not unique: the idea that both science and faith could be a path to the "truth," although regarded as heresy, was nonetheless in active circulation in sixteenth-century Italy—the example of Galileo, of course, springs immediately to mind—as was the suggestion that one's inner beliefs might differ from those professed outwardly.[86] Erculiani's theories caught the attention of the Church, however, and she was interrogated by the Inquisition on suspicion of heresy. Although the trial documents have been lost, a contemporary account of her questioning is given in the jurist Giacomo Menochio's *Consiliorum sive Responsorum,* a collection of legal opinions regarding cases and sentences.[87]

The principal question at stake in Erculiani's trial was, indeed, the relationship of science and faith—or rather, the separation of the two. Was Erculiani's argument purely philosophical, or were her opinions theological in nature? Menochio (1532–1607) maintained that she clearly separated the two, presenting her views only as scientific theory and acknowledging that theology offered the true explanation.[88] The substance of the interrogation seems to have been clearly anticipated by Erculiani and her interlocutors at the time of their correspondence. Not only does Erculiani seek to address this very question in her words to Berzeviczy, but Garnero does the same in his letter to Erculiani. Taking pains to underscore his own status as a good Christian, as well as Erculiani's, Garnero forcefully reiterates that theirs is a scientific debate, but also that Erculiani is completely orthodox in her beliefs. After first declaring his admiration for Erculiani's Christian spirit as well her scientific writing, Garnero declares his own spiritual rectitude and clarifies the nature of their discussion: "[you] deign to share with me the most lofty and profound treasures of your intellect and the generous gifts of the Holy Spirit you possess; I speak of the Holy Spirit as the professed Christian that I am, because all good and virtue comes from God, who contains all that is good, since He is the highest good, virtue, life, and truth; but leaving theology aside for now, I return to your arguments."[89] As indicated in the *consilium,* Garnero's attempt to place theology aside *(lasciando per hora da canto la Theologia)* lay at the heart of Erculiani's Inquisition interrogation. Menochio admits that Erculiani's statements about the flood and the natural imperative of death would be heretical were they meant to be interpreted through a theological lens, but maintains that this intent cannot be proved. He notes that Erculiani, in her own response to the charges, insisted that natural philosophy allows for such debate, while at the same time acknowledging the accepted theological position that both the flood and death are punishment for sin.[90]

Menochio's defense of Erculiani also demonstrates the degree to which gender, as scholars have long noted, played a major role in such trials.[91] In Erculiani's case, the old arguments of the *querelle des femmes* that questioned women's intellectual mettle—the very arguments she had set out to refute in her *Letters*—are used in her defense by Menochio. Repeatedly, Menochio makes the case that the views expressed by Erculiani should be discounted—or at the very least considered of lesser gravity—since they were expressed by a woman. A man's heretical speech has more potential to infect and influence, according to this reasoning, than that of a woman, in whom it is most often merely the result of ignorance, and may be corrected.[92] Even in the unwelcome legal arena of the Inquisition, the issue of women's intellectual capacities that Erculiani addressed in her scientific writings was still at the forefront. Menochio's defense of Erculiani furnishes an uncomfortable ending to a chapter in which Erculiani had sought so forcefully to place women's intellect and reason on an equal plane with that of men.

Erculiani's trial proceedings raise interesting questions about the function of apothecarial space in the early modern period. Not only was the pharmacy a locus for social interaction and the exchange of news, it was considered by the Inquisition to be a fertile breeding ground for heterodoxy, and as a result many apothecary shops were kept under official surveillance.[93] Given Erculiani's self-identification as an apothecary, together with the network of contacts she was able to build among philosophers and scientists in Padua and abroad, we might speculate that what heterodox tendencies she had were nurtured and developed within the walls of the Tre Stelle. Against such a backdrop, Erculiani's decision to seek publication in Kraków, rather than Padua, falls into sharper focus. Under Zygmunt I, Poland enjoyed a reputation for religious tolerance in sharp contrast to the increasingly repressive climate of the Counter-Reformation in Italy or the bloodshed in France during the Wars of Religion (1562–1598). Indeed, the Henrician articles enacted by the Union of Warsaw in 1573, and accepted by all Polish kings, including Anna Jagiellon and Stephen Báthory, included a pledge of religious tolerance. Although Anna and Stephen were Catholics, they accepted Protestantism, particularly among the nobility, as the cost of maintaining domestic stability.[94] Tellingly, Erculiani's dedicatory letter to Anna, along with its flattering inclusion of Stephen's merits as well, concludes with the observation that what the author seeks most urgently from her patron is protection from eventual detractors *(malivoli)*: "I know Your Majesty will be a most able defender of this work against malevolent critics."[95] If she refers, on one level, to those who would object to a woman publishing a work of natural philosophy, Erculiani's decision to choose this particular patron suggests that her motivations—and her concerns—went deeper: by publishing in Poland, Erculiani hoped both to extend her ties to a country in the midst of its own intellectual Renaissance, and to benefit from its more permissive political and religious policies.

Although Erculiani's efforts to protect herself from Inquisition scrutiny failed, her *Letters on Natural Philosophy* offered an opportunity to forge intellectual relationships across boundaries of gender and geography and enter into scientific debate with philosophers in Italy and abroad. Using the connections she had forged through her work in the Tre Stelle, and capitalizing on the patronage relationship that already existed between Anna Jagiellon and her subjects in Padua, Erculiani's efforts brought her recognition, if not the sort she may have intended. Her example, however, opens new avenues for thinking about the nature of scientific communities in early modern Europe, the role of women within those communities, and the transnational characteristics of such circles of knowledge.

Margherita Sarrocchi, the *Scanderbeide,* and Galileo

Orthodox in her views on matters both literary and religious, Margherita Sarrocchi (1560–1617) had no occasion to incur the suspicion of the Church as Erculiani did (indeed, her support of Galileo, whom she met in Rome, noticeably diminished once the scientist's theories came under increased scrutiny by the Roman Inquisition[96]). Sarrocchi, however, had a much higher profile and greater public presence than Erculiani, moving in both scientific and literary circles. As a woman renowned for her learning in literature and the sciences, she confronted other kinds of obstacles, including the hostility of some of her contemporaries (Giambattista Marino, once an admirer, notoriously dismissed Sarrocchi as a "magpie" in his poem *Adone* [1623], and others routinely criticized her ambition and drive as unseemly in a woman, or accused her of unchastity).[97] On occasion, Sarrocchi, like Erculiani, likened her role as a woman in the world of learning to a battle of sorts with male detractors. Yet—also like Erculiani—Sarrocchi had many supporters, and she participated in scientific culture in public, visible ways that shed important light on the power wielded by early modern women in such circles. Indeed, Sarrocchi's scientific as well as literary interests led her to engage with some of the most important intellectual figures of her day. Although Sarrocchi's extant works do not focus explicitly on natural philosophy, as Erculiani's *Letters* do, it is likely she wrote works of a scientific nature that are lost to us today, including a treatise on Euclidean theorems.[98] Even her most famous work, the epic poem *Scanderbeide,* harbors glimpses of Sarrocchi's scientific inclinations, particularly in its portrayal, in an early version, of the enchantress Calidora, depicted as a woman learned in natural philosophy (in a reflection of the Counter-Reformation climate in which the work was composed, a prefatory note to the reader clarifies that this "maga turca" is intended only as an entertaining imitation of other poets, and has no other meaning).[99] The dedicatory letter to the 1606 edition of the *Scanderbeide* (addressed to Costanza Colonna Sforza, marchioness of Caravaggio, and attributed to an unidentified academician), clearly states that knowledge of the sciences is critical to crafting an epic poem: "To compose an epic, one must be learned in every science and every art: this has been demonstrated by the authors of good poems, men of profound erudition. Such erudition is not lacking in Signora Sarrocchi, with respect to every science."[100] The letter further states that the remaining cantos of the poem, not included in the present edition, contain additional scientific material dealing with cosmology, astrology, and natural philosophy in general: "as for science, [they] deal with the heavens, astrological knowledge, [and] a study of very curious natural things, all appropriately and poetically explained."[101] (In fact, the narrative thread concerning Calidora, in which such themes figure most prominently, was excised from the 1623 version as a result of Sarrocchi's efforts to streamline the action of the poem around the bellic narrative). Some of Sarrocchi's occasional poetry ventures into the arena of natural philosophy, using astrology as a structuring theme or as a source of metaphor; and her presence as the dedicatee of scientific works penned by others (or as a contributor of paratextual material to such works), along with a constant refrain of admiration for her learning among her contemporaries, make Sarrocchi an important figure for thinking about the influence of women in scientific culture by the early seventeenth century.[102]

Originally from Naples, Sarrocchi was brought to Rome after the death of her father by her guardian, Cardinal Guglielemo Sirleto (1514–1585). Having placed his ward in the convent of Santa Cecilia in Trastevere, Sirleto arranged for her to receive rigorous instruction in the liberal arts and sciences.[103] Sarrocchi published her first poem by the age of fifteen, and initiated a correspondence with Tasso around 1583.[104] By 1588 she was married, and had taken the name Biraga; but it was her friendship with the mathematician Luca Valerio (1553–1618), who had been her tutor, that would remain her most sustaining relationship and intellectual bond. According to one seventeenth-century biographer, Valerio was Sarrocchi's "constant companion" (contubernalis perpetuus), and he shared her residence for a time; epistolary evidence shows that they regularly read and discussed the latest scientific discoveries together.[105]

Sarrocchi's contemporaries praised her erudition, singling out her excellence in the sciences in particular. She often appears as an *exemplum* in the continuing tradition of *querelle des femmes* literature, her reputation in the sciences now sounding a new note in the litany of women's accomplishments. One such catalog of famous and virtuous women, Cristofano Bronzini's *Della dignità e nobiltà delle donne* (On the dignity and nobility of women, 1622), devotes a substantial section to Sarrocchi, describing her as: "A real *virtuosa*, and famous for her learning in all the most difficult sciences and the major languages . . . accomplished not only in these languages and in poetry . . . but also in philosophy, theology, geometry, logic, astrology, and exceedingly well versed in many other most noble sciences and in the belles lettres."[106] In typical *querelle* fashion, Bronzini positions Sarrocchi within a tradition of learned women stretching back to Antiquity, noting that she merits mention "along with the Corinnas, the Sapphos, the Hypatias, and the worthiest women who ever lived" (in this, he echoes the praise of Sarrocchi offered by the dedicatory letter to the 1606 *Scanderbeide*).[107] He marvels that this "singular" woman is so exceptional that she was even offered a chair in Geometry and Logic at the venerable University of Bologna.[108] Another early biographer of Sarrocchi, Bartolomeo Chioccarelli, likewise singles out her erudition in both literature and philosophy, noting: "Margherita Sarrocchi was a brilliant poetess, famous in our day throughout Italy. This woman, admired and celebrated in our age, was not only learned in the study of poetry but also renowned for her distinction in philosophy and in nearly all the sciences and disciplines."[109]

Praise of Sarrocchi abounds in sonnets dedicated to her by her contemporaries and in letters addressed to her by other writers, in which her poetic gifts share center stage with her accomplishments in the scientific disciplines.[110] Sarrocchi's admirers generally regard her as both an exception and a marvel, in a conflation of baroque imagery and persistent stereotypes about the learned woman. Giulio Cesare Capaccio calls her a "marvel of the female sex"; Maurizio Cataneo, "a unique woman / and rare marvel / A marvel, certainly . . . / but a miraculous one sent from the Heavens."[111] Even her death certificate, recorded on October 29, 1617, emphasizes her exceptionality in both literature and science: "The renowned Margarita Sarochia, of Naples: through her intellect, wisdom, and through every literary genre raised above the rest of her sex; learned in the natural sciences, in theology, in mathematics."[112]

Sarrocchi was also well known for the *ridotto* she hosted in her home, where she forged ties to literary figures such as Tasso and Marino as well as to others in the scientific community, including Federico Cesi, founder of the Lincei, and to the academy's most famous member, Galileo. As Bronzini writes, her home regularly attracted "the most noble and virtuous spirits ever to live in or pass through Rome."[113] Galileo himself recalled the connections he made there, perhaps hoping they would help to lay the groundwork for the favorable reception of his theories: in a 1612 letter to Sarrocchi, he asks to be remembered to "S[ignor] Luca [Valerio] and the other gentlemen of letters [signori i literati] I met in your home."[114] While there are other examples of learned women hosting similar gatherings and functioning as cultural mediators and promoters (we might think of Vittoria Colonna in Naples, for example, or Lucrezia Gonzaga in Fratta Polesine[115]), Sarrocchi's salon is of particular interest for its scientific character, for the role it played as a locus of debate for new scientific theory, and for the influence her gatherings had beyond her palazzo. Finally, as we will see, not only did Galileo send copies of several works to Sarrocchi (as she sent her *Scanderbeide* to him): her reputation was such that correspondents from as far away as Perugia sought her opinion on scientific matters, including Galileo's discovery of Jupiter's satellites and his new design for the telescope.[116] Complimenting Sarrocchi on her erudition in the sciences, one such correspondent, Guido Bettoli, notes that her salon is a magnet for gifted intellectuals: "the meeting place and academy for the best minds in Rome."[117] The use of the term "academy" here is sugges-tive, indicative of regular gatherings that had achieved a certain intellectual stature and public reputation.

Sarrocchi also took part in formalized academic life beyond her own *ridotto*. One of a handful of women to participate as a regular member of an Italian academy, Sarrocchi earned the admiration of her peers through the lec-tures and orations she delivered at such gatherings (although some commented peevishly on what they perceived as her excessive need for public recognition).[118] Bronzini praises the quality of her contributions, remarking on her ability to respond spontaneously to questioning from her distinguished audience, which, on occasion, included Galileo:

> This exceptional woman was heard to hold forth, with great ability, on na-ture and the movements of the heavens, and other truly gainful and celes-tial things; [it was as if] with her words she nearly paralyzed the intellect of all who listened; so that anyone would, with great admiration, begin to pay her close attention and then to interrogate her—including Galileo of Tus-cany, to whose questioning the learned lady responded not just readily, and prudently, with erudite and well-reasoned arguments, but she raised such profound and important questions on his part that it kept him busy for quite a while.[119]

Underlined in Bronzini's account is not just Sarrocchi's erudition, but the agility and originality of her mind. It is tempting to speculate on the nature of Sarrocchi's discussions regarding the "movements of the heavens," which we can assume were Aristotelian in nature and must have given rise to interesting debate with Galileo, who had not yet been ordered to abandon the teaching of Copernicanism.

Similarly, Rossi draws on personal recollections of Sarrocchi's public per-formances and the applause with which they were routinely met, while Fran-cesco Agostino della Chiesa mentions Sarrocchi's vibrant presence in "the public Academies of Rome."[120] Sarrocchi was originally associated with the Accademia degli Umoristi in Rome, but after a few years her preference for a Tassian literary ethos over the more pronounced baroque taste that dominated the Umoristi—and perhaps her rift with Marino—led her to form the Accademia degli Ordinati, along with Cardinal Giovan Battista Deti and Giulio Strozzi.[121] The Jesuit Giovanni Domenico Ottonelli singled Sarrocchi out as the co-organizer of this academy, noting: "there lived in Rome that honored lady of the name of Sirocchia who held a literary academy in the palace of a leading lord, and many gentlemen used to relish participating in it, in order to hear her learned conversation and erudite discourse."[122] Sarrocchi's close ties to Valerio and Galileo, moreover, attest to her connections with the influential and exclusive Accademia dei Lincei, a scientific academy that also focused on literature and philology.[123] As we will see further on, some even blamed Sarrocchi's influence for Valerio's later decision to withdraw from the Lincei in protest of its support of Galileo's Copernicanism. It was Sarrocchi's major literary work, the *Scanderbeide,* however, that brought her recognition beyond Rome and formed a central component of her correspondence with Galileo, in the years just before his troubles with the Church would make a continued friendship untenable for the orthodox Sarrocchi.

The *Scanderbeide*

One of few epic poems authored by an early modern woman writer, the *Scanderbeide* recounts the story of the Albanian prince, George Kastrioti, or Scanderbeg (1442–1468), and his struggles against the Ottoman Turks.[124] Taken from his family by Murad II, Scanderbeg first converted to and later abjured Islam, leading an Albanian resistance supported by various Italian states and by the Pope. Sarrocchi's decision to set her poem in the Balkan context reflects historical anxieties regarding the ongoing threat of Turkish expansion: the war in Hungary, which pitted Christians against Turks, raged from 1593 to 1606, the period in which the *Scanderbeide* was composed. The tapestry of political partnerships and alliances among Albanians and Italians presented in the *Scanderbeide* thus mirrors a real conflict that would require the intervention and cooperation of Italy and its European allies to resolve.[125] Sarrocchi's poem adheres closely in many respects to the model of Tasso's *Gerusalemme liberata* (*Jerusalem Delivered,* 1581) in its organization around a single theme, the centrality of the hero's conversion, and the triumph of a divine Christian plan. That Sarrocchi admired Tasso is evident not just from her imitation of him in her epic, but from their earlier, mutually complimentary, correspondence and exchange of sonnets dating to 1583–1585.[126] Despite Sarrocchi's undisputed erudition and curiosity about all aspects of natural philosophy, the geographical contours of her narrative remain closely drawn, confined to the familiar, well-explored territory between Italy and the world to the immediate east. Absent from her poem, for example (in contrast to other examples of Seicento epic), is any mention of the New World.[127]

In its final form, the 1623 edition of the *Scanderbeide* consisted of twenty-three cantos; an earlier, incomplete edition, published in 1606, had only eleven. While scientific discourse is not the poem's primary narrative language, Sarrocchi's interest in natural philosophy and even medicine is evident at several points, including a key episode relating to the hero's conversion to Christianity, and the presentation of a central female figure in the 1606 edition, the *maga* Calidora. Critics have pointed to the influence of Sarrocchi's poem on at least one later epic by a woman writer, Lucrezia Marinella's *L'Enrico* (1635), particularly in Marinella's appropriation of the Calidora character and her learning in natural philosophy.[128] Calidora herself invites comparison to an earlier *maga* figure, Moderata Fonte's Circetta in *Floridoro* (1581). Taken together, these three figures—Circetta, Calidora, and Erina—create a triptych of female characters who function as new interpreters of science in the literary realm.[129]

Sarrocchi's interest in medicine is most apparent in cantos I and II, in which Alexander (based on the historical Scanderbeg) is poisoned by a corrupt servant. Sarrocchi's description in these stanzas of the course and symptoms of the poisoning and the efforts at treatment reflects some knowledge of medical culture and also suggests a familiarity with both the canonical and popular medical literature that proliferated in sixteenth-century Italy and often contained recipes for poisons and their antidotes. Poison was a real threat in Renaissance Italy: rumors abounded regarding the untimely deaths of politically inconvenient noblemen and women.[130] While its narrative purpose here is to instigate Alexander's conversion, Sarrocchi's rendering of the poisoning episode highlights the role of learned experts and printed texts in confronting medical crises. At first, the king does not notice the tasteless, odorless substance sprinkled over his dinner: "The hidden obnoxiousness did not manifest its deadly power right away, so Alexander ate, for the immediate pleasure of the food was not hampered and the taste gave him no warning."[131] Only later does he begin to feel its effects: an "aching tug at the heart," a "stabbing pain, the likes of which he had never experienced before."[132] Pallid and immobile with pain, he later tosses restlessly in bed as a "freezing cold spread all over his body."[133] Alexander's symptoms, as the poison slowly spreads through his veins, are consistent with arsenic poisoning.[134] Rushing to his aid, the king's loyal friends summon "several great experts in the medical arts" to find a remedy for their king's distress. These experts study the king's symptoms, using observation and reason to determine that he is indeed a victim of poisoning; they settle on liquid remedies as the best way to alleviate his distress (in fact, victims of arsenic poisoning commonly experience intense thirst).[135] Significantly, Sarrocchi specifies that the physicians try every prescription suggested by "either the old or the new books:" that is, the canonical medical texts as well as the vernacular, popular tradition of *libri di segreti,* or books of secrets, which contain numerous recipes for such liquid antidotes.[136] That Sarrocchi herself was familiar with such works is reiterated by her poetic contribution to Giulio Iasolino's *De remedii naturali* (On natural remedies, 1588), a medical manual centered around the healing properties and "secrets" of various baths around Ischia.[137] Her inclusion in such a work, which positions her alongside others who lend authority to the text, underscores Sarrocchi's reputation for medical-scientific knowledge. Although the king's friends are unable to cure Alexander, unaware that his food continues to be poisoned, their interventions lessen the poison's potency: "The hero was losing strength but was not in danger of his life, for as much as one potion was harmful, the other proved beneficial."[138] The physicians' failure to cure the patient, in fact, has less to do with skill than with Sarrocchi's desire to use this episode to mark Alexander's conversion. It sets the stage for a key speech by the wise Christian, Mohamet, about God's love and Christian charity, which sets in motion a true miracle: Alexander feels infused with a strength "far greater than he had ever enjoyed before, just as the rays of the sun shine more splendidly after a cloud has by unfortunate chance obscured it. Alexander soon rose feeling quite well, and . . . gave devout thanks to God."[139] While the episode acknowledges the importance of medical knowledge and pays close attention to the clinical observation of the body's reactions, this is subsumed—in keeping with Sarrocchi's poetic program—by a providential narrative of divine plan.

Present in the 1606 edition of the poem but not the 1623 version is a second major narrative thread involving Calidora, the wife of a Turkish guard, seduced and abandoned by the faithless Serrano.[140] In this episode too, learning—this time in the area of natural philosophy—plays a central role. When she first appears in canto III, Calidora is described (much like Sarrocchi herself) as "[a woman] learned in every liberal art, / and adorned with masculine valor."[141] The nature of her erudition, it is soon evident, revolves around an Aristotelian natural philosophy: "With acute intelligence she penetrates, from the principle causes, the secondary, which produce different effects in us all, whether sterile or fecund, depending on the variations in the heavens; they cause good fortune, or bad. And thus the light and movement of the stars and the workings of nature are all known to her."[142] Sarrocchi immediately establishes Calidora's particular interest in studying the heavens, which reflects the astrological conviction, common to early modern scientific thought, that the planets influence human actions and events. Calidora reserves special attention in this respect for Jupiter, whose movements can also foretell future events: "You, highest God of lofty and thunderous kingdoms, called Jove [*Giove*] because you help [*giovare*] others; your virtue imbues the lovely winged birds [with] the lofty secrets of your sacred breast. Reveal now, with voice and sign, the varied, innumerable events of humankind."[143] Wearing a diadem adorned with a golden sun and a silver moon, Calidora evokes not just astrological influence, but also the imagery of the alchemical wedding of male and female elements. Alchemical transmutation is commonly depicted in early modern illustrations as a union of *sol* and *luna:* it is likely that Sarrocchi would have been familiar with these images, which formed an integral part of early modern scientific exploration.[144] Learned in natural philosophy but also a woman in love, Calidora herself embodies such a union of attributes typically gendered male and female.

While Calidora's first appearance underscores her knowledge of astronomy and astrology, we soon learn that she is a *maga* who invokes the planets and the stars as well as the elements of the earth, in preparing her love magic. Cast off by Serrano, Calidora retreats to the forest to prepare a spell to win him back. Here, she demonstrates knowledge of a different kind: her magic is a pastiche of astrology, herbal medicine, and popular love recipes. After fashioning a wax figure in her lover's likeness, Calidora sets the spell in motion: "at the appointed hour, in a lonely spot, she sprinkles amber, myrrh, and other choice scents over a fire of ebony that burns with no smoke at all. . . . She lets down her golden locks in the wind and, removing the hearts from three live doves with a single knife, she writes Serrano's name upon the page in their blood."[145] The experiment is conducted at an astrologically propitious time ("at the appointed hour") and summons the power of traditional elements of recipes for love magic.

Sarrocchi's emphasis on natural philosophy and the influence of the heavens throughout her descriptions of Calidora ultimately serves the magical purpose of her character, but Sarrocchi's own interest in astrology is attested to by many sources, including her admirer Bronzini who singles out her belief in the "influence of the stars" *(influssi delle stelle)*, but takes pain to stress that her cosmological beliefs remain within the bounds of orthodoxy. Bronzini reports a conversation between Sarrocchi and a gentleman in which she asserts her belief that the stars "guide our bodies to good and evil, which man can always resist," but is careful to distinguish herself from heretical views with respect to free will.[146] Bronzini's clarification is noteworthy, given how Sarrocchi's interests in literature and astronomy would converge in her correspondence with Galileo—a connection that cooled as Galileo began to run afoul of the Church.[147] Sarrocchi herself was wary departing from orthodox doctrine: as we have seen, the prefatory letter to readers in the 1606 edition of the *Scanderbeide* takes pains to clarify that any "fantastical" episodes in the work are poetic fictions that do not reflect the poet's real views.[148] Despite Sarrocchi's innovative introduction of natural philosophy to the *maga* trope, the Calidora episode fell to the wayside as Sarrocchi worked to revise her poem and bring it more fully in line with her primary goal: a conversion story fully suited to her Counter-Reformation context. In this effort, she sought the help of Galileo himself.

Sarrocchi and Galileo

Galileo frequented Sarrocchi's salon during his stay in Rome in 1611, and they had a mutual friend in Valerio, whom Galileo had first met in Pisa and admired as a "new Archimedes" and with whom he also corresponded.[149] Sarrocchi's letters to Galileo, as well as the letters she exchanged with others regarding Galileo, show both the continuous efforts she made to seek the scientist's advice and support for her *Scanderbeide,* and the lengths she went to in return—for a time—to praise and defend Galileo's astronomical discoveries.

Galileo was a writer as well as scientist: his *Dialogue on the Two Chief World Systems* (1632) is structured not as a formal treatise, but as a vernacular dialogue, and he wrote several sonnets and even a comedy.[150] He also composed works of literary criticism, including two lectures he read before the Accademia Fiorentina "Circa la figura, sito e grandezza dell'Inferno di Dante" (On the shape, location, and size of Dante's hell). Most relevant for our discussion here is his *Considerazioni al poema del Tasso* (Considerations on Tasso's poem), in which he accuses Tasso's work of lacking in realism and clarity of style and expresses instead his strong preference for Ariosto. Crystal Hall argues that

Galileo "methodically and consistently incorporated" aspects of Ariosto's poem into philosophical arguments.[151] Inventories of Galileo's personal library show that he possessed numerous works of vernacular literature including several editions of *Orlando furioso* and *Gerusalemme liberata,* although the annotated edition of Tasso from which he worked is lost. He refers to this book in a 1614 letter to Francesco Rinuccini, the Tuscan Resident in Venice, describing how he made notes on blank pages bound into his copy of the *Gerusalemme* for this purpose and reasserting his opinion on the superiority of Ariosto over Tasso.[152] Although the *Considerations* was never published (whether due to the loss of his notes or to an awareness that his pro-Ariosto views were not in fashion), Galileo's thoughts on Tasso's epic were well known among his circle. As Hall writes, "[t]hough not widely accessible, these notes circulated around a sufficiently large group to make his ideas generally known and even to generate rumors that pro-Tasso literary critics needed to publish quickly to beat Galileo's *Considerazioni* into print."[153]

Given Galileo's anti-Tassian stance, it may seem surprising that Sarrocchi sought his literary advice regarding her decidedly Tassian poem, the *Scander-beide.* However, the correspondence between Sarrocchi and Galileo, which in addition to her requests for literary counsel includes her opinions about Galileo's own discoveries and works, demonstrates how issues of literary exchange, scientific community, and patronage were closely linked, overshadowing stylistic differences (and in fact, as we will see, Sarrocchi and Galileo shared certain literary convictions). Although our record of their correspondence is incomplete—we possess seven extant letters from Sarrocchi to Galileo and only one of his letters to her—their epistolary interactions suggest that each sought validation and support from the other, creating a web of mutual obligation between the scientist and the poet that transcended distinctions of gender as well as poetic camp.[154] Sarrocchi faced criticism of her poem, both because of her sex and in reaction to her own anti-Marino stance, which led her to abandon the Accademia degli Umoristi for the Ordinati. Galileo, for his part, needed to bolster support among the scientific community in Rome for his discoveries with the telescope. He was especially anxious to establish his ownership of Jupiter's four satellites: these were the "Medicean stars" named for his patron Cosimo II, in a brilliant gesture that catapulted Galileo to fame.[155]

Sarrocchi and her poem had been introduced to Galileo by letter as early as 1609. In a letter dated May 23, of that year, Valerio informs Galileo that he will be sending the scientist, together with a selection of his own works, a copy of the 1606 edition of Sarrocchi's *Scanderbeide*.[156] Certainly Sarrocchi had met Galileo in person by 1611, during his first sojourn in Rome. Galileo had not yet run into trouble with the Inquisition: on the contrary, his improvements to the telescope and subsequent discoveries were generating great curiosity and excitement. In visiting Rome, Galileo hoped to gain the approval of the Collegio Romano and continue to gather support for his ideas. Sarrocchi and Galileo may have first met at her salon, or at one of the numerous gatherings in which Galileo demonstrated the power of his perfected lens.

The first extant letter from Sarrocchi to Galileo is dated July 29, 1611, soon after Galileo had returned to Florence. Sarrocchi requests Galileo's advice about her *Scanderbeide:* she wants him to read the poem critically, without fear of offending her, for it is far worse to publish with errors: "The principal favor I desire from you is that you read my poem with the greatest diligence and with an enemy eye, so that you may note every little error, and believe me that I truly mean it, and will accept every criticism you make as a sign of your great kindness and great affection."[157] In other letters, she expresses her request in similar language, hesitating to send her manuscript until she has a clean version and asking "the favor of your most acute judgment" *(purgatissimo giudicio).*[158] In fact, it would take close to six months before she sent Galileo this clean copy, in January 1612.[159]

As mentioned earlier, it is interesting that Sarrocchi should ask a well-known anti-Tassian to critique a Tassian poem. Sarrocchi herself expresses a measured relief upon learning that Galileo has agreed to read the *Scanderbeide:* "Signor Spinello [Benci] wrote me about your kindness in favoring me with a revision of my poem, which made me very happy, although I was never in doubt of it."[160] However, many writers sought Galileo's literary advice and opinions, including Girolamo Magagnati, Cesare Marsili, and Lodovico Cardi, a friend to both Galileo and Sarrocchi. Sarrocchi observes, "Who could doubt the courtesy of my signor Galileo, who is adorned with so many virtues and so great an admirer of men of letters [*letterati*]?"[161] Much of what Galileo objected to in Tasso, in addition to a lack of verisimilitude, had to do with stylistic affectation. Although Sarrocchi's poem is Tassian in structure—and her style cannot be characterized as devoid of baroque flourish—she did, nonetheless, take a stand against the more extreme baroque trend epitomized by poets such as Marino, aligning her views more closely, perhaps, with Galileo than might initially appear.[162] Certainly, she subscribed to the view that Tuscan was the most appropriate literary language, and part of what she desired from

Galileo—who shared this view—was a revision of her poem to polish its language. In a later letter, dated January 13, 1612, she turns to Galileo not just as a respected author, but as a Tuscan one, who can help her refine her poem according to the current dictates of poetic style, in which *la lingua toscana* still reigned. Sarrocchi explains that she has labored to make her language as Tuscan as possible, but asks Galileo to correct errors in wording or spelling:

> In addition, do me the favor of looking over and correcting the language, because I would like it to be as Tuscan as possible, at least in the expressions—so long as it does not diminish the exalted style, being that Tuscan sounds so sweet. This is why, where you are accustomed to dropping the *r*, I sometimes leave it, as, for example, where in Tuscany one would say *trincea* [trench], I said *trincera*, and so on. . . . I would like you to look it over again for spelling. You will find many changes and altered lines.[163]

In addition to seeking Galileo's linguistic expertise, Sarrocchi also turned to him as a guide to the complexities of securing a patron for her work. As historians have highlighted, Galileo understood the patronage system, and used it to convert his scientific discoveries into a more stable and lucrative position at the Medici court in Florence.[164] Galileo's own contemporaries turned to him for advice on how to establish their own networks of patronage. Lorenzo Ceccarelli, for example, asked Galileo's assistance in parlaying two sonnets written for the marriage of Ferdinand II into financial reward.[165] Likewise, Sarrocchi, in search of a new patron for her revised poem, asks Galileo to give some thought to the work's eventual dedication, perhaps casting around the Medici court for possibilities: "Once you have read it over, and if it seems appropriate, you may do as you like with the dedication, for I will submit, completely, in all things to your wise counsel."[166]

Sarrocchi never asks without offering something in return, however, and her comments here are an illuminating window onto the close links between patronage strategies and writing practices. Her poem, she notes, will contain a passage listing the names of all the various Italians who come to the aid of the work's Albanian hero, Scanderbeg. Sarrocchi has left much of this list blank—or rather, inserted temporary names—until she can, at a later date, rework it by inserting the names of her friends and supporters (including, ideally, "a prince or two"). Among these, she assures her correspondent, she will be pleased to put members of Galileo's family, and even a reference to Galileo himself: "It's true I haven't finished the list of Italians who help Scanderbeg, since I haven't completely determined whom to include, and also in order to leave space to praise a prince or two. If you send me some names from your family, I will honor my pages with your ancestors; and, if I can, I will mention you, too, as someone yet to be born."[167] Thus, in exchange for reading Sarrocchi's epic, Galileo may expect to find his own name in a place of honor within its pages. Secure in her talents as a writer, Sarrocchi hastens to assure Galileo, who has been ill, that such an adjustment will not take long, and can wait until he is feeling well enough to turn his attention to her poem: "Waiting to finish the list doesn't matter, since for someone who is experienced in such things, as I am, it will be a matter of only fifteen or twenty days [to complete it]."[168] In another letter to Galileo, Sarrocchi says the list will take only eight or ten days to complete.[169] A catalog of Italian family names is integrated into the 1606 edition of the poem, composed prior to Sarrocchi's acquaintance with Galileo (VII, 13). Although similar references to Italian figures and families as Sarrocchi mentions in her letter are incorporated throughout the 1623 edition, Galileo's name does not appear.[170] It is probable that, given the turn in Galileo's fortunes after 1616 and Sarrocchi's own position regarding Copernicanism, she thought better of her promise.

Despite Sarrocchi's clear recognition that the support of a well-known figure like Galileo would lend authority to her poem, her letters to him establish the two on nearly equal footing. It is the exchange of work between them—Sarrocchi's *Scanderbeide,* and Galileo's scientific works, including the *Discorso intorno alle cose che stanno in su l'acqua (Discourse on Floating Bodies,* 1612) and the *Istoria e dimostrazioni intorno alle macchie solari (Letters on Sunspots,* 1613)—as well as the promise of mutual endorsement, that help to maintain this even plane. References to gender are oblique in Sarrocchi's letters, therefore, but not absent. Her comment to Galileo that she continues work on *Scanderbeide* despite bouts of illness and problems at home ("domestic troubles and . . . constant illness") recall, for example, Erculiani's comments in her *Letters on Natural Philosophy* regarding the effect of domestic duties on her writing—a problem frequently remarked upon by women writers.[171] Sarrocchi does not explicitly highlight her identity as a woman writer, nor does she position herself as a champion of women, as Erculiani does; but her profound concerns about errors in the *Scanderbeide,* as well as language and style, reflect a heightened awareness of the criticism she faces as a woman in publishing an epic poem of such magnitude. Indeed, we know from Valerio that Sarrocchi encountered hostility and ridicule from male detractors: "the adolescent opposition, that even now those hoarse and scornful naysayers move against her," a reference to reactions to her ambitious *Scanderbeide* and likely her falling out with Marino.[172] Sarrocchi expresses concern about the quality of her copyist ("it is very difficult to find someone to transcribe it correctly") and about possible errors in her work.[173] As she writes to Galileo, she realizes that "just as print can demonstrate one's erudition, so it sometimes shows one's poor judgment. For this reason, not wishing to commit such a mistake, *in propria causa advocatum quero* [I seek an advocate on my behalf]."[174] In Galileo, Sarrocchi seeks both guide and protector.

Finally, Sarrocchi is concerned about the structure of her poem. She was deeply interested in the debates over poetics that unfolded following the publication of Tasso's *Gerusalemme liberata,* the heroic style of which she sought to imitate. The *Scanderbeide* underwent significant revisions from its first incarnation in 1606, as Sarrocchi sought to streamline the narrative according to a single dominant storyline, as her model demanded. To this end, she also asks Galileo's advice about how to reorganize the cantos in a way that will allow her to adhere to established poetic principles:

> I would also like you to do me the favor of dividing this poem, as you see fit, into more cantos, because these seem too long to me. I will also tell you that I have labored to follow the rules of Aristotle, Valerius, Hermogenes, Longinus, and Eustachius, who advise drawing all plotlines into one; and so I labored in the verse to imitate things closely, so for things regarding war, I tried to make the style elevated, and for things regarding love, I tried to make it sweet.[175]

In fact, as noted earlier, several episodes, including that of Calidora, would be left out of the 1623 edition of the *Scanderbeide,* in the interest of maintaining focus on Scanderbeg's rebellion. Sarrocchi's request to Galileo here is in keeping with his own expressed preference for a clear, unaffected style and greater verisimilitude in epic poetry.

Although Sarrocchi was keenly interested in Galileo's literary advice, their exchange was not only literary in nature. Her requests for help with her manuscript are accompanied by her promises to offer her opinion on Galileo's work and to promote and defend him in Rome and anywhere else she can. From Sarrocchi's letters to Galileo as well as other letters written to her or in which she is mentioned, we see that this was an ongoing endeavor for her until at least 1613, after which point her support for him seems to have waned (at least, there is no evidence of further correspondence between the two after this date). The attention with which Sarrocchi follows Galileo's discoveries and writings and also the fact that others turn to her for her opinion on these matters offer compelling testimony of the reputation Sarrocchi enjoyed in both scientific and literary circles.

In the earliest letter we have from Valerio to Galileo, which predates Galileo's stay in Rome and induction into the Accademia dei Lincei, Valerio praises one of Galileo's works, noting: "the signora Margarita Sarrocchi also read it and holds the same opinion of its author as I do" (ever a supporter of Sarrocchi's literary ambitions, Valerio combines this sentiment with a request that Galileo read Sarrocchi's *Scanderbeide*).[176] Later letters by both Valerio and Sarrocchi show that Galileo continued to send his works to the two of them, including a copy of his *Discourse on Floating Bodies* dispatched directly to Sarrocchi, for which she thanks him with a letter of her own, writing: "Signor Luca and I will read it with affection and admiration."[177] While Galileo circulated his works widely, seeking the support of prominent figures including Federico Cesi, Giuliano de' Medici, Monsignor Piero Dini, and numerous others, Sarrocchi seems to have been one of few female recipients of Galileo's works (with the exception of his daughter, Suor Maria Celeste, to whom he was very close).[178] In addition to admiring Sarrocchi's erudition and her position in Roman intellectual society, Galileo may have also hoped to benefit from her connections to Church bureaucracy: Sarrocchi's former guardian, Cardinal Sirleto, had been a respected scholar of Counter-Reformation theology and Vatican librarian whose stature would have reflected positively on his ward.

The epistolary exchanges featuring Sarrocchi and Galileo are primarily concerned with Galileo's observations of Jupiter's four "Medicean stars" (actually satellites, or moons), observed with his telescope and described in *Sidereus nuncius,* or *Starry Messenger* (1610). The discovery had enormous implications, for it toppled the prevailing Aristotelian view that all celestial bodies orbited the earth. Almost immediately, in May 1610, Valerio and Sarrocchi were reading and praising *Sidereus nuncius,* as evidenced by a letter in which Valerio reminds Galileo, on their behalf, not to completely abandon the study of terrestrial movement: "I beg you not to let yourself be so distracted by the stars that you neglect the different movements of the earth; the signora Margherita Sarrocchi begs the same, having become no less your supporter than I am."[179] Despite Galileo's enormous success, the scientist was anxious about cementing his reputation. He worried that some, using the telescope incorrectly, might not be able to observe the Medicean stars. More important, he still needed to establish their orbital periods. Galileo continued to study and observe the satellites during his 1611 trip to Rome, recording his observations during a banquet hosted for him by Cesi and at a gathering in his honor at the Collegio Romano.[180] He rushed his newest observations into print in 1612 with the *Discourse on Bodies in Water,* which he sent to Valerio, Sarrocchi, and many others.

Galileo's haste was warranted. In August 1611, Cardi reported a comment by Giovanni Antonio Magini that the discovery of satellites was itself of little consequence, and that the real praise would go to whoever established their periods.[181] Likewise, Cesi urged Galileo to publish a "supplement" to the *Sidereus nuncius* describing the rings of Saturn and the phases of Venus.[182] Galileo's concern about safeguarding his discovery of the Medicean stars and establishing support among the academic community in Rome is evident in a letter of 1611 from Valerio to Marc'Antonio Baldi. Here—addressing Galileo's own apprehensions—Valerio assures Baldi that he has observed the moons himself, and it is not a trick of the telescope lens. Swearing on his identity as a philosopher who loves the truth above all, Valerio describes the experience of seeing Jupiter's "stars" in vivid detail. He asks his correspondent to share what he has written in order to support Galileo's discoveries, "so that, if the occasion presents itself, you may, with this letter written by my own hand, assure some of those stubborn people who like to spread rumors that I am not of any other opinion than that which I express in writing as my own."[183] Galileo kept a copy of Valerio's letter, marked "Attestation of signor Luca Valerio."[184]

The suspicions against which Valerio defended Galileo were fueled by a letter from Cosimo Sassetti of Perugia to Monsignor Dini voicing concern about the veracity of Galileo's discoveries.[185] Although later letters circulated protesting that Sassetti's accusations did not speak for the Studio di Perugia, the incident sparked much epistolary debate. Many still found it difficult to accept Galileo's discoveries, despite eventual corroboration from figures like Johannes Kepler (1571–1630) and Christopher Clavius (1537–1612). Like Valerio, Sarrocchi also found herself in the position of supporting Galileo's discoveries of Jupiter's satellites and the phases of Venus, responding to epistolary requests for her opinion on the matter. That others wrote to Sarrocchi directly to seek her judgment is, again, strong testimony to the reputation and influence she had garnered among the scientific community. Guido Bettoli of the University of Perugia, acknowledging her erudition in this area, begs her to offer her view: "The marvelous discoveries one keeps hearing about regarding signor

Galileo Galilei's telescope, or shall we say lens, which continually inspire everyone to offer their opinion, spur me to the presumption of writing to greet you, and to ask you to favor me with your opinion; you being so perfectly learned in every science, I hope for a perfect account of the truth, since you too must have seen it for yourself a thousand times."[186] Bettoli goes on to explain the controversy over Sassetti's letter, anxious that Sarrocchi should know that it does not reflect the opinions of the Studio di Perugia. As we can see, both Sarrocchi's endorsement and her good opinion carry weight:

> With this digression, I wished to clarify things to you, in case that letter or opinion, to which signor Galileo attempts to respond, should reach your discerning ears or those of virtuous men. Here no one knows anything about it, except that it involves signor Galileo, something that is more than a little upsetting; I don't know how they will manage it. I know how magnanimous and virtuous you are, and how you defend the virtuous, so I will not say any more about this, only that I will await your response and that you must honor me with your commands.[187]

Bettoli's words make clear both the influence of Sarrocchi's opinion in matters of science and her equal intellectual status to the "virtuous men" of Rome's scientific community, as well as the degree to which her connection to Galileo was widely recognized.

In a series of letters to Bettoli and to Galileo himself, Sarrocchi proceeds to defend Galileo against the allegations from Perugia. Sarrocchi, like Valerio, insists:

> Everything they say about signor Galileo's discovery of the stars is true: that is, that there are four unfixed stars around Jupiter that move independently, always maintaining an equal distance from Jupiter, but not from one another; and I saw these myself through signor Galileo's telescope, and showed them to some friends, as the whole world knows. Saturn has two stars on one side and one on the other, and they almost touch it. One can see how Venus, when it joins with the sun, becomes illuminated and, like the sun, horned, until it can be seen to be full. . . . Many prominent mathematicians, particularly Father Claudio and Father Grienberger, denied this at first, but then reversed themselves, having assured themselves of the truth, and held public lectures about it.[188]

Demonstrating her comprehension of the importance and implications of Galileo's observations (although without following them through to their Copernican conclusions), Sarrocchi writes elegantly and clearly of the position of the satellites and the phases of Venus as well as Saturn's composite nature, while also demonstrating that she is following the latest developments in the controversy.

Loyal to Galileo on a both a personal and intellectual level—a bond attested to by her willingness to entrust him with the *Scanderbeide*—Sarrocchi kept him apprised of the grumblings from Perugia, sending him a copy of her letter to Bettoli so that he might see "everything that is going on," and also copies of correspondence with a friar from Perugia who had engaged her in debate on the same subject.[189] As Sarrocchi tells Galileo, an Augustinian friar at Santa Maria Novella in Perugia, called Innocenzio, had asked for her opinion: she defended Galileo's discoveries, only to be much offended by her interlocutor's response: "and having written him in defense of the truth of [your discovery of] the stars, and praised your genius . . . he responded with another letter, which greatly offended me, so I replied as I saw fit. . . . He answered, as you will be able to see for yourself, because I am sending you both of his letters."[190] In a second letter regarding this episode, Sarrocchi's identity as a woman comes to the forefront, as she goes on to describe her epistolary "battle" with Innocenzio in gendered terms:

> It seems the case that that friar has it in for me and wants to challenge me with words, by asking me to explain the meaning of certain terms as I was trying to use astrology to discuss the newly discovered stars, as if to say that the discovery of these stars isn't real; but I've set straight better people than him, and so I hope to do the same with him, even though I am a woman and he is a learned friar.[191]

It is interesting to note that Sarrocchi engaged in the more contested vein of astrology known as "judicial astrology" by casting and interpreting natal charts. Despite any problematic implications associated with this practice, genitures were commonplace in early modern Italy, often sought out by the elite, given as gifts, or offered in exchange for patronage.[192] Sarrocchi makes reference in these letters to such a request by the friar himself, noting, "I did him the favor he asked with respect to the natal chart, and he asked for another one for a young girl to whom an incredible thing had happened."[193] Clearly, Sarrocchi applied her knowledge of astrology in a variety of contexts. Here, however, her reputation in astrology is manipulated by the friar in an attempt to diminish and dismiss both the branch of scientific inquiry, and her as its practitioner,

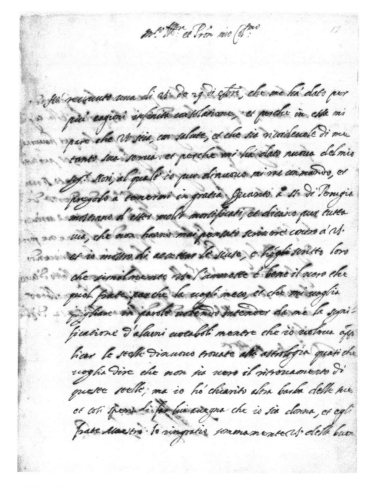

(Left and Right): Letter of Margherita Sarrocchi to Galileo Galilei, October 12, 1611 (Biblioteca Nazionale Centrale, Firenze, Mss Gal 23, cc. 12 r-v)

even as he seeks her expertise in a personal matter. Vexed, Sarrocchi accuses him of challenging her for the sake of establishing masculine authority.

In fact, in a letter to a colleague in Perugia, Innocenzio speaks dismissively of Sarrocchi, attributing her defense of Galileo's discoveries to the fondness she has for him rather than to her scientific opinion: "I understand that her affection for signor Galileo is such that perhaps she allows it to get in the way of the truth."[194] The accusation that Sarrocchi is governed by emotion rather

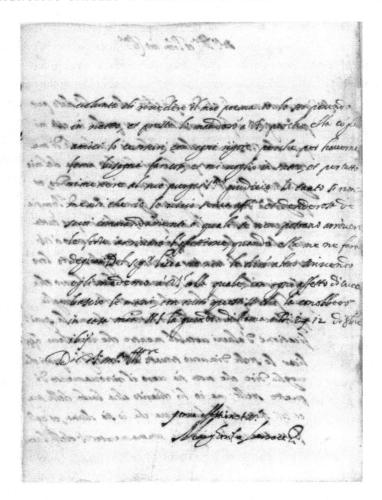

than reason is surely meant to play on gendered tropes, diminishing the intellectual import of any opinion she may offer. Innocenzio further insinuates that Sarrocchi's disposition is a vindictive one, writing to the same colleague, "I would have answered your most welcome letter sooner, were it not for fear of offending Sign[o]ra Margarita."[195]

The claim that Sarrocchi had a difficult personality seems to have constituted a common point of attack. Rossi, for example, accused her of reveling in her public enmities and disputes: he attributes her rift with the Accademia degli Umoristi to her pride, rather than her poetics.[196] Such characterizations demonstrate, more than anything else, the degree of influence Sarrocchi truly wielded in seventeenth-century intellectual culture in Rome and indeed the discomfort this stature caused among her contemporaries. Sarrocchi's relationship with Valerio, for example, was regarded with suspicion: not so much for its intimacy, but for the indecorous sway she was thought to hold over the mathematician's opinions. The Tuscan painter Lodovico Cardi, or Cigoli (1559–1613), a friend of both Valerio and Galileo, writes disparagingly of the power Sarrocchi was thought to exert over her former tutor, suggesting that Valerio is so enthralled with Sarrocchi that he neglects everything else, to the amusement of others.[197] When Valerio resigned from the Accademia dei Lincei in 1616, distancing himself from Galileo's Copernicanism, some, including Cesi, blamed Sarrocchi. Speculating after Sarrocchi's death in 1617 that Valerio might now contemplate a return to the Academy for which he had once served as *censore,* Cesi wrote to Faber da Acquasparta, "I hear that the infamous muse of our age has been relieved of earthly cares; what do you suppose will happen to her companion? Might he not return to us, now that he is freed of her extreme influence?"[198]

Some modern critics go so far as to attribute the waning of Valerio's support for Galileo to Sarrocchi's disillusionment with Galileo's failure to edit or promote the *Scanderbeide,* which in this analysis would have caused her to withdraw her support and to encourage Valerio to do the same.[199] Ugo Baldini and Daniele Napolitani cite a "coolness" in Valerio's letters to Galileo after 1612 (noting one instance in which Valerio does not send Galileo greetings from Sarrocchi), and in Sarrocchi's requests that Galileo return the draft she had entrusted to him.[200] However, nothing in Sarrocchi's request suggests anything but affection and a punctilious desire to be certain that only the most updated and correct copies of her draft are in circulation: "As for my poem," she writes, "please do me the favor of returning it, because I have made many changes, so your copy is no longer good. I will have it copied again, and will send it to you."[201] While it is true that this is the last letter we have from Sarrocchi to Galileo, that nevertheless does not mean it is the last one she wrote. Finally, as we know from Galileo's surviving letter to Sarrocchi, he was not well, a circumstance she clearly recognizes here, expressing her hope that he will soon return to perfect health.[202] Galileo's explanation that, although he is too ill to read her work, he wishes to reassure her of its safe arrival, is warmly genuine and entirely plausible. Galileo writes:

> Therefore, if I am brief in my reply to your most courteous letter, and in
> rendering the necessary thanks for your continued favor, you will excuse

my impotence, which does not allow me to exercise the mind, let alone the pen, without great damage. But to keep you from worrying whether the poem has safely arrived, I wished to write you these few lines, in which I also remind you of my service [to you]. . . . I kiss your hands with all affection, and pray God for your happiness.[203]

The epistolary testimony gives little reason to suspect that Sarrocchi was angered or offended by Galileo's treatment of her manuscript: more likely, she understood that distance and health were taking a toll on their agreement of exchange and support.

Favaro's extensive inventory of Galileo's library does not show that he held any works by Sarrocchi (whether the *Scanderbeide* or her poetry), although it does reflect his deep interest in vernacular literature, including romance and epic, as well as an interest in the so-called woman question. Among his books was a copy of Boccaccio's *Famous Women,* for example, and he possessed two works by the seventeenth-century poet and dramaturge Margherita Costa (like Galileo, closely affiliated with the Medici court).[204] Several of Costa's works seem to make reference to Galileo's discoveries and innovations, suggesting that the admiration went both ways.[205] He also had copies of scientific works by Piccolomini, who argued for making scientific texts more accessible to women; as well as two works by Valerio.[206] In the absence of further epistolary evidence, we can only speculate as to whether Galileo did eventually return Sarrocchi's manuscript to her. Her last letter to him also promises to send him, along with a revised copy of the *Scanderbeide,* some of her lyric poetry to revise.[207] The fate of these compositions is also unclear. Nonetheless, the correspondence between the poet and the scientist clearly reflects the reciprocal bonds of obligation among networks of literary and scientific communities that were fundamental to making an intellectual career in early modern Italy. Literary patronage was a valuable tool for authorizing scientific discoveries, while the support of a scientific luminary such as Galileo could surely lend credit to a poet (at least prior to 1616). Indeed, Sarrocchi's request that Galileo revise her lyric compositions along with her *Scanderbeide* is accompanied by a promise to read and comment upon his *Letters on Sunspots:* "When I send you my poem, I will beg you to review my lyric compositions. In the meantime I will read your treatise, and then I will write you at greater length [about it]."[208] For his part, Galileo was soliciting poetic contributions in praise of his discoveries for a planned (but never executed) second edition of

Sidereus nuncius, including an offering by Valerio; it is quite plausible that he sought Sarrocchi's support as well.[209] An epigram by Valerio had already appeared in the first edition of the *Letters on Sunspots* (1613), along with a pair of elegies also thought to be his.[210] Clearly, Galileo recognized the value of literary support.

Against such a backdrop of exchange, collaboration, and reciprocal support among scientists, mathematicians, and poets, therefore, Sarrocchi's relationship with Galileo becomes clearer. If, in cultivating a friendship with Galileo, Sarrocchi hoped to receive literary advice and guidance in finding a patron for her work among the Medici, Galileo recognized Sarrocchi's position of influence among Rome's scientific community and the potential support she could offer him with her endorsement of his discoveries and likely with her poetic compositions as well.

The examples of Camilla Erculiani and Margherita Sarrocchi reveal how, by the early seventeenth century, women had assumed an increasingly public presence in scientific culture. Unlike early sixteenth-century examples such as Caterina Sforza, whose experiments with alchemy and medicine were undertaken in a primarily private context—recorded in manuscript but not intended for broad circulation—Erculiani and Sarrocchi sought to contribute in more formal ways to the rich scientific debates afoot, taking the early empiricism espoused by Sforza and expanding it to the study of the stars and the world around them. As women writers, both Erculiani and Sarrocchi confronted opposition, but they also found support and esteem among their male counterparts. Indeed, as can be surmised from the presence of women as dedicatees of scientific treatises, in addition to their own intellectual contributions, scientific education for women was no longer to be considered beyond the scope of a woman's mind. Instead, it added a new dimension to the centuries-old *querelle* over women, erudition in the sciences now taking its place as a quality to be praised alongside catalogs of women learned in poetry and literature.

Gender receives different emphasis in Erculiani's work than in Sarrocchi's. For Erculiani, science takes center stage as a crucial step for women to achieve the recognition and status they deserve. As Erculiani writes in the dedication to her *Letters on Natural Philosophy,* she means to use her own scientific learning on behalf of the reputation of all women. Sarrocchi is less overt in her com-

mentary on gender and science, both in her literary works and in her epistolary exchanges, but when it comes to her intellectual debate with men such as
Padre Innocenzio, she, too, casts the battle in gendered terms.

Regardless of the degree to which gender is foregrounded by these writers,
the engagement of Erculiani and Sarrocchi with early modern scientific culture
brought them recognition far beyond their own immediate circles. Erculiani's
activity at the Tre Stelle apothecary and her *Letters on Natural Philosophy* carried her name as far afield of Padua as Kraków (and earned her the attention
of the Inquisition), while Sarrocchi's legacy comprises not only her impressive
literary accomplishments but also her contributions to scientific culture through
her own salon and her championing of Galileo's early discoveries. In many
respects, the examples of Erculiani and Sarrocchi lay the groundwork for the
efforts of women in the following decades not only to become learned in the
sciences but also to pass on their knowledge to other women.

Epilogue

The women discussed in these pages by no means constitute the full extent of women's participation in early modern scientific culture. They serve to illustrate something of the variety of forms that participation could take, but much work remains to be done to fully recover and recognize the activity and contribution of women to early modern science. Scholars have begun to fill in some of the missing pieces of women's contributions to scientific and medical culture (or, to use Pamela H. Smith's term, "techno-medico-science").[1] As we continue to stretch and challenge the borders of our investigation, we must also take into account issues such as cross-gender collaboration, the nature of transnational intellectual networks and communities of ideas, and the blurring of boundaries between disciplinary areas and technologies. Carol Pal, arguing for a reformulation of the way we understand the early modern Republic of Letters, observes that by the mid-seventeenth century the Republic of Letters was a "surprisingly inclusive, heterogeneous, and dually gendered" network of writers, scholars, philosophers, and scientists. An investigation of women in scientific culture must consider the myriad ways in which women participated in the production and circulation of scientific knowledge, not just the obstacles to their activity; and position them in connection to the vast and varied communities of knowledge in which they participated, in person and on the printed page, in Italy and beyond.

That women in early modern Italy participated in scientific culture in a variety of ways is abundantly evident from the literary as well as the archival record. Some, such as Caterina Sforza, pursued an interest in alchemy and experiment, while others, including Moderata Fonte and Lucrezia Marinella, found ways to integrate science into literary narrative. By the seventeenth century, women participated in scientific inquiry and debate in increasingly public ways: Camilla Erculiani, the apothecary from Padua, circulated her *Letters on Natural Philosophy* as far afield as Poland; Margherita Sarrocchi, author of the epic poem *Scanderbeide,* served as a cultural mediator, endorsing and disseminating Galileo's discoveries with the telescope. The empirical culture that encouraged clinical observation and practical experiment at the dawn of the so-called Scientific Revolution spilled over in increasing measure to the pages of literary works, informing not just scientific treatises but vernacular dialogues, chivalric romance, and epic poetry. Despite the chilling effects of the Counter-Reformation in Italy, women continued to find ways to engage with the new science, if not always to the same extent as their counterparts elsewhere in Europe.

While I have focused here on women who were interested in various aspects of natural philosophy—from alchemy to astronomy—in their own right, women's encounters with scientific inquiry took many forms, often crossing boundaries of gender as well as geography. Collaboration and dialogue between men and women were more common than is often recognized, and women made fundamental contributions to the promotion and circulation of new scientific concepts through their own ideas and works as was well as through their activity as patrons of the new philosophy. Women also formed an important audience for the dissemination and authorization of scientific theory, as can be seen in their growing presence as the dedicatees of scientific treatises on a variety of subjects throughout the sixteenth and seventeenth centuries. Before publishing his *Dialogue on the Two Chief World Systems,* Galileo, for example, turned to Christina of Lorraine, mother of his patron Cosimo II, to lay out his Copernican views. Prompted by a conversation in which she had questioned heliocentrism on the basis of scripture, the *Letter to the Grand Duchess Christina of Tuscany* (1615) attempts to work out an acceptable solution for the conflict between science and faith that Galileo hoped to avoid. Although Galileo's decision to address the Grand Duchess stems primarily from issues of theology and power at court, the prominent presence of a female figure in such a work is striking. On the one hand, Christina is a *donna schermo*, a proxy for the theologians of the Roman Curia with whom Galileo wished to engage.[2] On the other hand, as an interlocutor interested, if not expert, in natural philosophy but also well known for her piety, Christina stands in for a nonspecialist audience to whom Galileo can address a broad, persuasive defense of his position.[3] In this, his earliest comprehensive defense of Copernicanism, Galileo specifically enlists the protection and participation of a woman and, perhaps by extension, of women readers more generally.

Even when science was not a primary focus, the worlds of science and literature collided in intriguing ways among early modern men and women. The Benedictine nun Arcangela Tarabotti (1604–1652), author of several protofeminist and polemical works, was the daughter of an alchemist and corresponded from within the walls of a Venetian convent with the famous French astronomer Ismaël Boulliau.[4] As Lynn Westwater has shown, despite the vast gulfs of nationality, language, and gender separating their two worlds, the nun and the scientist shared a network of relationships that stretched from Italy to France and Holland, as well as a mutual intellectual appreciation and respect.[5] That Boulliau, a French Copernican who supported the views of Galileo, should have corresponded with a cloistered nun in Counter-Reformation Venice at all is astonishing. What common interests brought these two figures from such disparate worlds into contact? Tarabotti did not address natural philosophy or scientific discovery in her works, which are primarily concerned with condemning forced enclosure and defending women; and her epistolary exchanges with Boulliau are not scientific in nature. Rather, they concern the astronomer's dedicated efforts on Tarabotti's behalf to publish her most controversial work, *La tirannia paterna (Paternal Tyranny [or Innocence Deceived])*, abroad. As Westwater notes, however, the correspondence between Tarabotti and Boulliau suggests a "confluence of radical scientific and social views," one that opens up new ways of understanding the circulation of unorthodox views in early modern Europe.[6]

Although this study has concentrated on women's engagement with science in Renaissance Italy, women studied, practiced, and wrote about science throughout Europe. In France, the example of Marie Meurdrac serves as a telling illustration of women's presence in the sciences, despite their exclusion there from formal academic life. Meurdrac was the author of a Paracelsian chemistry manual published in 1666 and directed specifically at a female readership. Titled *La Chymie charitable et facile, en faveur des Dames* (Charitable and easy chemistry for women), the work is dedicated to a female patron and clearly positions itself within the *querelle des femmes* in a prefatory letter.[7] Admitting that she hesitated to publish a treatise that had been undertaken "for my sole satisfaction," in the knowledge that "teaching was not the profession of a woman," and that she must protect her reputation, Meurdrac finally concludes that "mind has no sex," and, given the same instruction and attention as men, women can "become [men's] equal."[8] Public lectures on medicinal chemistry were held in Paris beginning in the early seventeenth century, sometimes attended by women.[9] Yet women had little opportunity for more extensive instruction, an obstacle Meurdrac seeks to remedy with her handbook. At over 300 pages, her manual contains extensive explanations and illustrations of alchemical vessels and procedures, along with numerous medicinal and cosmetic recipes. Book 6, in particular, containing beauty secrets similar to those found in sixteenth-century books of secrets, is addressed specifically to women. Acknowledging that many women lack access to a laboratory and equipment, Meurdrac—who seems not to have shared this problem—offers her assistance: "For those ladies who are content to know, without desiring to make the effort to perform the operations which they judge to be necessary, because of the time required, or the different sort of vessels, and of other utensils which are needed, or who fear not to succeed, I shall personally explain . . . and I shall take care to perform myself whatever one may want me to teach."[10] Meurdrac's example illustrates a key moment in which scientific learning was moving out of the private realm and into the public for women, who are now recognized as a clear and eager audience for such a chemistry handbook.

In England, Margaret Cavendish (1623–1673) furnishes another important example, attracting increased attention from historians of science. The first Englishwoman to write at length on natural science and philosophy, Cavendish was a Cartesian who participated fully in the scientific debates of her day and was associated with the influential Newcastle Circle society.[11] She was also the only woman in a period of centuries to visit the Royal Society of London, in what has been interpreted as a protest of the exclusion of women from such institutions.[12] Cavendish was part of a network of scientists and philosophers that included William and Charles Cavendish, Thomas Hobbes, and Descartes. Some of her works, such as *Philosophical and Physical Opinions* (1653), included commentary on women's lack of access to education, although she did not formulate a consistent feminist position. While she eventually came to oppose the new experimental science, she benefited from the opportunities it presented for practice outside of the institutional bounds of the university or even the academy.

By the eighteenth century, Italy saw a handful of women formally recognized for their accomplishments in science. In 1732, Laura Bassi became the first woman in Europe to be offered a teaching position at a university. Although Bassi did not publish widely, her academic status allowed her, as Paula Findlen notes, to achieve recognition for her scientific work "through her actions rather than her pen," using the patronage system and her own role as a promoter of the new science to establish a place for herself in the scientific community.[13] The mathematician Maria Gaetana Agnesi and the anatomist and wax modeler Anna Morandi Manzolini also received university degrees and obtained teaching positions by the 1750s.[14] A dissector as well as a sculptor, Morandi's accurate and precise wax models integrate the scientific with the aesthetic, continuing to illustrate the multivalent nuances of practice and representation in early modern scientific culture.

Although no woman in early modern Italy achieved the canonical stature of Galileo, their contributions to the shaping of scientific culture and to its reflections in literary discourse were vital. As practitioners, patrons, writers, and teachers; in courts, academies, and in the home, women influenced and embraced the full range of the new philosophy. Daughters of alchemy, women sought knowledge and understanding of the natural world through study and experiment, independently and in collaboration with others. By recasting our gaze to encompass a wider range of contexts for scientific activity, we can begin to integrate the influence of women into new narratives of the era of the "Scientific Revolution."

Abbreviations

ASF	Archivio di Stato di Firenze
BNCF	Biblioteca Nazionale Centrale Firenze
Med. a. Pr.	Mediceo Avanti il Principato
Mss Gal	Manoscritti Galileiani

Notes

Introduction

1. Moderata Fonte, *Floridoro. A Chivalric Romance* (Chicago: University of Chicago Press, 2007), IV, 3 (*Tredici canti del Floridoro,* IV, 3: "L'oro che sta nelle minere ascoso, / Non manca d'esser or, benché sepolto, / E quando è tratto e se ne fa lavoro / È così ricco e bel come l'altro oro").

2. For two overviews of the shifts in thinking about the so-called Scientific Revolution, see the introduction by Lorraine Daston and Katharine Park in *The Cambridge History of Science;* and Pamela H. Smith's essay, "Science on the Move: Recent Trends in the History of Early Modern Science," *Renaissance Quarterly* 62 (2009): 345–375. See also Diana Robin, "Women on the Move: Trends in Anglophone Studies of Women in the Italian Renaissance," *I Tatti Studies in the Italian Renaissance* 16, nos. 1/2 (2013): 13–25. At the origins of such shifts, see Thomas S. Kuhn, *The Structure of Scientific Revolutions* (Chicago: University of Chicago Press, 1962); and Rachel Carson, *Silent Spring* (Boston: Houghton Mifflin, 1962). On women, gender, and science, see Londa Schiebinger's *The Mind Has No Sex? Women in the Origins of Modern Science* (Cambridge, Mass.: Harvard University Press, 1989) and *Nature's Body: Gender in the Making of Modern Science* (New Brunswick, N.J.: Rutgers University Press, 2004).

3. See Susan Broomhall, *Women's Medical Work in Early Modern France* (Manchester: Manchester University Press, 2004); Gianna Pomata, "Medicina delle

monache. Pratiche terapeutiche nei monasteri femminili di Bologna in età moderna," in *I monasteri femminili come centri di cultura fra Rinascimento e Barocco,* eds. Gianna Pomata and Gabriella Zarri (Rome: Edizioni di storia e letteratura, 2005), 331–363; Sharon Strocchia, "The Nun Apothecaries of Renaissance Florence: Marketing Medicines in the Convent," *Renaissance Studies* 25 (2011): 627–647; Alisha Rankin, *Panaceia's Daughters; Noblewomen as Healers in Early Modern Germany* (Chicago: University of Chicago Press, 2013); Sara Pennell, "Perfecting Practice? Women's Manuscript Recipes in Early Modern England," in *Early Modern Women's Manuscript Writing,* ed. V. Burke and J. Gibson (Aldershot: Ashgate, 2004), 237–258; Elizabeth J. Spiller, "Introduction," in *Seventeenth-Century Women's Recipe Books: Cooking, Physic and Chirurgery in the Works of Elizabeth Talbot Grey and Aletheia Talbot Howard* (Aldershot: Ashgate, 2008), ix–li.

4. See, for example, Penny Bayer, "Madame de la Martinville, Quercitan's Daughter and the Philosopher's Stone: Manuscript Representations of Women Alchemists," in *Women and Scientific Discourse in Early Modern Europe,* ed. Kathleen Long (Aldershot: Ashgate, 2010), 165–182; Tara Nummedal, "Alchemical reproduction and the career of Anna Maria Zieglerin," *Ambix* 48 (2001): 56–68. See also Robin L. Gordon, *Searching for the Soror Mystica: The Lives and Science of Women Alchemists* (Lanham, Md.: University Press of America, 2013).

5. Virginia Cox, *Women's Writing in Italy, 1400–1650* (Baltimore: Johns Hopkins University Press, 2008), 160.

6. The *Trotula* designates a group of three medical texts (*Book on the Condition of Women; Treatments for Women;* and *Women's Cosmetics*) that originated out of southern Italy in the twelfth century. The history of the texts and their author is complex, but scholars now maintain that a historical Trota of Salerno authored the *Treatments for Women* (see Monica Green, "In Search of an 'Authentic' Women's Medicine: The Strange Fates of Trota of Salerno and Hildegard of Bingen," *Dynamis* 19 [1999]: 25–54; and *The Trotula: An English Translation of the Medieval Compendium of Women's Medicine* [Philadelphia: University of Pennsylvania Press, 2002]). John Benton calls the *Trotula* "the most widely circulated medical work on gynecology and women's problems" of the thirteenth to fifteenth centuries (see "Trotula, Women's Problems, and the Professionalization of Women's Medicine in the Middle Ages," *Bulletin of the History of Medicine* 59 [1985]: 30–53, 35).

7. See Green, "In Search of an 'Authentic' Women's Medicine," 33. Green adds that the editor of this sixteenth-century work, Johannes Schottus, suggests that the "femaleness" of Trota and Hildegard was "itself a contribution to their experiential knowledge and their authority" (ibid., 35).

8. Carol Pal, *Republic of Women: Rethinking the Republic of Letters in the Seventeenth Century* (Cambridge: Cambridge University Press, 2012), 5.

9. See Paula Findlen, *Possessing Nature: Museums, Collecting, and Scientific Culture in Early Modern Italy* (Berkeley: University of California Press, 1996), 51–52. From

the time of John of Rupecissa—who sought the curative *quintessence*—to that of Paracelsian iatrochemistry, alchemy and medicine were closely linked (see Lawrence M. Principe, *The Secrets of Alchemy* [Chicago: University of Chicago Press, 2012], 71; and Leah DeVun, *Prophecy, Alchemy and the End of Time: John of Rupecissa in the Late Middle Ages* [New York: Columbia University Press, 2009], 71).

10. Smith, "Science on the Move," 358.

11. On the problem of terminology, see ibid., 345n1.

12. Nummedal, "Alchemical Reproduction," 57. See also Alisha Rankin, "Becoming an Expert Practitioner: Court Experimentalism and the Medical Skills of Anna of Saxony (1532–1585)," *Isis* 98, no. 1 (March 2007): 23–53.

13. See Lynette Hunter and Sarah Hutton, eds., *Women, Science and Medicine, 1500–1700: Mothers and Sisters of the Royal Society* (Stroud: Sutton, 1997), 2; and in the same volume, Lynnette Hunter, "Women and Domestic Medicine: Lady Experimenters, 1570–1620," 95.

14. On the gendered connotations of alchemical images see, for example, Sally G. Allen and Joanna Hubbs, "Outrunning Atalanta: Feminine Destiny in Alchemical Transmutation," *Signs: Journal of Women in Culture and Society* 6, no. 1 (1980): 210–219; Penny Bayer, "From Kitchen Hearth to Learned Paracelsianism: Women's Alchemical Activities in the Renaissance," in *"Mystical Metal of Gold": Essays on Alchemy and Renaissance Culture,* ed. Stanton Linden (Brooklyn: AMS Press, 2007), 365–386; Lyle Massey, "The Alchemical Womb: Johann Remmelin's *Catoptrum Microcosmicum,*" in *Visual Cultures of Secrecy in Early Modern Europe,* eds. Timothy McCall, Sean E. Roberts, and Giancarlo Fiorenza (Kirksville: Truman State University Press, 2013), 208–228.

15. See H. M. E. De Jong, *Michael Maier's* Atalanta fugiens. *Sources of an Alchemical Book of Emblems* (Leiden: Brill, 1969), 176–177.

16. Ibid., 177.

17. Ibid., 69.

18. Ibid., 65–66.

19. See Carolyn Merchant, *The Death of Nature: Women, Ecology, and the Scientific Revolution* (San Francisco: Harper and Row, 1980). For a critique of interpretive schools regarding alchemy since the eighteenth century, see Lawrence M. Principe and William R. Newman, "Some Problems with the Historiography of Alchemy," in *Secrets of Nature: Astrology and Alchemy in Early Modern Europe,* eds. William R. Newman and Anthony Grafton (Cambridge, Mass: MIT Press, 2001), 385–432.

20. For Principe, such sexualized and gendered imagery reflects the use of analogy to "explain the unknown by means of the known"; the male-female binary providing an "ever-ready means of labeling or classifying anything that falls into two subcategories" (Lawrence M. Principe, "Revealing Analogies: The Descriptive and Deceptive Roles of Sexuality and Gender in Latin Alchemy," in *Hidden Intercourse: Eros and*

Sexuality in the History of Western Esotericism, eds. Wouter J. Hanegraaff and Jeffrey J. Kripal [Leiden: Brill, 2008], 215).

21. See Kathleen P. Long, "Odd Bodies: Reviewing Corporeal Difference in Early Modern Alchemy," in *Gender and Scientific Discourse in Early Modern Culture,* ed. Kathleen P. Long (Aldershot: Ashgate, 2010), 63–86.

22. Scholars such as Lawrence M. Principe stress the "interplay between theory and practice" in early modern alchemy; such an interplay must be said to extend to other branches of scientific knowledge, such as medicine and astronomy, as well (see Principe, *The Secrets of Alchemy,* 108).

23. See Pier Desiderio Pasolini, *Caterina Sforza* (Rome: Loescher, 1893), 3:601.

24. The secrets book par excellence, well known to early modern readers, the pseudo-Aristotelian *Secretum secretorum* circulated widely through the seventeenth century (see W. F. Ryan and Charles B. Schmitt, eds., *Pseudo-Aristotle: The Secret of Secrets. Sources and Influences* [London: Warburg Institute, 1982]). As William Eamon points out, this text was "among the most widely read of the medieval and Renaissance periods" (*Science and the Secrets of Nature: Books of Secrets in Medieval and Early Modern Culture* [Princeton, N.J.: Princeton University Press, 1994], 45).

25. David Gentilcore, *Medical Charlatanism in Early Modern Italy* (Oxford: Oxford University Press, 2006); M. A. Katritzky, *Women, Medicine and Theatre 1500–1700: Literary Mountebanks and Performing Quacks* (Aldershot: Ashgate, 2007); Tara Nummedal, *Alchemy and Authority in the Holy Roman Empire* (Chicago: The University of Chicago Press, 2007).

26. On the gendered implications of secrecy in contexts ranging from early modern understandings of anatomy to books of secrets, see Monica Green, "From Diseases of Women to 'Secrets of Women': The Transformation of Gynecological Literature in the Later Middle Ages," *Journal of Medieval and Early Modern Studies* 30 (2000): 5–39; Katharine Park, *Secrets of Women: Gender, Generation, and the Origin of Human Dissection* (New York: Zone Books, 2006, rpt. 2010); Elaine Leong and Alisha Rankin, eds., *Secrets and Knowledge in Medicine and Science, 1500–1800* (Aldershot: Ashgate, 2011); and Timothy McCall, Sean E. Roberts, and Giancarlo Fiorenza, eds., *Visual Cultures of Secrecy in Early Modern Europe* (Kirksville, Mo.: Truman State University Press, 2013).

27. See Londa Schiebinger, "Maria Winkelman at the Berlin Academy: A Turning Point for Women in Science," *Isis* 78 (1987): 175. Despite the obstacles faced by women scientists, Schiebinger points out that between 1650–1720, some 14 percent of German astronomers were women (ibid., 177). Similarly, in Italy there were likely many more women like Sarrocchi and Erculiani who took an interest in science and the new discoveries, but of whom we have no record.

28. See Virginia Cox, *The Prodigious Muse: Women's Writing in Counter-Reformation Italy* (Baltimore: Johns Hopkins University Press, 2011), 19–31.

1. Caterina Sforza's Experiments with Alchemy

1. "El talcho è stella de la terra et [h]a le scaglie lucide e se trova ne l'isola di Ciprij et il suo colore è simile al cetrino et guardandolo essendo insieme in massa dimostra verde e vedendolo verso l'aria dimostra come cristallo et [h]a le infracripte virtù senza le altre che non sonno in libro noctate quale seria el desiderio de li alchimisti saperlo: Prima per fare le donne belle e levarsi omni segno o machia del viso de sorte che se una donna de sesanta annj la farà parere de vinti. . . . Ancora dicta acqua de talcho o vero polvere de esso chi ne bevesse in vino biancho guarisse uno che fusse avenenato et chi in quel giorno ne havesse preso in vino biancho serà sicuro de veneno et de omni morbo e peste. . . . Ancora se fa mentione che dicta acqua fa de lo argento oro, et de le zoie false le fa perfecte et fine" ([Caterina Sforza], *Experimenti de la Ex[ellentissi]ma S[igno]ra Caterina da Furlj Matre de lo inllux[trissi]mo S[ignor] Giovanni de Medici,* in Pier Desiderio Pasolini, *Caterina Sforza* [Rome: Loescher, 1893], 3:617–618). I have modernized accents and simplified punctuation; all translations from the *Experiments* are my own. Talc is a soft mineral, commonly used for cosmetic purposes as talcum powder.

2. As Lawrence M. Principe notes, the two principal goals of European alchemy were the transmutation of metals and the preparation of medicines (see *The Secrets of Alchemy* [Chicago: University of Chicago Press, 2013], 71).

3. See Julia Hairston, "Out of the Archive: Four Newly-Identified Figures in Tullia d'Aragona's *Rime della Signora Tullia di Aragona et di diversi a lei* (1547)," *MLN* 118 (2003): 259. On Caterina's *Experiments* see, in addition to Pasolini, *Caterina Sforza,* vol. 3: Umberto Foschi, "Fantasia e superstizione delle ricette di Caterina Sforza," *Bollettino economico. Organo ufficile della C.C.I.A.A. di Ravenna* 43, no. 2 (1988): 31–36; Natale Graziani and Gabriella Venturelli, *Caterina Sforza* (Milan: Dall'Oglio, 1988), 141–149; Mario Tabanelli, "Ricette di medicina dal libro *Degli experimenti* di Caterina Sforza," *La Piê* 43 (1970): 195–198; and *Ricettario di bellezza,* ed. Luigi Pescasio (Castiglione delle Stiviere, 1971). See also the essays contained in the exhibition catalog edited by Valeria Novielli, *Caterina Sforza: Una donna del Cinquecento* (Imola: Mandragora, 2000).

4. See Machiavelli, *Discorsi* 3, 6, modified in *Le istorie fiorentine,* 8, 34; *Arte della guerra,* 6; and *Il principe,* chap. 20. Caterina Sforza was by no means universally admired: she was also feared, and many opposed her designs to regain power after being divested of her claim on Imola and Forlì. For Caterina's political biography, see Pasolini, *Caterina Sforza;* Ernst Breisach, *Caterina Sforza. A Renaissance Virago* (Chicago: University of Chicago Press, 1967); and Graziani and Venturelli, *Caterina Sforza.*

5. On alchemy at the Medici court, see, for example, Suzanne Butters, *The Triumph of Vulcan: Sculptors' Tools, Porphyry, and the Prince in Ducal Florence* (Florence: Leo S. Olschki, 1996); Alfredo Perifano, *L'alchimie à la cour de Côme Ier de Médicis: Savoirs, culture et politique* (Paris: Honore Champion, 1997); Valentina Conticelli,

Guardaroba di cose rare et preziose: lo studiolo di Francesco I de' Medici: arte, storia, e significati (Lugano: Lumieres Internationales, 2007).

6. "A dare gran peso a uno scudo, o Ducato de oro senza carigo de conscientia, et se fosse meno tre giulij verrà a justo peso *Secondum Cosimum*" (Pasolini, *Caterina Sforza*, 3:620, italics mine). Cosimo was the patron of Marsilio Ficino, who dedicated a translation of Hermes Trismegistus to Cosimo; the Latin translation was immediately translated into Italian. An alchemical work attributed to Cosimo, *Cosmus de Medicis servus fidelis manu propria scripsit Pio paper* (1475), is preserved in the Bibliotoeca di San Marco in Venice; Giovanni Carbonelli suggests that Caterina may have known this work (see Giovanni Carbonelli, *Sulle fonti storiche della chimica e dell'alchimia in Italia* [Rome: Istituto Nazionale Medico Farmacologico, 1925], x).

7. "Questo è un rimedio da guarire omne sorta de febre provata per Cosimo De Medici" (Pasolini, *Caterina Sforza*, 3:680).

8. On the ambiguity of early modern uses of the terms "experiment" and "experience" see Charles Schmitt, "Experience and Experiment: A Comparison of Zabarella's View with Galileo's in *De Motu*," *Studies in the Renaissance* 16 (1969): 80–138. See also Alisha Rankin, "Becoming an Expert Practitioner: Court Experimentalism and the Medical Skills of Anna of Saxony (1532–1585)," *Isis* 98 no. 1 (March 2007): 52; and Jole Agrimi and Chiara Crisciani, "Per una ricerca su '*experimentum-experimenta*': Riflessione epistemologica e tradizione medica (secoli XIII-XIV)," in *Presenza del lessico greco e latino nelle lingue contemporanee: Ciclo di lezioni tenute all'Universita di Macerata nell'a.a. 1987/88,* ed. Pietro Janni and Innocenzo Mazzini (Macerata: Universita degli studi, 1990) 9–49.

9. Monica H. Green highlights the importance of women's activity in collecting and preserving "family remedies" (*Women's Healthcare in the Medieval West* [Aldershot: Ashgate, 2000], 20); see also T. Adolphus Trollope, *A Decade of Italian Women* (London: Chapman and Hall, 1859), 1:264. In early modern Germany, Alisha Rankin has demonstrated that noblewomen were admired for their knowledge and expertise with medicinal remedies and secrets, often setting them down in "little handwritten books of medicine, or books of art" (*Panaceia's Daughters: Noblewomen as Healers in Early Modern Germany* (Chicago: The University of Chicago Press, 2013), 61. See also the essays collected in *Renaissance Studies* 28, no. 4 (2014), edited by Sharon T. Strocchia, including Rankin, "Exotic Materials and Treasured Knowledge: The Valuable Legacy of Noblewomen's Remedies in Early Modern Germany," 533–555.

10. Biblioteca Nazionale Centrale di Firenze (BNCF), Fondo Magliabechiano, XI; some of these are cited in Raffaele Ciasca, *L'arte dei medici e speziali nella storia e nel commercio fiorentino dal secolo XII al XV* (Florence: Olschki, 1927), 350–353. While many of these are anonymous and therefore their female authorship cannot be established with certainty, their close attention to women's health and beauty suggests, at the very least, a female audience. Many more similar works are listed in the Biblioteca Nazionale's inventories: see, for example, in Fondo Palatino, 1023, 1024, and 1026.

11. The *Ricettario galante* is transcribed from a manuscript in the Biblioteca dell'Università di Bologna (1352), titled *Libro per farsi bella—Secreti sul principio del secolo XVI—Secreti Cod. Ms. Saec. XV—Ex Bibliotecha Io. Iacobi Amadei Bononien. Canonici S. Mariae Majoris* (see *Ricettario galante. Del principio del secolo XVI,* ed. Olindo Guerrini [Bologna: Gaetano Romagnoli], 1883). Magliabechiano XI, 73, no. 2 (BNCF), contains recipes attributed to "Hypolita Sforza, Duchess of Calabria."

12. On women's recipe books, including in early modern England, see Jayne Elizabeth Archer, "Women and Chemistry in Early Modern England: The Manuscript Recipt Book (c. 1616) of Sarah Wigges," in *Gender and Scientific Discourse in Early Modern Culture,* ed. Kathleen P. Long (Aldershot: Ashgate, 2010), 191–216; Elaine Leong and Sara Pennell, "Recipe Collections and the Currency of Knowledge in the Early Modern 'Medical Marketplace,'" in *Medicine and the Market in England and Its Colonies, c. 1450–c. 1850,* ed. Mark S. R. Jenner and Patrick Wallis (New York: Palgrave, 2007), 133–152; Lynette Hunter, "Women and Domestic Medicine: Lady Experimenters 1570–1620," in *Women, Science, and Medicine 1500–1700: Mothers and Sisters of the Royal Society,* ed. Lynette Hunter and Sarah Hutton (Stroud: Sutton, 1997), 89–107; Sara Pennell, "Perfecting Practice? Women's Manscript Recipes in Early Modern England," in *Early Modern Women's Manuscript Writing,* ed. V. Burke and J. Gibson (Aldershot: Ashgate, 2004), 237–258; Linda Pollock, *With Faith and Physic: The Life of a Tudor Gentlewoman, Lady Grace Mildmay, 1552–1620* (New York: St. Martin's Press, 1993); Elizabeth J. Spiller, "Introduction," in *Seventeenth-Century English Recipe Books: Cooking, Physic and Chirurgery in the Works of Elizabeth Talbot Grey and Aletheia Talbot Howard* (Aldershot: Ashgate Publishing, 2008), ix–xxxi.

13. For an overview of alchemical activity at early modern European courts, see Didier Kahn, "Alchimia," in *Storia della scienza. Vol. 5: La rivoluzione scientifica* (Treccani, 2002), 389–398.

14. See Daniel Jütte, "Trading in Secrets: Jews and the Early Modern Quest for Clandestine Knowledge," *Isis* 103, no. 4 (Dec. 2012): 671.

15. See Bruce T. Moran, "Courts and Academies," in *The Cambridge History of Science. Vol. 3: Early Modern Science,* ed. Lorraine Daston and Katharine Park (Cambridge: Cambridge University Press, 2006), 253. On the role of the court in the development of the new scientific culture, see also Mario Biagioli's fundamental study of Galileo, *Galileo Courtier: The Practice of Science in the Culture of Absolutism* (Chicago: University of Chicago Press, 1993), 2.

16. William Eamon, "Court, Academy, and Printing House: Patronage and Scientific Careers in Late Renaissance Italy," in *Patronage and Institutions: Science, Technology, and Medicine at the European Court,* ed. Bruce T. Moran (Rochester, N.Y.: Boydell Press, 1991), 39.

17. For this term, see Rankin, "Becoming an Expert Practitioner," 23. See also Tara Nummedal's work on Anna Maria Zieglerin, a female alchemist in a Northern European ducal court: "Alchemical Reproduction and the Career of Anna Maria

Zieglerin," *Ambix* 48 (2001): 56–68 and "Anna Zieglerin's Alchemical Revelations," in *Secrets and Knowledge in Medicine and Science, 1500–1800,* ed. Elaine Leong and Alisha Rankin (Aldershot: Ashgate, 2011), 125–142.

18. In addition to a number of Latin recipes that were likely acquired from authoritative textual sources, a handful of secrets are attributed to specific people, for example, King Robert of Naples (1277–1343); see Pasolini, *Caterina Sforza,* 3:690.

19. On Cristoforo de Brugora (about whom little is known), see Anna Laghi, *Cristoforo de Brugora speziale della duchessa Bona e della corte sforzesca,* in *Atti del nono convegno culturale e professionale dei farmacisti dell'alta Italia* [Pavia, 1959], 141–147); see also Graziani and Venturelli, *Caterina Sforza,* 309–310n4. Archival documents show that Brugora sought—and won—permission to establish a private medicinal garden near the Basilica di S. Ambrogio to be used for the cultivation of "herbs, roots, flowers, seeds, and other things" to tend to the health of Bona and members of her court ("desiderando di fare uno horto in dicto teragio per tenirgli et in continente havere per uso et salute de quilli de caxa de vostra Signoria de quelle cose che cum fatica si tienano. Prega et richiede di gratia a la vostra Signoria che li voglia donare chel possa fare uno orto in dicto teragio . . . et liberamente che niuno che li abia a confinare non li possa tenere [uscii] aperti per intrare in dicto orto aciò possa havere le erbe, radice, flori, somenze et altri cose etiamo pure et nete senza alcuno suspecto de alchuni li possa praticare") (cited in Laghi, *Cristoforo de Brugora,* 144–145).

20. On Caterina's garden at Ravaldino, see Joyce de Vries, *Caterina Sforza and the Art of Appearances: Gender, Art and Culture in Early Modern Italy* (Aldershot: Ashgate, 2010), 103. Raffaella Fabiano Giannetto, *Medici Gardens: From Making to Design* (Philadelphia: University of Pennsylvania Press, 2008).

21. Conferred on the Sforza in 1470, Imola and Forlì were of strategic importance due to the position of the Romagna between northern and southern Italy, as well as for their agricultural economy and commerce with the East (see Graziani and Venturelli, *Caterina Sforza,* 17).

22. The major biographies of Caterina Sforza remain Pasolini, Breisach, and, most recently, de Vries. See also the essays collected in Novielli, *Caterina Sforza.* A more speculative account of Caterina's life is in Elizabeth Lev, *The Tigress of Forlì: Renaissance Italy's Most Courageous and Notorious Countess, Caterina Riario Sforza de' Medici* (New York: Houghton Mifflin, 2012).

23. See Machiavelli, *Discorsi* 3, 6, modified in *Le istorie fiorentine* 8, 34. Machiavelli, *Il principe,* chap. 20, cites Sforza as a negative example with reference to the usefulness of fortresses to rulers under seige. Sforza exchanged letters with Machiavelli and, as late as 1503, borrowed money from him (see Alessandro Sarti to Caterina Sforza, October 30, 1503, ASF, Med. a. Pr., f. 78, c. 302 and November 3, 1503, ASF, Med. a. Pr., f. 78, c. 306; in Pasolini, *Caterina Sforza,* 3:482–483).

24. Specifically, Caterina's lifting of her skirts appears to have originated with Machiavelli. See Julia Hairston, "Skirting the Issue: Machiavelli's Caterina Sforza," *Re-*

naissance Quarterly 53, no. 3 (2000): 687–712. On the incident at the Rocca di Ravaldino and Machiavelli's reworking of it, see also John Freccero, "Medusa and the Madonna of Forlì: Political Sexuality in Machiavelli," in *Machiavelli and the Discourse of Literature,* ed. Albert Russell Ascoli and Victoria Kahn (Ithaca: Cornell University Press, 1993), 161–178; Francesco Bausi, "Machiavelli and Caterina Sforza." *Archivio storico italiano* (1991), disp. IV: 887–892. For an early disavowal of the episode see Pier Desiderio Pasolini, "Nuovi documenti su Caterina Sforza," *Atti e Memorie: Deputazione di Storia per la Romagna* 15 (1897): 91–94.

25. See Machiavelli, *Il principe,* chap. 20, and *Arte della guerra,* 1:670–671.

26. See Graziani and Venturelli, *Caterina Sforza,* 71–72.

27. "Deliberò dita Madona de volerla adotare de qualque sove zentileze, zoè di de fare una casina come uno barco, overo zardine, come per ora se costuma ala cità di Feraria e in multe alter loco, da poter tener hogna soa salvadichume overe animale inracionale de hogne altra rasone; e qui lore pore a piantare hogne suova frascha frotifera." (Bernardi, *Cronache forlivesi,* vol. 1, part 2, 122–123; cited in de Vries, *Caterina Sforza,* 104).

28. Graziani and Venturelli, *Caterina Sforza,* 49. On Isabella d'Este and her *buffona,* a female jester, see Alessandro Luzio and Rodolfo Renier, "Buffoni, nani e schiavi dei Gonzaga ai tempi di Isabella d'Este," *Nuova antologia* 34 (1891): 618–650; 25 (1891): 112–146.

29. de Vries, *Caterina Sforza,* 35. On the court of Caterina and Girolamo as the "reflection of a second-tier signoria" (*lo specchio di una signoria di second'ordine*), see Graziani and Venturelli, *Caterina Sforza,* 73.

30. de Vries, *Caterina Sforza,* 129–175.

31. "se Franzosi biasmano la viltà de li homini, almeno debeno laudare lo ardire e valore de le donne Italiane," Isabella d'Este (Gonzaga) to the marchese of Mantua, January 14, 1500 (Archivio di Stato di Mantova, Archivio Gonzaga, Copialettere della marchesa, doc. LXXI; cited in Pasolini, "Nuovi documenti," 105n1).

32. "before a notary public she renounced the titles to her states in her own name and on behalf of her children" ("per mano di publico Notareo renuntiò a li stati in suo nome proporio et come tutricie de' sua figlioli"; letter of Francesco Fortunati to Ottaviano and Cesare Riario, July 8, 1501, in Pasolini, *Caterina Sforza,* 3:455).

33. See Kate Lowe, *Nuns' Chronicles and Convent Culture in Renaissance and Counter-Reformation Italy* (Cambridge: Cambridge University Press, 2003), 176 and n128. Sforza, like other noblewomen of the period, was granted visitation privileges to the enclosed convent of Le Murate under the pontificate of Sixtus IV (1471–1484) (see Sharon T. Strocchia, *Nuns and Nunneries in Renaissance Florence* [Baltimore: Johns Hopkins University Press, 2009], 169).

34. See Sister Giustina Niccolini, *The Chronicle of Le Murate,* ed. and trans. Saundra Weddle (Toronto: Iter Inc., Centre for Reformation and Renaissance Studies, 2011), 142–143: "She was always very affectionate toward and devoted to our convent,

and they say she made the first cell located beyond the *sala,* which gives onto the large courtyard with the dormitory above."

35. Ibid., n161. On Sforza's patronage of this and other institutions see Joyce de Vries, "Casting Her Widowhood: The Contemporary and Posthumous Portraits of Caterina Sforza," in *Widowhood and Visual Culture in Early Modern Europe,* ed. Allison Levy (Aldershot: Ashgate, 2003), 88.

36. Letters exchanged among Sforza's sons Cesare, Galeazzo, and Ottaviano lament that Caterina herself is the main obstacle to such a return (ASF, Med. a. Pr., f. 77, n. 129; see Pasolini, *Caterina Sforza,* 3:474); this opinion is echoed by the Florentine commissioner general of the Dieci di Libertà and the Balia, Giovan Battista Ridolfi, who cautions that Sforza is "strongly hated and feared" by her former subjects (*lei vi è forte odiata et temonla assai*) (ASF, Dieci di Balia, Carteggio, Responsive, Registro 74, c. 135; see Pasolini, *Caterina Sforza,* 3:473–474).

37. Sforza's grave is no longer visible but Pasolini transcribes the inscription from her tombstone: CATERINA SFORTIA / MEDICES / COMITISSA ET DOMINA / IMOLAE FORLIVII / OBIIT IV KAL. IUNII / MDIX (Pasolini, *Caterina Sforza,* 3:587).

38. Caterina's testament is in ASF, Med. a. pr., Cart. Priv., f. 99, n. 12; it is transcribed in Pasolini, *Caterina Sforza,* 3:537–547. She distributed her holdings outside of Florence among her first five sons, while her Medici son, Giovanni, inherited those within Florence. She also made provisions for Ottaviano's daughter, Cornelia, and Galeazzo's daughter, Giulia. Caterina's daughter, Bianca, is not named in the will (perhaps because she had received a dowry upon her marriage in 1503).

39. [Sforza], *Experimenti de la Ex[ellentissi]ma S[igno]ra Caterina da Furlij Matre de lo inlux[trissi]mo S[ignor] Giovanni de Medici.* I am grateful to the curent owner of the archive, who wishes to remain anonymous, for allowing me access to the manuscript during my 2010–2011 National Endowment for the Humanities fellowship year for research on this book. On Lucantonio Cuppano, see Carlo Coppi da Gorzano, "Il Conte Lucantonio Coppi detto Cuppano, ultimo condottiero delle Bande Nere e dimenticato Governatore Generale di Piombino (1507–1557)," *Rivista araldica* 3 (1960): 87–105. Cuppano appears as an interlocutor in the *Rime* of Tullia d'Aragona (see Hairston, "Out of the Archives," 259).

40. "Questo pergamena è una procura di Gio. de' Medici . . . per l'eredità della Caterina Sforza Riario di lui madre 1514" ([Sforza], *Experimenti*). Pasolini assigns a date of circa 1525 to the manuscript on the basis of the letter by Giovanni dalle Bande Nere regarding the collection, discussed shortly (see Pasolini, *Caterina Sforza,* 3:609, 614n1). A note in a much later hand designates the manuscript as part of the library of Giovan Battista Mannajoni, a well-known seventeenth-century physician and member of the Società botanica fiorentina who traveled with the botanist and natural philosopher Pier'Antonio Micheli (see Giovanni Targioni Tozzetti and Adolfo Targioni-Tozzetti, eds., *Notizie della vita e delle opere di Pier'Antonio Micheli* [Florence: Le Monnier, 1858], 193n1). It was then acquired by Pietro Bigazzi, a Florentine book collector,

in a sale of Mannajoni's medical books. It passed through several more hands before being purchased by Pasolini in October 1887 (Pasolini, *Caterina Sforza*, 3:609).

41. Even the name of the copyist is partially encoded. The key, which proves generally accurate when applied to the encrypted passages in the manuscript, reads:

b   f   h   p   x

A   E   I   O   U

42. "Ci trovamo manco nelli forzeri da Roma uno libro scritto a mano di ricette di più et varie cose operate: che sensa fallo nisuno lo ritrovamo, ché in ogni modo lo volemo" (Jovanni de' Medeci [Giovanni dalle Bande Nere] to Francesco Suasio, December 29, 1525, transcribed in Carlo Milanesi [raccolte da Filippo Moisé], "Lettere di Giovanni de' Medici, detto delle Bande Nere," *Archivio Storico Italiano*, Nuova serie, v.IX, part 2 [Firenze: Vieussieux Editore, 1859], 127). As Rankin shows in a study of Anna of Saxony (known throughout Europe for her medical-alchemical recipes and especially her white and yellow *acqua vitae*), it was not uncommon for mothers to pass recipes on to their sons in this fashion (see Rankin, "Becoming an Expert Practitioner," 35). For another example, see the case of Isabella d'Este and her son Federico, discussed shortly.

43. Pasolini, *Caterina Sforza*, 3:601.

44. "In nome de Dio in questo libro se noteranno alcuni experimenti cavati da lo originale de la illux.ma madonna Caterina da Furli Matre de lo illux.mo S.re joann de medici mio S.re et patrone et per essere lo originale scripto de man propria de dicta madonna . . . non me curarò durare fatiga a rescriverli" (*Experimenti*, iv.).

45. [Sforza], *Experimenti de la Ex.ma S.ra Caterina da Furlj, Matre de lo Inllux.mo Signor Giovanni de Medici* (Imola: Ignazio Galeati e Figlio, 1894), x. The copy is executed in at least two hands: that of Cuppano and a second hand that made additional entries.

46. "Il vero modo de calcinarlo sie questo, como usava madama da Furlì" ([Sforza], *Experimenti*, 2r-v; see Pasolini, *Caterina Sforza*, 3:618). The question of transcription and transmission is a complex one, complicated by the very nature of such manuscript recipe collections, which were often added to over time. Because the Cuppano manuscript is held in a private archive, and because Pasolini's 1894 edition is rare, page references for citations from the *Experiments* are provided for Pasolini's far more accessible, although not comprehensive, 1893 edition, where possible and unless otherwise noted.

47. Trollope's speculation that Caterina compiled the whole of her *Experiments* only during her final years in Florence seems unlikely, although she was certainly collecting recipes until the year before her death (see Trollope, *A Decade of Italian Women*, 1:264). Pescasio surmises that the recipes were compiled over the course of her life (*Ricettario di bellezza di Caterina Sforza*, 5).

48. "se deve extimare tucti li altri essere veri per essere experimentati da cusi alta madonna" (Pasolini, *Caterina Sforza*, 3:617).

49. In her study of cookery and "household physick" books in early modern England, Elizabeth Tebeaux emphasizes the role of technical writing as a memory aid meant to supplement women's existing knowledge: "The fact that these books . . . gave less emphasis to exact quantities than to ingredients and placed little emphasis on precise instructions for the making and then using [of] foods and folk medicines implies that women readers had a basic knowledge of cooking which they learned through oral transmission and from demonstration" ("Women and Technical Writing, 1475–1700: Technology, Literacy, and Development of a Genre," in *Women, Science and Medicine 1500–1700: Mothers and Sisters of the Royal Society,* ed. Lynette Hunter and Sarah Hutton [Stroud: Sutton, 1997], 37).

50. "Per non essere le altre sue virtù notate in dicto originale si non per infinite, lassarò in questo volume lo spatio a causa si mai persona arivasse a questa cognitione si degni comunicarlo" (Pasolini, *Caterina Sforza,* 3:618). The Cuppano transcription includes more than a dozen blank pages. Regardless of how much blank space was available, readers were likely to make their own notes in the margin regarding the success of the recipe or any adjustments that should be made to it.

51. "è mortale peccato tenere ascoso tanto tesoro" (ibid.). Highlighting both the secrecy and the utility of Sforza's secrets, Cuppano again adds that the recipes were "written down in [Sforza's] own hand, and so as not to keep these marvelous secrets hidden I now leave a record of them ("experimentati da cusi alta madonna e scripti de sua propria mano et perché questi mirabili secreti non siano ascosi per me se ne tira memoria", [ibid., 617]).

52. On the traditions and strategies of secrecy associated with alchemical texts, see Pamela O. Long, *Openness, Secrecy, Authorship: Technical Arts and the Culture of Knowledge from Antiquity to the Renaissance* (Baltimore: Johns Hopkins University Press, 2001); on the culture of secrecy more generally, see Karma Lochrie, *Covert Operations: The Medieval Uses of Secrecy* (Philadelphia: University of Pennsylvania Press, 1999).

53. For example, recipes for increasing sexual desire or for determining—or simulating—virginity. On this aspect of Sforza's compilation, see Meredith K. Ray, "Impotence and Corruption: Sexual Function and Dysfunction in Early Modern Italian Books of Secrets," in *Cuckoldry, Impotence and Adultery in Europe (15th–17th Century),* ed. Sara Matthews-Grieco [Aldershot: Ashgate, 2014], 125–146).

54. Pasolini, *Caterina Sforza,* 3:669.

55. "Hec omnia pulveriza soctite quam polverim commisceas cum antimone et tanto vini albi quantum tu ipse judicaveris oportare postea destilato hoc lambico er ignem lentum et efficietur aqua clarissima" (ibid., 689).

56. "fa regiovenire la persona et de morto fa vivo ciò e si una persona fusse tanto gravata de infermitate che li medici l'abandonassino per incurabile e morta la reduce a sanità" (ibid., 639).

57. Ibid.

58. See William Newman, "The Homunculus and His Forebears: Wonders of Art and Nature," in *Natural Particulars: Nature and the Disciplines in Renaissance Europe,* ed. Anthony Grafton and Nancy Siraisi (Cambridge, Mass: MIT Press, 1999), 321–345.

59. "la seconda è meglio che la prima ma la terza è de color de sangue, e sopra tutte" (Pasolini, *Caterina Sforza,* 3:622),

60. "A far dormire una persona per tal modo che porai operare in cirurgia quelche vorai e non sentirà est probatum" (ibid., 769).

61. See ibid., 3:679, and also 672, 673. Poison was a real threat: Sforza herself was accused of having attempted to poison Pope Alexander IV by sending him letters contaminated with plague, although it is widely agreed that this accusation was a pretext on Alexander's part for confining her to Castel Sant'Angelo. Among the recipes for protecting written documents in the *Experiments* is one for an "Acqua da cancelare le lettere scritte" and another "Acqua da scrivere che non si veda," both offered in Latin (ibid., 784, 790).

62. See Nancy Siraisi, *Medieval and Early Renaissance Medicine: An Introduction to Knowledge and Practice* (Chicago: University of Chicago Press, 1990), 147–148; see also Anna Maria Guccini, "L'arte dei Semplici: Alchimie e medicina naturalistica tra conoscenza e credenza all'epoca di Caterina," in *Caterina Sforza,* ed. Novielli, 131–138.

63. See Marsilio Ficino, *Three Books on Life: A Critical Edition and Translation with Introduction and Notes,* ed. Carol V. Kaske and John R. Clark (Binghamton, N.Y.: The Renaissance Society of America, 1989); see also Peter J. Forshaw, "Marsilio Ficino and the Chemical Art," in *Laus Platonici Philosophi: Marsilio Ficino and His Influence,* ed. Stephen Clucas, Peter J. Forshaw, and Valery Rees (Leiden: Brill, 2011), 198–249. Caterina may well have known Ficino's *De vita libri tres* (1480–1489), especially given Ficino's links to Cosimo de' Medici, to whom Caterina's father had ties.

64. In some cases, the affinity is merely linguistic, as in "corallo" (coral) to treat the "core" (heart) (see Pasolini, *Caterina Sforza,* 3:612–613). On the vast bibliography on Paracelsus, see, inter alia: Paracelsus, *Selected Writings,* ed. Jolande Jacobi, trans. Norbert Guterman (New York: Pantheon Bollingen Series 28, 1958); Philip Ball, *The Devil's Doctor: Paracelsus and the World of Renaissance Magic and Science* (New York: Farrar, Straus and Giroux, 2006), 144–145; Allen Debus, *The Chemical Philosophy: Paracelsian Science and Medicine in the Sixteenth and Seventeenth Centuries* (New York: Science History Publications, 1977); Walter Pagel, *Paracelsus: An Introduction to Philosophical Medicine in the Era of the Renaissance,* 2nd ed. (Basle: Karger, 1982). On Della Porta's doctrine of signatures, see William Eamon, *Science and the Secrets of Nature: Books of Secrets in Medieval and Early Modern Culture* (Princeton, N.J.: Princeton University Press, 1994), 214–215.

65. See Pier Andrea Mattioli, *I discorsi nei sei libri di Pedacio Dioscoride Anzarbeo della materia medicinale* (Venice, 1712), cited in Emmanuela Renzetti and Rodolfo Taini, "Le cure dell'amore: Desiderio e passione in alcuni libri di segreti," *Sanità scienza e storia: Rivista del Centro italiano di storia sanitaria e ospitaliera* (1986): 72.

66. Pasolini, *Caterina Sforza*, 3:625. On the relationship between alchemy and astrology see William R. Newman and Antony Grafton, eds., *Secrets of Nature: Astrology and Alchemy in Early Modern Europe* (Cambridge, Mass.: MIT Press, 2006). On humoral theory in Renaissance medicine, see Siraisi, *Medieval and Early Renaissance Medicine*.

67. "non mi parse conveniente a notificarlo nequiquam parlarne, mi essere vero creditore dela prefata bona memoria di Madonna mia illustrissima de fiorini 587 et ultra in magiore somma per resto de robbe date a prefate sua llustrissima Signoria in Forlì como appare uno clarissimo conto per li mei libri. Item de la bona memoria del magnifico Ziovann di Medici per robbe de la mia botega date a sua Magnificentia" (Lodovico Albertini to Francesco Fortunati, ASF, Med. a Pr., unnumbered [see Pasolini, *Caterina Sforza*, 3:551, doc. 1364]). In her testament, Sforza left to Francesco Fortunati, her agent and confessor, the disposition of her books, papers, and letters: "Item jure legati reliquit et legavit dicto domine Francisco sine prejudicio tamen supra vel infra dispositorum per dictam Illus.am dominam omnes et singulos suos dicte Ill. me Domine testatricis libros scripturas et literas et alias quascumque scripturas publicas vel privatas et omnes et quoscumque libros scripturas et literas quascumque penes eam quomodo libet existentes cujuscumque et cujusvis qualitatis vel condicionis existerent ita quod possit de illis prefatus dominus Franciscus disponere et facere" (ibid., 541, doc. 1355]). She also made Fortunati guardian of her son Giovanni (see de Vries, *Caterina Sforza*, 231).

68. On *speziali* and their shops, see Adriano Galassi and Romano Sarzi, *Alla Syrena: Spezeria del '600 in Mantova* (Mantua: Editoriale Sometti, 2000); see also Sharon Strocchia, "The Nun Apothecaries of Renaissance Florence: Marketing Medicines in the Convent," *Renaissance Studies* 25 (2011): 627–647; James Shaw and Evelyn Welch, *Making and Marketing Medicine in Renaissance Florence* (Amsterdam: Rodopi, 2011).

69. See Graziani and Venturelli who surmise that Sforza kept a laboratory: "un vero e proprio laboratorio di alchimia con calderoni, storte ed alambicchi in grado di estrarre e distillare vegetali, sostanze organiche ed inorganiche per poi comporre medicamenti, crème e lozioni di bellezza" (*Caterina Sforza*, 143). On Sforza's renovations and decoration of the Paradiso, see de Vries, *Caterina Sforza*, 74–127.

70. The chronicler of Le Murate, Giustina Niccolini, recalled Sforza's charity in sending grain to the sisters during the food shortage of 1498 and the gift she made to the convent in her will; while Sforza's letters to the abbess make note of the donations she regularly made to the convent (*Chronicle*, 142–145).

71. "Ho con summo piacere lecte le lettere vostre vedendo che non ve ne siate domenticata de me: et che ne le oratione vostre ne faciate di continuo qualche commemoratione. . . . Le pome granate et altri fructi dell'orto vostro ne sonno state gratissime per multi respecti" (Pasolini, *Caterina Sforza*, 3:398, 402).

72. See Strocchia, "Nun Apothecaries." Convent apothecaries were generally staffed by a skilled, expert practitioner and an apprentice, an "arrangement that par-

alleled the transmission of other craft skills" (637). For similar activity by nuns elsewhere in Italy, see, for example, Gianna Pomata, "Medicina delle monache. Pratiche terapeutiche nei monasteri femminili di Bologna in età moderna," in *I monasteri femminili come centri di cultura fra Rinascimento e Barocco,* ed. Gianna Pomata and Gabriella Zarri (Rome: Edizioni di storia e letteratura, 2005), 331–363. Given Caterina's close connection to the convent of Le Murate, it is interesting to note that when, decades later, its pharmacy was badly damaged in a flood, it was Caterina's descendant Ferdinando de' Medici who personally financed its restoration (see Strocchia, "Nun Apothecaries," 646).

73. On the scriptorium at Le Murate, see Lowe, *Nuns' Chronicles,* 288–299.

74. As Graziani and Venturelli note, the production of cosmetics was part and parcel of the Renaissance's "scientific meditation on the nature of man" and interest in achieving a harmonious balance between microcosm and macrocosm (*Caterina Sforza,* 143).

75. "A fare un acqua mirabile a conservare il viso contra ogni macula: . . . fa stilare ultra modo de acqua rosata la prima e quasi argento la seconda oro la terza e quasi balsino." (Pasolini, *Caterina Sforza,* 3:621).

76. "La prima acqua manda via lentiggine del volto . . . la seconda acqua . . . spegne le fistole. La terza, ch'è più di tutte forte . . . rode il ferro . . . raunando insieme queste tre acque, fa bello et biondo il capo come fila d'oro da la mattina a la sera" (*Ricettario galante,* 7). When combined, the three waters cause the hair to turn a permanent golden color, a result also prized in Sforza's text, which offers more than a dozen recipes for lightening the hair.

77. Pasolini, *Caterina Sforza,* 3:627, "Ad idem."

78. Ibid., "A fare la faccia bianchissima et bella et lucente, et colorita."

79. Ibid., 630–631.

80. See, for example, a clustering of recipes for beauty waters "a far bella" in ibid., 630–638.

81. Firenzuola, *On the Beauty of Women,* 45.

82. "Piglia cinabro, zaffrano, et solfo et fa destillare queste cose per lanbicco et quando te hai lavato el capo pettinate al sole et bagnia el pettine spesso in questa acqua stillata et cusí te asciutta al sole et verrà bella come oro" (Pasolini, *Caterina Sforza,* 3:657).

83. "Piglia semenza de ortiga et falla bollir in la lissia che fai con la tua cenere al solito et lavate et veneranno bellissimi" (ibid., 653).

84. Firenzuola, *On the Beauty of Women,* 63.

85. "Piglia succo de cicuta, ogne dí fatte et se ben fossino grandi deventaranno piccole et avvertise voi giovani che non sete in eta perfetta et che sete putte, se voi ogne dí le[vate] vostre tecte con ditto onzo non ve creseranno più et restaranno in quelle bellezza e quella durezza" (*Experimenti,* ed. Pasolini [1894], 180). Highly poisonous, hemlock was used as a sedative and to reduce swelling but had to be handled

with care. Pasolini's 1893 transcription of the *Experiments* does not include this cluster of recipes relating to the breasts.

86. "se una donna fuse de sesanta anni la farà parere de vinti" (Pasolini, *Caterina Sforza*, 3:617–618).

87. "A cavare l'acqua del talcho: Piglia el talcho calcinato nel modo dicto de sopra et ponilo alambicho in questo modo farai u[n] fornello da vento e mectivi dentro un catinecto de ramo che sia copoluto infondo e sempre se vada allargando incima e mectili dentro . . . et benserato intorno che non possa vaporare; poi li darai focho lentamente de carbone (ibid., 619).

88. "Est quippe Caterina inter mulieres nostri seculi formosissima, et eleganti aspectu, ac per omnis corporis artus mirifice decorata est" (J. F. Foresti da Bergamo, *De plurimis claris selectisque mulieribus* [Ferrara: Lorenzo Rosso da Valenza, 1497], 561).

89. On the possible identification of Caterina Sforza as the subject of Lorenzo di Credi's *Portrait of a Young Woman with Jasmine* (1485–1490), see de Vries, *Caterina Sforza*, 259.

90. Sforza had to be practical about her luxury acquisitions, sometimes pawning them in times of financial strain or using them as collateral (de Vries, *Caterina Sforza*, 156–159).

91. See Joyce de Vries, "Caterina Sforza's Portrait Medals: Power, Gender, and Representation in the Italian Renaissance Court," *Women's Art Journal* 24, no. 1 (2003): 24. On inventories of Caterina's possessions, see also de Vries, *Caterina Sforza*, 130–166, 209–210, 230–232.

92. On Rudolf II and alchemy. see Robert J. W. Evans, *Rudolf II and His World: A Study in Intellectual History, 1576–1612* (London: Thames and Hudson, 1997), and Peter Marshall, *The Magic Circle of Rudolf II: Alchemy and Astrology in Renaissance Prague* (New York: Walker, 2006). For the court of Philip II, see William Eamon, "Masters of Fire: Italian Alchemists in the Court of Philip II of Spain," in *Chymia: Science and Nature in Early Modern Europe (1450–1750),* ed. Miguel Lopez Perez and Didier Kahn (Cambridge: Cambridge Scholars Publishing, 2010), 138–156.

93. The *Experiments* includes a medicinal recipe for a potent emetic powder attributed to Maximilian (see Pasolini, *Caterina Sforza*, 3:670). On experiments at the Medici court, see Luciano Berti, *Il Principe dello Studiolo: Francesco I dei Medici e la fine del Rinascimento fiorentino* (Florence: Edam, 1967); Giulio Cesare Orlandi Lensi, *L'arte segreta: Cosimo e Francesco de' Medici alchimisti* (Florence: Convivio, 1978); Valentina Conticelli, *Guardaroba di cose rare et prezioso: Lo studiolo di Francesco I de' Medici: arte, storia e significati* (La Spezia: Lumieres Internationales, 2007). On perceptions abroad of the Italian aristocracy's fascination with alchemy, for example, in England, see Michael G. Brennan, "The Medicean Dukes of Florence and Friar Lawrence's 'Distilling Liquor' (*Romeo and Juliet*, IV.i.94)," *Notes and Queries* 38, no. 4 (1991): 473–476. On the Medici *fonderia* founded by Cosimo I and continued by Francesco I, see

Valentina Conticelli, *L'alchimia e le arti: la fonderia degli Uffizi da laboratorio a stanza delle meraviglie* (Livorno: Sillabe, 2012).

94. On the alum trade, see Charles Singer, *The Earliest Chemical Industry: An Essay in the Historical Relations of Economics and Technology Illustrated from the Alum Trade* (London: The Folio Society, 1948); Raymond de Roover, *The Rise and Decline of the Medici Bank, 1397–1494* (Washington, D.C.: Beard Books, 1999), 152–164. On the production and importance of saltpeter, see Bert S. Hall, *Weapons and Warfare in Renaissance Europe: Gunpowder, Technology, and Tactics* (Baltimore: Johns Hopkins University Press, 1997). A sixteenth-century letter in the Archivio di Stato of Florence, offers a typical example of this intertwining of alchemy with commerce and military defense: it describes a "gold tincture" (*tintura d'oro*) that may prove of great value, especially in wartime: "Vi scrissi alli [di] passati che m'era venuto per amicitia per le mani una tintura d'oro, et che me ne saria dato un saggio, pensando ch'io sia huomo da farmi partito, credendo forse ch'io habbia vinto, dove ho perduto. Finalmente me l'hanno dato come vedrete qui incluso, lo potrete far' mirare, et quant' a me si promette mantenere el segreto chiaro secondo questo saggio, quale t[occa] di 22 [carati], sta all'effusione . . . perche mi dicono che d'una libbra d'argento ne raveranno in sei g[missing text] . . . messa d'argento, et messa d'oro, la cosa saria d'imp[missing text, possibly *importanza*] et massime per valersene in un bisogno di guerra, o d'altro simil. . . . Qui non bisogna dubitare . . . poi che non habbiamo da sborsare un carlino del nostro si primo non habbiamo fatto mille esperientie del fatto (Lattanzio Riccolini to Ugolino Grifoni, Rome, 28 ottobre, 15[?] [ASF, Med. a. Pr., 2, f. 140 1]).

95. For this reason, as Lawrence M. Principe puts it, "Alchemical knowledge was potentially as dangerous [in the early modern period] as the knowledge of nuclear weapons is today" ("Revealing Analogies: The Descriptive and Deceptive Roles of Sexuality and Gender in Latin Alchemy," in *Hidden Intercourse: Eros and Sexuality in the History of Western Esotericism*, ed. Wouter J. Hanegraaff and Jeffrey J. Kripal [Leiden: Brill, 2008], 218–19). In fact, most laws against alchemy focused on the production of counterfeit coinage.

96. "Questa è una acqua mirabile che dessolve el ferro et tutti li altri metalli et congela el mercurio et si pone in loco de balsamo" (Pasolini, *Caterina Sforza*, 3:777); see also the recipes, 781–782.

97. "Aqua rubea fixa ad rubeum qua tingit omnia coloris auris sic fit 24 karatarum" (ibid., 782).

98. "Aqua que redit denarios novos argenti, veteres in aspettu" (ibid., 621, 780).

99. "A dare gran peso a uno scudo, o ducato de oro. . . . Quando voli dare peso alli scudi ponili drento et lassali stare un poco de poi cavali et vedi si è al peso se non mettili di novo et fa cusí finché serà al peso" (ibid., 620).

100. "A dare gran peso a uno scudo, o ducato de oro senza carigo de coscientia" (ibid.). On alchemical fraud, see Tara Nummedal, *Alchemy and Authority in the Holy Roman Empire* (Chicago: The University of Chicago Press, 2007). Forlì was authorized

by papal bull to coin money (see Trollope, *A Decade of Italian Women*, 1:265). On the iconography of the coins Caterina had minted during her regency, see de Vries, *Caterina Sforza*, 50–59.

101. As de Vries notes, Caterina slowly "lost not just the material objects, but also the physical proof of the position her family had so recently enjoyed and . . . their social ties" (*Caterina Sforza*, 157). On pawning luxury goods and using jewels as loan collateral, see 156–159; see also Evelyn S. Welch, *Shopping in the Renaissance: Consumer Cultures in Italy 1400–1600* (New Haven: Yale University Press, 2005), 196–203.

102. "Io mi trovo alle spalle 24 boche: 5 cavalli et tre muli et a tucti ho ad fare le spese, et non ho uno soldo; et qui non ho trovato bene alcuno, né persona me ha voluto subvenire pure da uno bicchiere d'acqua" (Caterina Sforza to Ottaviano Riario, July 22, 1502, in Pasolini, *Caterina Sforza*, 3:467–468).

103. See de Vries, *Caterina Sforza*, 156–159.

104. Over a thousand letters and documents pertaining to Caterina Sforza are recorded, and many transcribed, in Pasolini's invaluable *Caterina Sforza*, vol. 3. For an assessment of Pasolini's archival findings and suggestions for further work, see Cecil H. Clough's critique in "The Sources for the Biography of Caterina Sforza and for the History of Her State During Her Rule, with Some Hitherto Unpublished Letters Illustrative of Her Chancery Archives," *Atti e Memorie, Deputazione di storia patria per le provincie di Romagna*, 15–16 (1963): 57–143.

105. In a letter to Florimonte Brognolo, dated October 12, 1504, Isabella asks for "the recipe for the tooth powder, because we have finished ours" (*la recetta de la polvere de denti, perché la havemo finita*), cited in Alessandro Luzio and Rodolfo Renier, "Il lusso di Isabella d'Este, Marchesa di Mantova: accessori e segreti della toilette," *Nuova antologia di scienze, lettere ed arti* [1896]: 675. In a much later letter, she inquires after cosmetic powder from Venice: "We wish to acquire some Cyprus powder of the best quality; therefore you will do us a great favor by looking for it in Venice, if there is some good [powder] to be had; and if so, please send it to us" (*Haveriamo desiderio d'havere della polvere di Cipri che fosse de tutta excellentia; però ci farete cosa gratiossima a far cercare lí in Venetia se n'è di buona, et trovandosene vi piacerà de mandarcene*) (letter of Isabella d'Este, April 7, 1532, p. 674). On Isabella d'Este's letters, see Deanna Shemek, "In Continuous Expectation: Isabella d'Este's Epistolary Desire," in *Phaethon's Children: The Este Court and its Culture in Early Modern Ferrara*, ed. Dennis Looney and Deanna Shemek (Tempe: Arizona Center for Medieval and Renaissance Studies, 2005), 269–300 and Sarah Cockram, "Epistolary Masks: Self-Presentation and Dissimulation in the Letters of Isabella d'Este," *Italian Studies* 64, no. 1 (2009): 20–37.

106. See the letter of Isabella d'Este to her son Federico, May 18, 1516: "We send a small box, with three little containers of *compositione*: the crystal one with the gold lid is for the lady Queeen, those of horn are for madame the Queen Mother and her sister, the duchess of Lansone" (*Mandiamo una scatoletta, nela quale sono tre busoletti*

di compositione, quel di cristalo col coperto d'oro per la S.ra Regina, quelli di corno, l'uno per madama matre dil re, l'altro per la duchessa di Lansone sua sorella, cited in Luzio and Renier, "Il lusso di Isabella d'Este," 678–679).

107. "che noi sapiamo et possiamo fare, perchè in componer questi odori non cederessimo al miglior profumero del mondo, e però supplicate S[ua] M[aestà] non cambiare la bottega" (ibid.).

108. See the letter of Galeazzo Bentivoglio to Isabella d'Este, May 5, 1514: "Ho combatuto cum quanti profumieri ha questa città et cum quante signore co sono cusí spagnole come italiane che V[ostra] Ex[cellenza] fa et adopera la più excellente mistura et compositione che si trovi al mondo" (cited in ibid., 678).

109. Ibid., 682.

110. Isabella d'Este to Federico d'Este, May 18, 1516. See Welch, *Shopping in the Renaissance,* 270 for a discussion of this phrase. The recipient of the gift, whom Isabella hopes will not "change shop," was the queen of France. A 1508 letter from Isabella d'Este to Taddeo Albano listing the numerous ingredients the marchioness used to create her "compositioni" reflects her direct involvement in production (see in Luzio and Renier, "Il lusso di Isabella d'Este," 676).

111. "La recepta megliore del ovo et del zaferano non la voglio mandare per esser cosa tropo mirabile et de vera substantia" (Luigi Ciocha to Caterina Sforza, March 23, 1502 [ASF, Med. a. Pr., f. 77, n. 85 r-v, 85 bis r-v; see Pasolini, *Caterina Sforza,* 3:606]). I have modernized accents and punctuations in Caterina's correspondence but have retained the original orthography.

112. Caterina Sforza herself worked diligently to forge ties to the Mantuan court, declaring her devotion to Federico Gonzaga in several letters and seeking his support on her behalf abroad (see, e.g., Pasolini, *Caterina Sforza,* 3:458–459). Ciochi also wrote to Caterina to apprise her of the Marchioness's fashion choices and hairstyles, for which she was famous: in one letter he laments that it is simply too difficult to describe Isabella d'Este's clothing and hair in writing—one must see it in person: "Regarding sending you the fashions . . . in clothing and hairstyle worn by our illustrious madame marchesa, I can only say that there are not enough artists or canvases in the world because every day she does something new. But if a person expert [in such things] were to come here, he would see it for himself in four days and could more easily explain to you personally in an hour what I could neither write nor an artist paint in a month" (*"Circha el mandarvi li desegni de . . . vestimente come di testa che porta la n[ost]ra Ill[ustrissim]ma M[adam]a Marchesana dico che non ce bastariano quanti pictori et carte siano al mondo perchè ogni dì fa cose nove. Ma solo uno homo experto che venesse qua in quattro giorni vederia con l'ochio, et poteria molto meglio referire a bocha in una hora, ch'io non potria scrivere né uno pictore designare in uno mese"*) (Alovisius Ciocha to Caterina Sforza, December 1, 1502, ASF Med. A. Pr., f. 78, c. 181r-v and 181 bis r-v). The letter also contains a reference to a hand lotion ("unto") Ciochi is trying to procure for Caterina at Mantua (*Circha al unto per le mane dico che vederò di haverne da queste madone*).

113. "Ma zuro a Vostra Exellentia di portarvela se ben dovesse venire aposta perchè io anchora mi voglio trovar presente a tanto experimento et a una tanta satisfactione . . . et anche Vostra Exellentia non trovaria mai homo simile a mi perché c'è bisogna animosità, zoe non temere spiriti, fede, zoe credulo, secreto, zoe non se scoprire con homo del mondo et haver li istrumenti necessarii a tanta opera che né in el studio di Bologna né in Ferrara né in Parigi né a Roma non furono mai simili a li mei" (Luigi Ciocha to Caterina Sforza, March 23, 1502 [ASF, Med. a. Pr., f. 77, n. 85; see Pasolini, *Caterina Sforza,* 3:606]).

114. "un servitor fedele, e secreto, e buono d'anima" (Cortese, *I secreti,* 32).

115. "Madona Costanza prega V[ostra] Ex[ellen]tia a mandargli qualche profumi et polvere di Cipri et io la prego a compiacerla perchè anchora le ha de le gentileze de recambiare Vostra Signoria" (Luigi Ciocha to Caterina Sforza, May 5, 1502; ASF, Med. a. Pr., V doc. 1168; see Pasolini, *Caterina Sforza,* 3:607).

116. "servitore et partigiano fidelissimo a farne tutti li esperimenti del mondo" (ibid.).

117. "El s[ignor]e Marchese mi ha detto due volte se le pilole zovano a V[ostra] Ex[ellentissima] . . . mi potrà avisare se le pilole zovano, o no" (Luigi Ciocha to Ottaviano Sforza, May 31, 1500 [ASF, Med. a. Pr., f. 77, c. 11]).

118. Anna Hebrea to Caterina Sforza, March 15, 1508: "El quale unguento lo ponirete la sera et tenetelo fino ala matina e poi ve laverete con l'acqua pura di fiume, poi bagnerete el viso con l'acqua ne la quale è scripta *Acqua da Canicare.* Poi ponirete un poco di questo unguento bianco; poi prenderete manco di un granello di cece di questo sulimato e lo distemperarete con quest'acqua dove e scripto *Acqua dolce,* e poneretila nel viso, et ogni cosa ponirete suttilissimamente perchè è meglio. Lo unguento nero vale carlini quattro l'oncia, et l'acqua da canicare vale carlini quattro la foglietta. La ceretta, cioè l'unguento bianco, vale carlini octo l'oncia. Lo sulimato vale un ducato d'oro l'oncia, et l'acqua dolce vale un ducato d'oro la foglietta. Se Vostra Illustrissima Signoria ne adoperarà io mi rendo certo che continuo ce mandarete" (ASF, Med. a Pr., f. 125, n. 228; see Pasolini, *Caterina Sforza,* 3:608). Anna adds a postscript cautioning that the black unguent, in particular, is quite bitter in taste, but not poisonous, deriving as it does from the aloe plant (*Lo unguento negro è amaro, abattendose ad andare in bocca, sappiate non essere cosa cattiva e la amaritudine vene perchè c'entra lo aloe,* 609).

119. See Ariosto, *Satire,* 62. Francisco Delicado's *Retrato de la Lozana andaluza* offers several depictions of Jewish women selling cosmetics and medicinal remedies: one character states that they "go throughout the city of Rome repairing lost virginities and hawking treatments for the face" (Delicado, *Portrait of Lozana,* 29). In addition to their association with the production of cosmetics, Jews were associated more specifically with alchemy, beginning with ancient alchemical texts citing "Maria the Jewess" as an alchemical adept. Caterina Sforza had contact with Jews in other arenas as well, for example, intervening in 1490 to have Guglielmo d'Aia establish a bank in

Forlì following Girolamo Riario's assassination (John Larner, *The Lords of Romagna: Romagnol Society and the Origins of the Signorie* [Ithaca, N.Y.: Cornell University Press, 1965], 135).

120. See Jütte, "Trading in Secrets," 671, 685.

121. "mando a la vostra S[ignoria] tre fiaschete de aqua celeste per il compagno mio Don Piero dove che io stago in cassa sua a Fiorenzuola perchè è messo fidato de la quale fiaschete et ge ne una per il malle de testa e del stomaco. L'altra si è per el figato ma el ge bisogna metere dentro onze j.ª de diarodon abbatis. L'altra si è contra peste pigliandone uno poceto la matina per quelo zorno non piarà peste e così è scrito e ligato uno breve al collo per cadauna fiasceta" (Frate Bernardino di Gariboldi to Caterina Sforza, January 21, 1504, ASF, Med. a Pr., f. 125, n. 19; see Pasolini, *Caterina Sforza*, 3:607).

122. "De li oley non o potuto lavorare, ma de curto la vostra Signoria li haverà" (ibid.).

123. "Prego V.S. me voglia fare due fiaschi d'acqua de pigne et io manderò per Cechino una soma delle nostre fructe de qua" (Bianca de' Rossi to Caterina Sforza, April 28, 1508, ASF Med. a. Pr., f. 125, n. 243).

124. "De la rizetta io ve la darrò a la venutta vostra" (Caterina Sforza to unnamed correspondent, unsigned, 1504 (ASF, Med. a. Pr., f. 125, n. 71; see Pasolini, *Caterina Sforza*, 3:503–504).

125. "prego che la Exellentia Vostra me voglia attendere ad quel che la me promisse, cioè de mandarme quella ricetta da faro oro da xviiij carati, et mandarmela intieramente et al più presto" (unknown correspondent to Caterina Sforza, undated [1504?], ASF, Med a Pr, f. 125, c. 10; see Pasolini, *Caterina Sforza*, 3:607–608).

126. "sperando che la Ex[cellen]tia V[ostra] et m[esser] Jacomo insiem cum mi pigliasseno piacere de lui che se reputa in questa arte grande philosopho et molto intelligente (Lorenzo de Mantechitis to Caterina Sforza [ASF, Med. a. Pr., f. 125, n. 202; see Pasolini, *Caterina Sforza*, 3:603–605]). As Pasolini notes, the letter is preserved among other documents from 1504, but the internal reference to Giacomo Feo, dates it to between 1490 and 1495.

127. "perchè non intendesse quello che se lavorasse, como da la Signoria Vostra hebbi in commissione; pure a l'ultimo li disse che non era nulla, immo non credeva fusse possibile ma che faceva per dare ad intendere al maestro, el quale era pertinace che non se potesse fare arzento né fusse possibile affarne per questa via alchimica, dove che monstrandoli questo argento calcinato et in sua presentia redurlo in corpo cum sapone negro, salnitrio overo borace, che in simile forma pare cenere, che lui remanerà stupefacto a vedere che quella polvere sia reducta in arzento" (ibid., 603).

128. "et non seria quasi in extrema necessità como sonno se havesse tale vertù" (ibid.).

129. "Quando ancora la S[ignoria] V[ostra] o vero m[esser] Jacomo [Feo] vorano che se faza alcuna prova de quelle cose che io ho, quale tengo per bone, se farà; et

maxime circa quanto se contiene in quello libretto . . . perché ultra di quelle mai più intendo sopra ad ziò affaticarmi. Quanto che non vorano se lasserà stare et non se farà se non tanto quanto comandarà V[ostra] S[ignoria] et il prefato m[esser] Jacomo" (ibid.). Pasolini speculates that this "little book" could be a leather-bound volume found among Caterina's inventory (see Pasolini, "Nuovi documenti," 113).

130. "Li disse ancora che io era conducto cum lo Re Maximiano, cosa incredibile che uno mio paro havesse modo et via de condurse cum cussí alto s[igno]re" (Pasolini, *Caterina Sforza,* 3:604). Maximilian I (1459–1519) was king of the Romans and then Holy Roman Emperor; he married Caterina's sister Bianca Maria Sforza in 1494.

131. In his *Life of Judah,* Leone Modena recounts his son Mordecai's death after experimenting with transforming lead into silver, speculating that the vapors and fumes of the "arsenics and salts" used in his efforts were to blame (Mark R. Cohen, *Autobiography of a Seventeenth-Century Venetian Rabbi: Leon Modena's Life of Judah* [Princeton, N.J.: Princeton University Press, 1988]); see also Raphael Patai, *The Jewish Alchemists* [Princeton, N. J.: Princeton University Press, 1995], 399–401).

132. "Mandatice palle tre de vetro tondo habiano il buco piccolo et che tengano doi bucali de mesura et xii cipolle marine che si chiamano schille" (Caterina Sforza to Francesco Fortunato, November 2, 1499, ASF, Med. a. Pr., f. 70, n. 87; Pasolini, *Caterina Sforza,* 3:606). A *bucalo* (*boccale*) was a wine measure equal to approximately one pint (see Ronald Zupko, *Italian Weights and Measures from the Middle Ages to the Nineteenth Century* [Philadelphia: American Philosophical Society, 1981], 30). Sforza and Fortunati were in regular epistolary contact, and Fortunati also wrote frequently to Sforza's children, especially her son Ottaviano. Sforza's testament shows that upon her death she left her "books, writings, and letters, both public and private" to Fortunati (see note 67). Alchemy was not all Caterina had time for during the siege of Ravaldino. Contemporary chronicles recount that during the siege, which lasted several weeks, at night, "those inside [i.e., Caterina and her supporters], show that they have little fear of those outside [i.e., Cesare Borgia and his men] . . . and do not hesitate to make music and dance" (*mostrano quelli che sono dentro avere poca paura de quelli che sono fora . . . et non mancano de sonare et ballare*) (Cristoforo Poggio to the Marchese of Mantua, December 31, 1499, cited in Graziani and Venturelli, *Caterina Sforza,* 327n29).

133. See Spratt, *Flora medica,* 2 vols. (London: Callow and Wilson, 1830), 1:134–136.

134. Hoping to create a poisonous or at least soporific gas, Gonzaga ordered that a mixture of antimony and poppy and sunflower seeds be inserted into artillery casings; as Roberto Navarrini notes, "[g]li ingredienti che compenevano la 'mistura' ci danno la misura della ingenuità e della credulità di Vincenzo. Alcuni sono innocui, altri, sebbene nocivi, adoperati ed impiegati male" ("La guerra chimica di Vincenzo Gonzaga," *Civiltà mantovana* 4 [1969]: 46). I thank Valeria Finucci for directing me to this reference.

135. Giacomo Feo is mentioned as participating in Caterina's experiment in Lorenzo de Mantechitis's letter cited in notes 125–129. While I have not found a similar reference to Caterina's last husband, Giovanni de' Medici (nor to Girolamo Riario,

although Girolamo's son Scipione is named in Lorenzo's letter), it seems likely that Giovanni would have been interested in Caterina's medicinal secrets at the very least. A letter from Giovanni to Caterina dated September 2, 1498, finds him at Bagni, taking the medicinal waters, along with Caterina's daughter, Bianca; he died ten days after writing the letter: "send me one or two of my black berets, so I can change during the day when I sweat, and also the other two large reddish ones . . . I have already taken the waters two days running in the women's bath here and by the grace of God may it do me some good" (*mi mandi una o due berrette delle mie nere, per poter mutarmi il dì quando sudo, et così due altre rosate doppie e grandi . . . io mi sono già' bagnato due dì nel bagno delle donne qui per grazia di Dio mi fa bene ogni cosa* (Giovanni de' Medici to Caterina Sforza, 2 settembre 1498, ASF Med. a. Pr., f. 78, n. 63; cited in Pasolini, *Caterina Sforza*, 3:848).

136. Suggesting a continued and very real interest in recipes and secrets on the parts of both Cuppano and his Medici employers, Cosimo and his wife, Eleonora of Toledo, sent medicinal remedies to him as he was dying. See da Gorzano, "Il conte Lucantonio Coppi," 105.

137. Stefano Rosselli's recipes were widely copied and circulated by his contemporaries in Florence (see Shaw and Welch, *Making and Marketing Medicine*, 304–307. For these manuscripts, see Rosselli, *Mes secrets: A Florence au temps des Medicis, 1593: patisserie, parfumerie, medecine*, ed. Rodrigo de Zayas (Paris: J. M. Place, 1966).

138. "Attorno a queste sopra dette cose spende quasi tutto il tempo ed ha un luogo che lo chiama il Casino ove, in guisa di piccolo arsenale in diverse stanze, ha diversi maestri che lavorano in diverse cose e quivi tiene i suoi lambicchi ed ogni suo artificio" (cited in Lensi, *L'arte segreta*, 88–90).

139. See P. F. Covoni, *Don Antonio de' Medici al Casino di San Marco* (Florence: Tipografia cooperativa, 1893), 202, 243.

140. "mai non [ho] sentito el magior dollore insemo con Bastiano e tuta la famiglia mia et mai più non vivrò contento perch'io ho perso la mia dolce patrona e più me dolle ch'io non me so' ritrovato a la fine a li servitie" (Lodovico Albertini, June 3, 1509, ASF, Med. a. Princ., f. 125, n. 258; see Pasolini, *Caterina Sforza*, 3:549). Breisach's view is that Albertini's grief stemmed mostly from the loss of Caterina's financial patronage (*Caterina Sforza*, 137).

2. The *Secrets* of Isabella Cortese

1. On the evolution of scientific patronage and the diffusion of scientific culture from the universities to the courts, academies, and ultimately printing houses, see William Eamon and Françoise Peheau, "The Accademia Segreta of Girolamo Ruscelli: A Sixteenth-Century Scientific Society," *Isis* 75 (1984): 327–342; and William Eamon, "Court, Academy, and Printing House: Patronage and Scientific Careers in Late Renaissance Italy," in *Patronage and Institutions: Science, Technology and Medicine at the European Court*, ed. Bruce T. Moran (Rochester, NY: Boydell Press, 1991),

25–50. See also Jo Wheeler, *Renaissance Secrets: Recipes and Formulas* (London: V & A Publishing, 2009), 9–10. For a succinct discussion of the divergence of the "new" science and its emphasis on experimentation from the Scholastic tradition, see Eamon, *The Professor of Secrets: Mystery, Medicine, and Alchemy in Renaissance Italy* (Washington, D.C.: National Geographic, 2010), 112–114. On early modern empirical culture more generally, see Gianna Pomata and Nancy G. Siraisi, eds., *Historia: Empiricism and Erudition in Early Modern Europe* (Cambridge, Mass: MIT Press, 2005); and Anthony Grafton and Nancy G. Siraisi, eds., *Natural Particulars: Nature and the Disciplines in Early Modern Europe* (Cambridge, Mass.: MIT Press, 1999).

2. See Eamon and Peheau, "The Accademia Segreta," 333. There was, however, a great deal of ambiguity in the usage of terms such as "experiment" and "experience" throughout the sixteeenth century (on this issue, see Charles Schmitt, "Experience and Experiment: A Comparison of Zabarella's View with Galileo's in De Motu," *Studies in the Renaissance* 16 [1969]: 80–138). On books of secrets, see William Eamon's pivotal study, *Science and the Secrets of Nature: Books of Secrets in Medieval and Early Modern Culture* (Princeton, N.J.: Princeton University Press, 1994); Eamon, "Arcana Disclosed: The Advent of Printing, the Books of Secrets Tradition, and the Development of Experimental Science in the Sixteenth Century," *History of Science* 22 (1984): 111–150; and Eamon, "How to Read a Book of Secrets," in *Secrets and Knowledge in Medicine and Science 1500–1800,* ed. Elaine Leong and Alisha Rankin (Aldershot: Ashgate, 2011), 23–46. See also Rudolph Bell, *How to Do It: Guides to Good Living for Renaissance Italians* (Chicago: University of Chicago Press, 2000); Allison Kavey, *Books of Secrets: Natural Philosophy in England 1550–1600* (Urbana: University of Illinois Press, 2007); and David Gentilcore, *Healers and Healing in Early Modern Italy* (Manchester: Manchester University Press, 1998), 96–124.

3. See Giovanni Battista Zapata, *Maravigliosi secreti di medicina e chirurgia* (Turin: appresso gli heredi del Bevilacqua, 1586), 2: "Truly, I believe that you [who are] poor, who have no wealth, worry greatly about your health . . . [but] even if you don't possess gems, gold, or precious stones, as the rich and powerful do, to fend off the maladies mentioned here (medicaments that are truly useless, and of no worth), at least you will have simple remedies, created by wise Nature for your benefit, which, even though they are just inexpensive simples, they will be . . . of the same usefulness and efficacy as those grand and expensive ones." *(Credo veramente voi poverelli, che sete privi di ricchezze, vi ritroviate in gran pensieri nelle vostre infermità . . . se non havrete gemme, oro, e pietre preziose, come i ricchi e potenti per discacciar detti mali [medicamenti che veramente sono vani, e di niun profitto] havrete almeno rimedij facili, che la sagace natura ha fatto, e prodotto in util vostro, i quali se ben saranno semplici e di vil prezzo, saranno . . . di tanta utilità e efficacia, quanto quei magistrali di gran valore.)* Not only questions of audience but also della Porta's theorization of natural magic and its workings, set the *Magia naturalis* apart from the practical texts examined in this chapter. Similarly, Marsilio Ficino's *De Vita libri tres* (1489) of the preceding

century—a fundamental contribution to the development of early modern natural philosophy, also composed in Latin—injects a cosmic element into its ruminations on the health and psychology of the intellectual that differentiates it from the later works studied here. On della Porta, see Eamon, *Science and the Secrets of Nature,* 195–233, and Luisa Muraro, *Giambattista della Porta mago e scienziato* (Milan: Feltrinelli, 1978); for an overview of della Porta in his Neapolitan context, see Arianna Borrelli, "Giovan Battista Della Porta's Neapolitan Magic and His Humanistic Meteorology," in *Variantology 5: Neapolitan Affairs. On Deep Time Relations of Arts, Sciences and Technologies,* ed. by Siegfried Zielinski and Eckhard Furlus in cooperation with Daniel Irgang (Cologne: Verlag Der Buchhandlung Walter König, 2011), 103–130. On Ficino see, inter alia, Marsilio Ficino, *Three Books on Life: A Critical Edition and Translation with Introduction and Notes,* ed. Carol V. Kaske and John R. Clark (Binghamton, N.Y.: The Renaissance Society of America, 1989); Sebastiano Gentile and C. Gilly, eds., *Marsilio Ficino e il ritorno di Ermete Trismegisto/Marsilio Ficino and the Return of Hermes Trismegistus* (Florence: Olschki, 1999); Peter J. Forshaw, "Marsilio Ficino and the Chemical Art," in *Laus Platonici Philosophi: Marsilio Ficino and His Influence,* ed. Stephen Clucas, Peter J. Forshaw, and Valery Rees (Leiden: Brill, 2011), 198–249.

4. Evangelista Quattrami, *La vera dichiaratione di tutte le metafore, similitudini, & enimmi de gl'antichi filosofi alchimisti* (Rome: Appresso Vincentio Accolti, 1587), 26, 37.

5. Carlo Dionisotti argued that the linguistic openness of the period between 1545 and 1563 engendered by the emerging primacy of the vernacular in Italy fostered the entry into the literary arena of marginalized groups with limited access to formal education, including women ("Letteratura italiana all'epoca del Concilio di Trento"). Virginia Cox has problematized this interpretation, arguing that it overlooks women's involvement in manuscript publication and therefore underestimates the extent of their engagement in literary culture (Cox, *Women's Writing in Italy, 1400–1650* [Baltimore: Johns Hopkins University Press, 2008], xx–xxi). On literacy rates in this period, see Peter Burke, ed., "The Uses of Literacy in Early Modern Italy," in *The Historical Anthropology of Early Modern Italy: Essays on Perception and Communication* (Cambridge: Cambridge University Press, 1987), 110–131; Paul Grendler, *Schooling in Renaissance Italy: Literacy and Learning, 1300–1600* (Baltimore: Johns Hopkins University Press, 1989). On rethinking our understanding of literacy itself, see Margaret W. Ferguson, *Dido's Daughters: Literacy, Gender, and Empire in Early Modern England and France* (Chicago: University of Chicago Press, 2003).

6. On the appropriation of medicine for women (such as gynecology and obstetrics) by male practitioners in the early modern period, see Katharine Park, *Secrets of Women: Gender, Generation, and the Origins of Human Dissection* (New York: Zone Books, 2006, rpt. 2010); see also Monica H. Green, "From Diseases of Women to 'Secrets of Women': The Transformation of Gynecological Literature in the Later Middle Ages," *Journal of Medieval and Early Modern Studies* 30 (2000): 5–39; see also Gianna

Pomata, "Was There a *Querelle des Femmes* in Early Modern Medicine?" *Arenal* 20, no. 2 (2013): 313–341. On early modern medical culture more generally, see Nancy Siraisi, *Medieval and Early Renaissance Medicine: An Introduction to Knowledge and Practice* (Chicago: The University of Chicago Press, 1990); Katharine Park, *Doctors and Medicine in Renaissance Florence* (Princeton, N.J.: Princeton University Press, 1985). On books of secrets and women readers, see Meredith K. Ray, "Prescriptions for Women: Alchemy, Medicine and the Renaissance *Querelle des Femmes,*" in *Women Writing Back/Writing Women Back: Transnational Perspectives from the Late Middle Ages to the Dawn of the Modern Era,* ed. Anke Gilleir, Alicia C. Montoya, and Suzan van Dijk (Leiden: Brill, 2010), 135–162.

7. See, for example, Matteo Palmieri's *La vita civile* (Florence: Sansoni, 1982), 17–20, which argues that mothers, by breastfeeding, pass their own good qualities to their children; see also Lodovico Dolce, *Dialogo della institutione delle donne* (Venice: Giolito, 1545). Henricus Cornelius Agrippa notes the positive properties of both breastmilk and menstrual blood in the *Declamatio de nobilitate et praecellentia foemini sexus* (see Henricus Cornelius Agrippa, *Declamation on the Nobility and Preeminence of the Female Sex,* ed. and trans. Albert J. Rabil, Jr. [Chicago: University of Chicago Press, 1996], 57–59). Agnolo Firenzuola linked female beauty to the beauty of the soul (see *On the Beauty of Women*).

8. See Laura Cereta, *Collected Letters of a Renaissance Feminist,* ed. and trans. Diana Robin (Chicago: The University of Chicago Press, 1997), 199–200.

9. Eamon and Peheau call the *Secreti di Alessio Piemontese* "one of the most extraordinary popular successes in the history of sixteenth-century literature" ("The Accademia Segreta," 330). Following its publication in 1555, the work was translated into Latin, English, French, Dutch, and German. It had fifty editions within fifteen years, and more than ninety editions by the end of the seventeenth century (see Eamon, "The *Secreti* of Alexis of Piedmont," *Res Publica Litterarum* 2 [1979]: 43–55). As Sandra Cavallo and Tessa Storey note, books of secrets share some commonalities with health regimens: handbooks for preventive care that also circulated widely in this period (see *Healthy Living in Late Renaissance Italy* [Oxford: Oxford University Press, 2013], 39–40).

10. Bairo's *Secreti medicinali* is based on his earlier Latin handbook *De medendis humanis corporis malis Enchidirion* (1512); see Eamon, *Science and the Secrets of Nature,* 163. As Giuseppe Palmero notes, an enormous quantity of material circulated in mansuscript and print material, in both Latin and the vernacular: it was from this "cultural crucible" that works such as *The Secrets of Alexis,* which inaugurated the "golden age" of the *libri di segreti* in print, and Bairo's *Medicinal Secrets* emerged (see Giuseppe Palmero, "La distillazione: I suoi prodotti ed i suoi usi nell'ambito della "letteratura dei segreti" tra Quatrrocento e Cinquencento," in *Grappa & Alchimia. Un percorso nella millenaria storia della distillazione,* Atti del convegno "Dall'Alchimia alla grappa: un percorso millenario nella distillazione," Greve in Chianti, 11 settembre 1999, ed. Allen J. Grieco [Rome: Agra Editrice, 1999], 17).

11. On Fioravanti, see Eamon, *The Professor of Secrets;* David Gentilcore, *Medical Charlatanism in Early Modern Italy* (Oxford: Oxford University Press, 2006), 211–294.

12. The term "poligrafo" has traditionally been used (often with negative implications) to indicate a professional writer who produces works in various genres depending on the demands of publishers and the editorial market. For a discussion of the *poligrafi* in early modern Italy, see Paul Grendler, *Critics of the Italian World* (Madison: University of Wisconsin Press, 1969). On Ruscelli's *Secreti nuovi,* see Eamon and Peheau, "The Accademia Segreta"; on Ruscelli as author of the *Secrets of Alexis,* see Eamon, *Science and the Secrets of Nature,* 143–154. Some have maintained that a real Alexis existed: see Mazzuchelli, *Gli scrittori d'Italia* (Brescia: G. Bossini, 1753), 1:465 and E. Erspammer, "I Secreti di Alessio Piemontese e la rivoluzione scientifica," in *Du Pô à la Garonne: Recherches sur les échanges entre Italie e la France à la Renaissance,* ed. A. Fiorato (Agen: Centre Matteo Bandello d'Agen, 1990), 376–377; see also G. Mombello, *Sur les traces d'Alexis Jure de Chieri. Le probleme des francisants piemontais au XVI siecle* (Geneva: Slatkine, 1984).

13. Some question remains as to whether this academy actually existed, or was instead a utopian vision. For a conceptualization of "virtual" intellectual communities in early modern Italy (with respect to women's salon culture), see Diana Robin, *Publishing Women: Salons, The Presses, and the Counter-Reformation in Sixteenth-Century Italy* (Chicago: The University of Chicago Press, 2007).

14. Eamon, "How to Read a Book of Secrets," 23; see also Elena Camillo, "Ancora su Donno Alessio Piemontese: Il libro di segreti tra popolarità e accademia," *Giornale storico della letteratura italiana* 162 (1985): 543.

15. See Claire Lesage, "La littérature des 'secrets' et *I secreti di Isabella Cortese,*" *Chroniques italiennes* 36 (1993): 146.

16. On the development of museums and collecting in the early modern period, see Paula Findlen, *Possessing Nature: Museums, Collecting, and Scientific Culture in Early Modern Italy* (Berkeley: University of California Press, 1996); see also Lorraine Daston and Katharine Park, *Wonders and the Order of Nature, 1150–1750* (New York: Zone Books, 2001).

17. Many scholars note the linked nature of alchemical and medical practice in the early modern period: see, for example, Chiara Crisciani, "From the Laboratory to the Library: Alchemy According to Guglielmo Fabri," in *Natural Particulars,* ed. Grafton and Siraisi, 295–319; and Katharine Park, "Natural Particulars: Medical Epistemology, Practice, and the Literature of the Healing Springs," in *Natural Particulars,* ed. Grafton and Siraisi, 347–367.

18. "A sublimare argento vivo, cioè a fare il solimato commune delle spetierie, che si adoprano da gli orefici, da gli alchimisti, dalle donne, & in molte cose di medicina" (*Secreti di Alessio,* 5:84). *Mercurius dulcis sublimatus* (as opposed to its lethal cousin, corrosive mercury sublimate) was considered useful as a purgative and in treating worms and venereal disease (see John Hill, *A History of the Materia Medica* [London: Longman, Hitch, and Hawes, 1751], 61–62).

19. "per ogni qualità di persona, così ricca & povera, dotta & indotta, & maschio & femina" (Ruscelli, *Segreti nuovi,* proem). Ruscelli explains that, among the few outsiders allowed to penetrate the secrecy of the *Accademia segreta* and examine its laboratory and grounds, were the wife and sister of its primary patron (a "prince" who is not otherwise identified in the text).

20. Eustachio Celebrino, *Opera nuova intitolata dificio delle ricette* (Venice: Giovanni Antonio et fratelli da Sabbio, 1528). On Celebrino, see Giovanni Comelli, *Ricettario di bellezza di Eustachio Celebrino, medico e incisore del Cinquecento* (Florence, 1960) and S. Morison, *Eustachio Celebrino da Udine Calligrapher, Engraver, and Writer for the Venetian Printing Press* (Paris: Pegasus Press, 1929); see also Eamon, *Science and the Secrets of Nature,* 127–130.

21. See Bruce Moran, *Distilling Knowledge: Alchemy, Chemistry, and the Scientific Revolution* (Cambridge, Mass.: Harvard University Press, 2005), 11.

22. "Quivi beletti son blanche e vermigli / Varie composition d'acque stillate" (Eustachio Celebrino, *Opera nuova piacevole . . . per far bella ciaschuna donna* (Venice: Bindoni, 1551).

23. Ibid.

24. Timoteo Rossello, *Della summa de' secreti universali in ogni materia* (Venice: Bariletto, 1561).

25. [Isabella Cortese], *I secreti della signora Isabella Cortese, ne' quali si contengono cose minerali, medicinali, arteficiose, & alchimiche, et molte de l'arte profumatoria, appartenenti a ogni gran Signora* (Venice: Bariletto, 1561). The 1584 edition of the *Secreti della signora Isabella Cortese* is available in a modern facsimile, edited by Chicca Gagliardo; the editor's introduction incorrectly gives the date of the first edition as 1584 rather than 1561.

26. See Eamon, *Science and the Secrets of Nature,* 136.

27. The secrets book par excellence, well known to early modern readers, the pseudo-Aristotelian *Secretum secretorum* circulated widely through the seventeenth century. See W. F. Ryan and Charles B. Schmitt, eds., *Pseudo-Aristotle: The Secret of Secrets. Sources and Influences* (London: Warburg Institute, 1982). As Eamon points out, this text was "among the most widely read of the medieval and Renaissance periods" (*Science and the Secrets of Nature,* 45). See also Lynn Thorndike, *A History of Magic and Experimental Science* (New York: Macmillan, 1923), 2:267; and Karma Lochrie, *Covert Operations: The Medieval Uses of Secrecy* (Philadelphia: University of Pennsylvania Press, 1999), 98. On the problem of secrecy more generally in medieval and early modern culture, see also Timothy McCall, and Sean E. Roberts, and Giancarlo Fiorenza, eds., *Visual Cultures of Secrecy in Early Modern Europe* (Kirksville, Mo.: Truman State University Press, 2013).

28. On artisans, craft knowledge, and scientific culture, see Pamela O. Long, *Artisan/Practitioners and the Rise of the New Science, 1400–1600* (Corvallis: Oregon State University Press, 2011).

29. "grandi huomini per dottrina"; "da povere feminelle, da artegiani, da conta-dini" (*Secreti di Alessio*, a2r).

30. Park, *Secrets of Women;* see also Park, "Dissecting the Female Body: From Women's Secrets to the Secrets of Nature," in *Crossing Boundaries,* ed. Jane Donawerth and Adele Seef (Newark, Del.: University of Delaware Press, 2000), 29–47.

31. On *Le segrete cose delle donne,* see Lochrie, *Covert Operations,* 93–134. On *De secretis mulierum,* see Helen Rodnite Le May, *Women's Secrets: A Translation of Pseudo-Albertus Magnus'* De Secretis Mulierum *with Commentaries* (Albany: Albany State University, 1992); and Dinora Corsi, Lada Hordynsky-Caillat, and Odile Redon, "Les se-crés des dames, tradition, traductions," *Médiévales* 14 (1988) 47–57.

32. Park, "Dissecting the Female Body," 34.

33. See Monica H. Green, ed. and trans., *The Trotula: A Medieval Compendium of Women's Medicine* (Philadelphia: University of Pennsylvania Press, 2001). The linkage of gender and medicine can be seen in other cultural arenas as well, for example, on the early modern professional stage, where M. A. Katritzky sees women actresses as closely connected to the "performative marketing" of medicinal and cosmetic reme-dies (*Women, Medicine and Theatre, 1500–1750: Literary Mountebanks and Performing Quacks* [Aldershot: Ashgate, 2007], 1).

34. Park, "Dissecting the Female Body," 34.

35. I borrow this term from Micaela di Leonardo, ed., *Gender at the Crossroads of Knowledge: Feminist Anthropology in the Postmodern Era* (Berkeley: University of Cal-ifornia Press, 1991).

36. See Lochrie, *Covert Operations,* 3.

37. Cereta, *Collected Letters,* 199.

38. [Ruscelli], *La prima parte de' secreti del reverendo donno Alessio Piemontese* (Pesaro: Appresso gli Heredi di Bartolomeo Cesano, 1562), 3r: "I always said that if secrets were known to everyone, they wouldn't be called secrets anymore, but rather public [knowledge]" (*Et sempre io diceva, che se i secreti si sapessero da ogn'uno, non si chiameriano più secreti, ma publici*).

39. "il revelare de' secreti fa perdere l'efficacia" ([Cortese], *Secreti della signora Is-abella Cortese,* 32).

40. On Garzoni's *Piazza universale,* see Paolo Cherchi, *Enciclopedismo e politica della riscrittura: Tommaso Garzoni* (Pisa: Pacini, 1981); and Paolo Cherchi and Bea-trice Collina, "Invito alla lettura della 'Piazza,'" in *Tomaso Garzoni, La piazza uni-versale di tutte e professioni del mondo,* ed. Paolo Cherchi and Beatrice Collina (Turin: Einaudi, 1996); see also George McClure, *The Culture of Profession in Late Renaissance Italy* (Toronto: University of Toronto Press, 2004).

41. Garzoni considered most alchemists to be frauds, and was ambivalent about the merits of secrets, which in his opinion could be useful or merely superstitious non-sense (see Tomaso Garzoni, *La piazza universale di tutte le professioni del mondo,* ed.

Giovanni Battista Bronzini, with Pina de Meo and Luciano Carcereri [Florence: Olschki, 1996], 191, 242).

42. "cosa oscura, velata, et occulta, la cui ragione non è talmente chiara, che debba a tutti esser nota, ma per natura a pochissimi manifesta" (ibid., 241).

43. Ibid., 241–242.

44. See Leong and Rankin, *Secrets and Knowledge in Medicine and Science,* 8.

45. Ibid., 98. On the function of the recipe, see also Eamon, for whom collections of recipes, which made it possible to compare, contrast, and discern among similar formulas, are another manifestation of an increased emphasis on praxis over theory (*Science and the Secrets of Nature,* 131).

46. See Kavey, *Books of Secrets,* 98.

47. Ibid., 98–99.

48. On books of cookery and domestic management in early modern England, see Michelle di Meo and Sarah Pennell, *Reading and Writing Recipe Books, 1500–1800* (Manchester, UK: Manchester University Press, 2013); see also Rebecca Laroche, *Medical Authority and Englishwomen's Herbal Texts, 1550–1650* (Aldershot: Ashgate, 2009); Elaine Leong, "Collecting Knowledge for the Family: Recipes, Gender and Practical Knowledge in the Early Modern English Household," *Centaurus* 55, no. 2 (2013): 81–103; and Leong, "Making Medicines in the Early Modern Household." *Bulletin of the History of Medicine* 82 (2008): 145–168.

49. Eamon, for example, devotes just passing consideration to the problem, surmising only that Isabella Cortese was likely a Venetian noblewoman (*Science and the Secrets of Nature,* 137); Virginia Cox briefly raises the question in *Women's Writing in Italy,* 317n5.

50. Massimo Rizzardini, "Lo strano caso della Signora Isabella Cortese, professoressa di *secreti,*" *Philosophia* 2, no. 1 (2010): 84.

51. The case for Ruscelli's authorship of the *Secreti di Isabella Cortese* is made most forcefully in Rizzardini, "Lo strano caso." Claire Lesage also addresses the question but concludes that style differences make shared authorship unlikely and that a historical Cortese could have existed ("La litterature des 'secrets,'" 170).

52. "Isabella Cortese, il cui nome si tiene esser mentito insieme con quel di don Alessio dal Ruscello" (Garzoni, *La piazza universale,* 242).

53. "ces quatre inventeurs, ces compilateurs de recettes, ces mirifiques distillateurs de quintessence ne sont, en somme qu'un seul et même homme" (Armand Baschet and Felix Feuillet de Conches, *Les femmes blondes selon les peintres de L'Ecole de Venise* [Paris: Aubry, 1885], 183). Cited in Lesage, "La littérature des 'secrets,'" 158.

54. Ragusa had a vibrant scientific culture and many connections with scientific communities in Italy, particularly Venice and Padua (see Chapter 4 here).

55. On Caboga, see Franceso Maria Appendini, *Notizie istorico-critiche sulle antichità storia e letteratura de' Ragusei* (Ragusa: Antonio Martecchini, 1803) and Renzo de' Vidovich, *Albo d'Oro delle famiglie nobili patrizie e illustri nel Regno di Dalmazia*

(Trieste: Fondazione Scientifico Culturale Rustia Traine, 2004). See also Rizzardini, "Lo strano caso," 83.

56. The documents are in the Archivio di Stato di Venezia, Senato terra, 1559–1560, R. 42, Speciales Personae, 168 (see Lesage, "La littérature des 'secrets,' " 166).

57. Vanuccio Biringoccio, *Li diece libri della pirotechnia* (Venice: Curtio Troiano Navò, 1558); the dedication is dated April 15, 1558.

58. Jo Wheeler, "The Ragusan Connection." On Curtio Troiano di Navò, see Corrado Marciani, "Troiano Navò di Brescia e suo figlio Curzio librai-editori del secolo XVI," in *La Bibliofilia* 73 (1971): disp.1, 49–60.

59. "non solamente l'huomo si contenta della investigatione, ma cerca in tutto & per tutto mettendo in opera, di farsi scimia della natura, anzi che superarla, mentre tenta di fare quello, che alla natura è impossibile, & che ciò sia vero, si può cavare da' secreti, che tutto il giorno si odono & veggono mettere in essecutione" (Cortese, *I secreti*, 2r-v).

60. "Tanto è cosa naturale il ricercare i secreti di natura che vediamo per le historie antiche, tutti gli huomini di giudicio haversi occupato in questo esercitio . . . quanto di bene habiamo cerca la cognitione delle cose naturali potiamo riconoscerlo da questo desiderio di investigare li secreti" (Rossello, *Della summa de' secreti*, s.n.).

61. "far diligentissima inquisitione & come una vera anatomia delle cose & dell'operationi della Natura" (Ruscelli, *Secreti nuovi,* proem).

62. Lesage, "La littérature des 'secrets,' " 162.

63. Cortese's dismissal of alchemical authorities mirrors the attitude taken in books of secrets toward medical authorities. As Camillo writes, "Nei libri di segreti l'appello al nuovo contatto fra sapere scientifico e sapere tecnico diventa uno strumento per polemizzare contro i medici accademici ed è forse un riflesso della più generale polemica tra i seguaci di Galeno e chi applicava metodi chimici, ossia tra gli ambienti accademici ed i seguaci di Paracelso" ("Ancora su Don Alessio Piemontese," 18).

64. "se vuoi seguire l'arte dell'Alchimia, & in quella operare, non bisogna che più seguiti l'opere di Geber, né di Raimondo, né di Arnaldo, o d'altri Filosofi, perchè non hanno detto verità alcuna ne i loro libri, se non con figure, & enigmati" (Cortese, *I secreti*, B2r).

65. "Et io ho studiato in tali libri più di trenta anni, e mai non ho trovato cosa alcuna buona" (ibid.).

66. See Lesage, "La littérature des 'secrets,' " 163–166.

67. On syphilis in the early modern period, see John Arrizabalaga, John Henderson, and Roger French, eds., *The Great Pox: The French Disease in Renaissance Europe* (New Haven, Conn.: Yale University Press, 1997); Laura McGough, *Gender, Sexuality and Syphilis in Early Modern Venice: The Disease That Came To Stay* (New York: Palgrave Macmillan, 2011).

68. Cortese, *I secreti*, 11: "Alle creste che vengono alle donne, per causa del parto, o per altra cagione."

69. "È nota quanto più questa cosa sarà distillata augmentarassi la virtù sua" (ibid., 26–27).

70. Ibid., 29 ("A guarir il mal della bocca per il mal francese").

71. Some of Cortese's recommendations resonate with Ruscelli's later description of the rules of the Accademia Segreta in Naples. Ruscelli offers a detailed picture of the society's membership, laboratory, and grounds. Like Cortese, he emphasizes that the laboratory must be remote and its secrets protected from outsiders, and the importance of having trusted helpers (Ruscelli, *Secreti nuovi,* proem).

72. On Caterina Sforza's epistolary exchanges with Luigi Ciochi see Chapter 1 here.

73. See in Cortese, *I secreti,* recipes for: "inchiostro che in quaranta dì sparisce e non si vede" (43); "reduttione d'argento" (56); "A indorar il ferro" (61); "A far il ferro frangibile da pestare" (63); "A far ottone bello" (68); "Oro potabile" (70). The presence of a recipe for potable gold suggests that the author of the *Secrets of Isabella Cortese* could have been familiar with a more theoretical work such as Ficino's *De vita.*

74. Ibid.: "A levar ogni macchia d'olio, e di grasso in panno" (92); "Pallotte di sapone per levar le macchie" (93); "Levar macchie d'ogni drappo e d'ogni colore" (93).

75. Ibid., 94: "A far drizzare il membro."

76. Ibid.: "Rossetto de scudellini per le donne" (101); "A conciar sollimato per le donne" (106); "pelatoio per donne" (163).

77. "Acqua che fa colorita la carne a chi è pallida. . . . Piglia due piccioni di penne bianche, e per otto dí siano cibati de pignoli overo per quindeci dí, poi squartagli e getta via la testa i piedi e le bidella, poi mettigli a lambicco a stillare con mezzo pane di polvere zuccarina et iv once d'argento fino, tre ducati d'oro, quattro molliche di pane buffetto bianco che sia stato sei giorni continui a molle nel latte caprino . . . tutte queste cose lambicca a lento fuoco, e n'uscirà acqua perfettissima per incolorir la carne pallida" (ibid., 196).

78. Ibid., 192. A *braccio* was a Venetian unit of length used for measuring fabrics (see Luca Mola, *The Silk Industry of Renaissance Venice* [Baltimore: Johns Hopkins University Press, 2000]).

79. "Acqua vita perfettissima: . . . metti a distillare in una boccia che habbia il collo longo un bracio e mezzo, nel bagno maria col suo capello, ben lutate le gionture, e quando vedrai che più non distillarà cosa alcuna sarà segno che lo spirito sarà uscito fuori, e veduto tal segno di subito leva via la boccia, e vuoterai fora detta acqua in un saggiolo piccolo di vetro, mettendone a volta per volta della detta acqua spirito. . . . Poi ritornare di nuovo con l'altro vino a cavarne per il simile, come la prima volta, e questo ordine si tenga per fine che ne haverai quanto ti piacerà" (Cortese, *I secreti,* 192).

80. [Falloppio], *Secreti diversi* (Venice: Zaltieri, 1588), 315–316. The empiric Leonardo Fioravanti (author of his own book of secrets) was likely the real author behind this work (see Eamon, *Science and the Secrets of Nature,* 166).

81. Books of secrets and their contents were the object of more general satire as well: for example, in Ariosto's *Herbolato* (Venice: Fratelli da Sabio, 1545), which takes aim at the fraudulent claims of medical charlatans, and on the commedia dell'arte stage. On charlatans, see Gentilcore, *Medical Charlatanism;* see also Katritzky, *Women, Medicine and Theater.*

82. See Alessandro Piccolomini, *La Raffaella, dialogo della bella creanza delle donne,* ed. Mario Cicognani (Milan: Longanesi, 1969). Piccolomini wrote scientific treatises and dedicated some of them to women (see Chapter 4 here). On Piccolomini's support of women and of one woman in particular, Laudomia Forteguerri, see Robin, *Publishing Women,* 124–159; and Cox, *Women's Writing in Italy,* 101.

83. Pietro Aretino, *Ragionamenti dialogo,* ed. Giorgio Barberi Squarotti and Carlo Forni (Milan: Rizzoli, 1998). On Aretino's dialogue, see also Rosenthal, "Epilogue," in *Aretino's Dialogues,* ed. and trans. Raymond Rosenthal (New York: Marsilio, 1994), 387–402.

84. See Alessandro Piccolomini, *La economica di Xenofonte, tradotta da lingua greca in lingua toscana . . .* (Venice: Comin da Trino, 1540). On the *Raffaella* and the *Economica* as "two sides of the same coin," see Diana Robin, "A Renaissance Feminist Translation of Xenophon's *Oeconomicus,*" in *Roman Literature, Gender, and Reception: domina illustris. Essays in honor of Judith Peller Hallett. Routledge monographs in classical studies, 13,* eds. Donald Lateiner, Barbara K. Gold, and Judith Perkins (London and New York: Routledge, 2013), 209. For Leon Battista Alberti's condemnation of makeup, see Alberti, *The Family in Renaissance Florence,* ed. and trans. Renée Neu Watkins (Prospect Heights, Ill.: Waveland Press, 1994), 84–85: "Her eyes were sunken and always inflamed, the rest of her face withered and ashen. . . . Thanks to makeup . . . she had been left in this diseased condition and seemed old before her time."

85. Francisco Delicado's *Dialogo del Zoppino* laments, "And they even apply poultices of wax, honey and figs to their bellies or try to remove wrinkles from the belly with pine waters. . . . And what of the foul color they apply to their lips, does that [not] smell?" ("E più si fanno alle lor pancie impiastri con cera, mele e fichi, o si discrespan la pancia con le sopraddette acque di pino. . . . E i putrefatti lisci che sui labbri si pongon, puzzano egli?" [Delicado, *Ragionamento del Zoppino,* 28]). Delicado (1480?–1534) was also the author of *La Lozana andaluza (Portrait of Lozana)* (Venice, 1528): this work features a female protaganist who prepares cosmetics and medicinal remedies for courtesans.

86. "Non sa che 'l liscio è fatto col salivo de le giudee che 'l vendon; né con tempre di muschio ancor perde l'odor cattivo. . . . Oh quante altre spurcizie a dietro lasso, di che s'ungono il viso" (Ariosto, *Satire,* ed. Cesare Segre [Turin: Einaudi, 1976], 62). Ariosto's association of Jews with the production of cosmetics is clearly negative, but elsewhere we find evidence of Jewish and Christian women exchanging cosmetic recipes as a matter of routine (see, for example, Caterina Sforza's letter to "Anna Hebrea" in her manuscript collection of recipes, *Experimenti,* discussed in Chapter 1).

The linkage of Jewish women to the production and sale of cosmetics is also evident in Delicado's *La Lozana andaluza* (see note 85). A number of prescriptive manuals warned against women embellishing their appearance with beauty waters, cosmetics, or fine clothing: for an early example, see Francesco da Barbarino's conduct manual, *Reggimento e costumi delle donne,* which cautions women not to rely on beauty waters and makeup but only on purity of soul (Francesco da Barbarino, *Reggimento e costumi delle donne,* ed. Giuseppe E. Sansone [Rome: Zauli, 1995], 73, 180).

87. See Annibal Guasco, *Discourse to Lady Lavinia His Daughter,* ed. and trans. Peggy Osborn (Chicago: The University of Chicago Press, 2003), 79.

88. See Robin, "A Renaissance Feminist Translation," 209–210.

89. The example of Margarita supports Tessa Storey's argument that the extent of ingredients required by many recipes would have been too much for nonexperts to manage: "many householders would have obtained their remedies, waters, oils and so on from small-scale vendors . . . who had the kitchen space and equipment to make such products" (see Tessa Storey, "Face Waters, Oils, Love Magic and Poison: Making and Selling Secrets in Early Modern Rome," in *Secrets and Knowledge,* 145).

90. Piccolomini, *La Raffaella,* 45: "basta che te ne farò ogni volta che vorrai."

91. Likely camphor (a white gum), often confused with camphire (henna) and used interchangeably.

92. Ambergris, a substance produced by sperm whales and traditionally used as a fixative by perfumers.

93. "Io piglio prima un par di piccioni smembrati, dipoi termentina viniziana, fior di gigli, uova fresche, mele, chioccioline marine, perle macinate e canfora; e tutte queste cose incorporo insieme, e mettolo dentro a' piccioni e in boccia di vetro a lento fuoco. Dipoi piglio muschio e ambra, e più perle e panelle d'argento, e macinate queste ultime cose al porfido sottilmente, le metto in un botton di panno lino, e legole al naso della boccia con recipiente sotto, e dipoi tengo l'acqua al sereno; e diviene una cosa rarissima," (Piccolomini, *La Raffaella,* 45). English translations taken from Piccolomini, *Raffaella of Master Alexander Piccolomini,* trans. John Nevison (Glasgow: Robert MacLehose, 1968), 34.

94. Piccolomini, *Raffaella,* 35–36 ("Si piglia argento sodo fino ed argento vivo passato per camoscio, e incoroporati insieme, si fan macinar per un dì per un medesimo verso con un poco di zuccaro fino; e dipoi il cavo del mortaio, e le fo macinare al porfido a un dipintore, e v'incorporo dentro panelle d'argento e perle; e di nuovo fo macinare al porfido ogni cosa insieme, e le rimetto nel mortaio, e le stempro la mattina a digiuno con saliva di mastice con un poco d'olio di mandorle dolci; e così liquido, rimenato un dì, stempro di nuovo il tutto con acqua di frassinella, e mettolo in un fiasco, e lo fo bollire a bagno marie; e così fatto quattro volte, gittando sempre l'acqua, la quinta la serbo . . . ed al fondo rimane il solimato," Piccolomini, *La Raffaella,* 46–47]).

95. Whereas the parodic elements of Piccolomini's dialogue complicate a reading of its presentation of women's secrets, Marinello—a physician—addressed his trea-

tise to "chaste young women" *(caste et giovani donne)* and defended himself (and his female readers) from criticism by celebrating the decorous use of cosmetic embellishment and cloaking his recipes in a Neoplatonic framework (Giovanni Marinello, *De gli ornamenti delle donne* [Venice: Valgrisio, 1574], a2).

96. On Piccolomini's works in defense and support of women, see Robin, *Publishing Women*, 136–159.

97. For a general discussion of Lando and the *Lettere di molte valorose donne*, see Ray, *Writing Gender*, 45–80; and for a brief preliminary discussion of the anthology's links to the *libro di segreti* model, 71–75. On the *Valorose donne*, see also Ireneo Sanesi, "Tre epistolari del Cinquecento," *Giornale storico della letteratura italiana* 24 (1894): 1–32; Novella Bellucci, "Lettere di molte valorose donne . . . e di alcune pettegolette, ovvero: di un libro di lettere di Ortensio," in *Le carte messaggiere: Retorica e modelli di communicazione epistolare: per un indice dei libri di lettere del Cinquecento,* ed. Amedeo Quondam (Rome: Bulzoni, 1981), 255–276; and Francine Daenens, "Donne valorose, eretiche, finte sante: note sull'antologia giolitina del 1548," in *Per lettera: La scrittura epistolare femminile,* ed. Gabriella Zarri (Rome: Viella, 1999), 181–207.

98. On Lando's authorship of the *Valorous Women,* see Meredith K. Ray, "Un'officina di lettere: le *Lettere di molte valorose donne* e la fonte della 'dottrina femminile,'" *Esperienze letterarie* (2001) no. 3, 69–91; and Ray, *Writing Gender*, 45–56.

99. See Park, *Secrets of Women;* see also Park, "Dissecting the Female Body."

100. Lando, *Lettere di molte valorose donne,* 113v, "non lo havrei dato a mio figliuolo."

101. Lando, *Valorose donne,* 58r, "tutto il male alli dí passati avenutovi nacque dal non haver voi potuto tener segreto quanto vi fu segretamente detto."

102. Ibid., "Saresti veramente scopiata se non partorivi questo poco di segretuzzo."

103. "non voglio . . . negare di non haver letto la parte mia" (ibid., 111r); "se verrete a bagni di Villa col vostro consorte, provederò che sappiate da cui di voi dua proceda; se mi accorgerò che in niuno di voi sia il difetto, desiderando d'haver un figlio maschio, pigliarò la matrice, & la natura della lepre qual farò seccare, & spolverizata la beret[e], con un poco di vino, & senza dubbio gravida rimarrete" (ibid., 111r-v). For hare as an ingredient in recipes to determine infertility, see [Ruscelli] *Secreti di Alessio,* 25v, "A sapere se una donna si potrà ingravedare, cosa verissima."

104. Lando, *Valorose donne,* 111r-v; compare Bairo, *Secreti medicinali,* 166v.

105. "perché siete solita di abortire, faretevi far dal vostro speciale, la presente polvere: semi d'apio, ameos, menta parte uguali dracme iij, mastiche, garophili, cardomomo, radici di rubea, maggiore parti uguali drac. iij, castorio zedoaria, ireos parte uguali dracme ii, zuccaro dracme y, pigliarete questa polvere col mele, et nel vino ne infunderete tre scruoppoli per volta et sarete sicura . . . di non sconciarvi mai" (Lando, *Valorose donne,* 111v–112r). Zedoary, or white turmeric, is a relative of ginger; castoreum is the yellowish secretion of the beaver's castor sac, used by perfumers. Lando's recipe seems to draw directly on similar instructions to protect against miscarriage in the *Trotula:* "take three drams each of wild celery, mint, and cowbane, three drams

each of cloves, watercress, and madder root, five drams of sugar and two drams each of castoreum, zedoary, and gladder [iris]. Let these be made into a very fine powder and let it be prepared with honey and let her be given three scruples of it with wine" (*Trotula,* 97).

106. See Marinello, *Delle medicine partenenti alle infermità delle donne* (Venice: Valgrisio, 1574), 222: "Percioché sono assai donne che disperdono senza poter rimediarvi." Like Riminalda's recipe, Marinello's calls for mastic, cardomon, and iris among other ingredients. See also Bairo, *Secreti medicinali,* 169.

107. On the role of the uterus in illness, see Laurinda S. Dixon, *Perilous Chastity: Women and Illness in Pre-Enlightenment Art and Medicine* (Ithaca, N.Y. and London: Cornell University Press, 1995).

108. See Lando, *Valorose donne,* 110r: "this relaxation of the uterus can be attributed only to an overabundance of cold humors" *(sappiate che d'altra cagione non procede questa relassatione de matrice, che dalla molta abondanza d'humori freddi);* compare *Trotula,* 220. If the uterus has descended, she should wash in a mixture of amber, balsam, and musk (see Bairo, *Secreti medicinali,* 179v–180, for similar advice). If it should wander out of the body, she provides an an herbal remedy to mix in wine; that this is offered in Latin is surely a sign that Lando was referencing a medical manual, many of which circulated in Latin as well as Italian (*Valorose donne,* 110r-v).

109. Lando, *Valorose donne,* 110v, "contadinelle da star al paragone con i più dotti phisici ch'oggidì sieno in Padova, o nella dotta Bologna."

110. On Lando's criticism of the learned professions, see Grendler, *Critics of the Italian World.* The issue is again raised where Genevra Malatesta solicits the opinion of a group of Ferrarese physicians regarding her friend's menstrual troubles; Genevra distinguishes between truly experienced doctors and those who "know less than their mules" ("ne sanno meno delle loro mule"; *Valorose donne,* 116r).

111. On Lando's involvement with the heterodox religious culture of sixteenth-century Italy, see Meredith K. Ray, "Textual Collaboration and Spiritual Partnership in Sixteenth-Century Italy: The Case of Ortensio Lando and Lucrezia Gonzaga," *Renaissance Quarterly* 62, no. 3 (Fall 2009): 694–747.

112. Lando, *Valorose donne,* 48r-v. On using egg yolk to keep the skin smooth, see Cortese, *I secreti,* book 4, 100; Celebrino, *Opera nuova piacevole,* 3v–4r; on hare's blood to remove spots from the face see Marinello, *Ornamenti delle donne,* 159.

113. Lando, *Valorose donne,* 116v.

114. Ibid., 113r, "Io voglio ricompensare il segreto, che alli dì passati mi mandaste, con un altro, di non minor virtù."

115. Crisciani, "From the Laboratory to the Library," 295.

116. "Con mio gran dispiacer ho risaputo, esser venuto a voi un scelerato Alchimista, il qual con false lusinghe v'ha pervertito il cervello & ha fatto intrare in humore, che tramutar si possino le sostanze de gli elementi, & di rame fare argento &

l'argento convertire in oro: l'è è pur una gran cosa che questi furfanti, mendichi & pidocchiosi, vogliono arrichir ogn'uno" (Lando, *Valorose donne,* 54r-v). On alchemical fraud, see Tara Nummedal, *Alchemy and Authority in the Holy Roman Empire* (Chicago: University of Chicago Press, 2007). Gentilcore examines the widespread phenomenon of medical fraud (which encompasses alchemy) in *Medical Charlatinism.* On Giulia Gonzaga, see Robin, *Publishing Women,* esp. 18–45.

117. "la più miracolosa acqua che mai né da huomo, né da donna sia stata fatta" (Lando, *Valorose donne,* 114r).

118. "Pigliate limatura d'argento, ferro, ramo, piombo, acciaio, oro, schiuma d'argento e schiuma d'oro. . . . Porrete dette cose per il primo giorno nell'urina di un fanciullo vergine, il secondo giorno in vino bianco caldo, il terzo nel succhio di fenocchio, il quarto giorno nel bianco dell'uova, il quinto giorno nel latte di femina che allatti un fanciullo" (ibid., 116r-v). Gold foam, as we learn in another letter, can be made by distilling wine, sulphur, salt, white tartar, and other ingredients over a series of days: a mere drop of the substance produced on the tenth day, when placed in a gilded goblet, will create a *schiuma d'oro,* or gold foam, that will turn anything it touches to gold (*Valorose donne,* 113r–114r). On progressive distillation, see Aromatico, *Alchemy: The Great Secret* (New York: Harry N. Abrams, 2000, 71); Allison Coudert, *Alchemy: The Philosopher's Stone* (London: Wildwood House, 1980), 43, 47–52.

119. "& quando n'havrete chiaramente veduto l'isperienza, imparate a credere a chi sa & per età & per isperienza più di voi, & far riverentia a fornelli, & a lambicchi" (Lando, *Valorose donne,* 114v).

120. See, for example, Bairo, *Secreti medicinali,* 1r–6r, for a potion that promises to "preserve youth, delay old age, and keep the body perpetually healthy and vigorous") *(conservar la gioventù, et ritardar la vecchiezza, et mantaner la persona sempre sana et vigorosa),* and Cortese, *I secreti,* book 4, 129 for a similar remedy. Another multifaceted potion is furnished by Argentina Rangona to promote health, get rid of freckles, clear up scabies and leprosy, cure gout, and freshen the eyes after crying, all subjects commonly treated in manuals (Lando, *Valorose donne,* 115r-v).

121. "i gran segreti della più segreta parte di Philosophia" (Lando, *Valorose donne,* 114v). See also 116r: "You have often mocked me because I spend my days having master Christopher's waters distilled: for my part, I have often laughed at the naiveté of you and your husband because you don't understand the value that is often found [in such potions]" *(Più volte avete riso di me, perché faccia tutto 'l giorno distillare acque da mastro Cristofero: io ho parimente riso della simplicità vostra e del vostro consorte che non sappiate quanta virtù spesso ci si trovi).* Lando's satire works on two levels: Isabella accuses her friends of the same naiveté of which she is also accused. That Isabella is Lando's mouthpiece for the spiritual treatise, *De la vera tranquillità dell'anima* (Venice, Manuzio: 1544), further underscores the element of critique.

3. Scientific Culture and the Renaissance *Querelle des Femmes*

1. "Che è al caso nostro, di grazia, il discorrer sopra cose tali? Siamo noi medici? Lasciateli parlar loro di siloppi, di empiastri e sí fatte pratiche, che è una vergogna che noi ne trattiamo" (Fonte, *Il merito delle donne* [Venice: Domenico Imberti, 1600], 125). Citations in the original Italian are taken from this 1600 edition. The Italian text is also available in a modern edition by Adriana Chemello, *Il merito delle donne* (Venice: Eidos, 1988). English citations are taken from Moderata Fonte, *The Worth of Women,* ed. and trans. Virginia Cox (Chicago: University of Chicago Press, 1997), 180.

2. By scientific knowledge, I mean knowledge relating to any kind of natural phenomena (meteorology, astronomy, botany, pharmacology, alchemy), as in the Renaissance sense. Discourses of science and medicine, furthermore, were closely linked. As Paula Findlen notes, an increasing interest in *materia medica* and the production of medicinal compounds led in the sixteenth century to the introduction of natural philosophy into the medical curriculum in Italian universities (see *Possessing Nature: Museums, Collecting, and Scientific Culture in Early Modern Italy* [Berkeley: University of California Press, 1994], 248–249).

3. "questa preminenza si hanno essi arrogata da loro" (Fonte, *The Worth of Women,* 59, 59n23; *Il merito delle donne,* 26).

4. On the debate over women as it relates to these works, see, for example, Constance Jordan, "Bocaccio's In-Famous Women: Gender and Civic Virtue in the *De Claris Mulieribus,*" in *Ambiguous Realities: Women in the Middle Ages and Renaissance,* ed. Carole Levine and Jeannie Watson (Detroit: Wayne State University Press, 1987), 25–47; Jordan, *Renaissance Feminism: Literary Texts and Political Models* (Ithaca, N.Y.: Cornell University Press, 1990), 35–50; Pamela Benson, *The Invention of the Renaissance Woman* (University Park: Pennsylvania State University Press, 1992), 9–31; see also Laura Cereta, *Collected Letters of a Renaissance Feminist,* ed. and trans. Diana Robin (Chicago: University of Chicago Press, 1997), 10–11.

5. The literature on the Renaissance *querelle des femmes* is extensive. See, inter alia, Ruth Kelso, *Doctrine for the Lady of the Renaissance* (Urbana: University of Illinois Press, 1956); Conor Fahy, "Three Early Renaissance Treatises on Women," *Italian Studies* (1956): 30–55; Francine Daenens, "Superiore perchè inferiore: il paradosso della superiorità delle donne in alcuni trattati italiani del Cinquecento," in *Transgressione tragica e norma domestica: esemplari di tipologie femminili dalla letteratura europea,* ed. Vanna Gentili (Rome: Edizioni di storia e letteratura, 1983), 11–50; Joan Kelly, "Early Feminist Theory and the *Querelle des Femmes,* 1400–1789," *Signs* 8, no. 1 (Autumn 1982): 4–28; Adriana Chemello, "La donna, il modello, l'immaginario: Moderata Fonte e Lucrezia Marinella," in *Nel cerchio della luna: Figura di donna in alcuni testi del XVI secolo,* ed. Marina Zancan (Venice: Marsilio, 1983); Benson, *Invention of the Renaissance Woman;* Cox, "The Single Self: Feminist Thought and the Marriage Market in Early Modern Venice," *Renaissance Quarterly* 48, no. 3 (Autumn 1995): 513–581.

6. On the the socioeconomic context fueling the debate in this period—including dowry inflation, a shift to land-based wealth, and a shrinking marriage market—see Cox, "The Single Self"; Jutta Sperling, *Convents and the Body Politic in Late Renaissance Venice* (Chicago: University of Chicago Press, 1999); and Brian S. Pullan, ed., *Crisis and Change in the Venetian Economy of the Sixteenth and Seventeenth Centuries* (London: Methuen, 1968).

7. Lucrezia Marinella, *Della nobiltà et eccellenza delle donne co' difetti et mancamenti de gli huomini* (Venice: G. B. Ciotti, 1600). The abridged English translation is edited by Letizia Panizza, *The Nobility and Excellence of Women and the Defects and Vices of Men* (Chicago: The University of Chicago Press, 1999).

8. Fonte's text was published, posthumously, just a few months after Marinella's. On this episode, see Beatrice Collina, "Moderata Fonte e il *Merito delle donne*," *Annali di italianistica* (1989): 145; see also Stephen Kolsky, "Moderata Fonte, Lucrezia Marinella, Giuseppe Passi: An Early Seventeenth-Century Feminist Controversy," *Modern Language Review* 96 (2001): 973–989.

9. See Passi, *I donneschi difetti*, chap. XV: "Delle donne maghe, incantatrici, malefiche o venefiche, fattochiere, streghe, o strigimache." Passi echoes a long tradition of associating women's knowledge of the natural world with witchcraft. He offers the example of Circe, for instance (who in Fonte's *Floridoro* [1581] becomes a symbol of the power of women's learning in natural philosophy), as a witch who "had four handmaidens to help gather herbs for her evil spells" (*Circe, la quale hebbe quattro ancelle per servitio di raccogliere quell'herbe, ch'ella nelle sue malie adoperava* [113]). Passi also condemns astrology as forbidden by Scripture and accuses women who engage in it of dealing with the Devil: "to know and reveal is a divine act. She who seeks to divine and know the things of the future, with the help of the devil, steals from God his own honor" (*il conoscere, et il revelare è atto divino. Quella donna dunque che cerca per mezzo del diavolo con l'aiuto, et arte sua d'indovinare, e sapere le cose future, toglie a Dio l'honor suo,* 140). Witchcraft was still a source of concern and subject of investigation in early modern Italy, but as Jonathan Seitz has shown, inquisitional investigation of cases of *maleficio* (using magic to cause harm) in 1550–1650 was relatively benign compared to other areas of Europe (see Jonathan Seitz, *Witchcraft and Inquisition in Early Modern Venice* [New York: Cambridge University Press, 2011]). On this subject see also Ruth Martin, *Witchcraft and the Inquisition in Venice 1550–1650* (Oxford: Blackwell, 1989); Brian P. Levack, *The Witch-Hunt in Early Modern Europe* (London: Longman, 1994); and Guido Ruggiero, *Binding Passions: Tales of Magic, Marriage and Power at the End of the Renaissance* (New York: Oxford University Press, 1993). For a general overview, see Ruggiero, "Witchcraft and Magic," in *A Companion to the Worlds of the Renaissance*, ed. Guido Ruggiero (Oxford: Wiley Blackwell, 2007), 475–490.

10. On Fonte's *Merito delle donne*, see the Italian edition and introduction edited by Chemello, and Cox's English translation (cited in note 1); see also Patricia LaBalme, "Women's Roles in Early Modern Venice: An Exceptional Case" in *Beyond Their Sex:*

Learned Women of the European Past, ed. Patricia H. LaBalme (New York: New York University Press, 1980); Beatrice Collina, "Moderata Fonte e il *Merito delle donne,*" *Annali di italianistica* (1989): 142–164; Malpezzi Price, "A Woman's Discourse in the Italian Renaissance: Moderata Fonte's *Il Merito delle donne,*" *Annali d'Italianistica* 7 (1989): 165–181; Malpezzi Price, *Moderata Fonte: Women and Life in Sixteenth-Century Venice* (Teaneck, N.J.: Fairleigh Dickinson University Press, 2003); and Malpezzi Price, "Venezia Figurata and Women in Sixteenth-Century Venice: Moderata Fonte's Writings," in *Italian Women and the City. Essays,* ed. Janet Levarie Smarr and Daria Valentini (Teaneck, N.J.: Fairleigh Dickinson University Press, 2003), 18–34. For Marinella's *Nobiltà,* see the English edition and translation by Letizia Panizza, cited in note 7. See also Malpezzi Price, "Lucrezia Marinella," in *Italian Women Writers: A Bio-Bibliographical Sourcebook,* ed. Rinaldina Russell (Westport, Conn.: Greenwood Press, 1994), 234–242; and Malpezzi Price and Christine M. Ristaino, *Lucrezia Marinella and the "Querelle des Femmes" in Seventeenth-Century Italy* (Teaneck, N.J.: Fairleigh Dickinson University Press, 2008); Stephen Kolsky, "The Literary Career of Lucrezia Marinella (1571–1653): The Constraints of Gender and the Writing Woman," in *Religion and Culture in the Italian Renaissance. Essays in Honour of Ian Robertson,* ed. F. W. Kent and Charles Zika (Turnhout: Brepols, 2005), 325–342. The work of Fonte and Marinella is often addressed together: see, for example, Malpezzi Price, "Moderata Fonte, Lucrezia Marinella and Their 'Feminist' Work," *Italian Culture* 12 (1994): 201–214; LaBalme, "Venetian Women on Women: Three Early Modern Feminists," *Archivio Veneto* 5, no. 117 (1981): 81–109; Adriana Chemello, "La donna, il modello, l'immaginario: Moderata Fonte and Lucrezia Marinella," in *Nel cerchio della luna: figure di donna in alcuni testi del XVI secolo,* ed. Marina Zancan (Venice: Marsilio, 1983), 95–170; Odorisio, *Donna e società nel Seicento* (Rome: Bulzoni, 1979); Kolsky, "Moderata Fonte, Lucrezia Marinella"; and Cox, "The Single Self."

11. With some exceptions: see Stephen Kolsky, "Wells of Knowledge: Moderata Fonte's *Il Merito delle donne,*" *Italianist* 13 (1993), 57–96; Claire Lesage, "Le savoir alimentaire dans *Il Merito delle donne* de Moderata Fonte," in *La table et ses dessous: Culture, alimentation et convivialité en Italie, XIV–XVIe siècles,* ed. Adelin Charles Fiorato and Anna Fontes Baratto (Paris: Presses de la Sorbonne Nouvelle, 1999), 223–234; Suzanne Magnanini, "Una selva luminosa: The Second Day of Moderata Fonte's *Il Merito delle donne,*" *Modern Philology* (2003): 278–296; see also Cox's introduction to Fonte, *The Worth of Women,* 9–10, and, briefly, Chemello in Fonte, *Il merito delle donne,* xix–xli.

12. Modern editions and in some cases translations of these works are available: for *Floridoro,* see Fonte, *Tredici canti del Floridoro,* ed. Valeria Finucci (Modena: Mucchi, 1995); Fonte, *Floridoro: A Chivalric Romance,* ed. Valeria Finucci, trans. Julia Kisacky (Chicago: University of Chicago Press, 2006). Marinella's *Arcadia felice* is edited by Françoise Lavocat (Florence: Olschki, 1998). The Italian edition of *L'Enrico* is edited by Maria Galli Stampino (Modena: Mucchi, 2011); as is the English translation, *Enrico; or, Byzantium Conquered: A Heroic Poem* (Chicago: University of Chicago Press, 2009). Citations throughout are taken from these editions.

13. As Virginia Cox notes, the terms "pro-woman" and "protofeminist" can be unwieldy. Like Cox, I use "pro-woman" to indicate works, themes, or figures that fall on that gynephilic side of the debate over women, while reserving "protofeminist" to indicate those with more directly politicizing implications (see Cox, *Women's Writing in Italy, 1400–1650* [Baltimore: Johns Hopkins University Press, 2008], xxvii).

14. As Cox points out, the strength of Fonte's reputation can be inferred from her inclusion as the sole female contributor to a poetic anthology in honor of Stephen Báthory, king of Poland, as well as from Niccolò Doglioni's claim that she was invited to contribute to a volume in memory of Gian Tommaso Costanzo, son of the famous Venetian condottiere Scipio Costanzo (Fonte, *The Worth of Women,* 39n21).

15. "Hebbe cosi profonda memoria, che udita una Predica, per lunga, che fosse, la recitava quasi tutta di parola in parola: Nel proporre, e rispondere ne' ragionamenti, fu non meno pronta, che arguta" (Cristofano Bronzini, *Della Dignità & Nobiltà delle Donne. Dialogo di Cristofano Bronzini D'Ancona. Diviso in Quattro Settimane, e ciascheduna di esse in Sei Giornate* [Florence: Zanobi Pignoni, 1625], 116). Bronzini singles out Fonte's *Il merito delle donne* for particular praise as a "most beautiful and learned discourse in prose" (*bellissimo, e dotto discorso in prosa,* 117).

16. Doglioni, in Fonte, *The Worth of Women,* 35–36. In addition to the *Worth of Women,* Fonte penned a *rappresentazione* titled *Le feste* in honor of the Venetian doge Nicolò da Ponte (1581); the chivalric romance *Floridoro* (1581); and the verse narratives *La passione di Cristo* (1582) and *La resurettione di Gesù Cristo* (1592); as well as occasional poetry and perhaps additional *rappresentazioni.* Fonte's *Le feste: Rappresentazione avanti al Serenissimo Prencipe di Venetia Nicolò da Ponte il giorno di S. Stefano, 1581* (Venice: D. and G. B. Guerra, 1582) is translated by Courtney Quaintance: *Le Feste, Written by Moderata Fonte.* For the modern editions and translations of *Floridoro,* see note 12.

17. Cox notes that Doglioni's *Vita* exists in various manuscript copies, "suggesting that it was distributed independently after Fonte's death" (Fonte, *The Worth of Women,* 31n1). A modern edition of Doglioni's biography of Fonte is in Moderata Fonte, *Il merito delle donne,* 3–10; it is translated in Fonte, *The Worth of Women,* 31–40. For recent contributions to Fonte's biography, the entry by Virginia Cox in *The Italian Women Writers Project,* s.v. "Fonte, Moderata." On Fonte's education and schooling, see Sarah Gwyneth Ross, *The Birth of Feminism: Woman As Intellect in Renaissance Italy and England* (Cambridge, Mass.: Harvard University Press, 2009), 194–209.

18. This was a relatively advanced age for a Venetian woman of Fonte's class to marry; the delay may have been tied to questions regarding her dowry (see Fonte, *The Worth of Women,* 37n19). Fonte's husband, Filippo Zorzi, worked for Venice's *Ufficio delle Acque* (ibid.).

19. Doglioni, *Vita,* in Fonte, *The Worth of Women,* 36. See also Stephen Kolsky, "Per la carriera poetica di Moderata Fonte: Alcuni documenti poco conosciuti," *Experienze letterarie* (1999): 10. Among Fonte's occasional poetry is a composition

dedicated to Doge Niccolò da Ponte (for whom she wrote *Le feste*) and several poems published in an anthology for Stephen Báthory of Poland, to which she is the only female contributor (*Del giardino de' poeti in lode del Serenissimo Re di Polonia . . . libro secondo* [Venice: Fratelli Guerra, 1583], in *Viridiarium poetarum . . . in laudes Serenissimi atque potentissimi D.D. Stephani Regis Poloniae . . . in duos libros divisum*, ed. Ippolito Zucconello [Venice, ad Signum Hyppographi, 1583]).

20. "gentil huomo suo grande amico" (Fonte, *Il merito delle donne*, 73; Fonte, *The Worth of Women*, 133); "Gran saper, c'huom mortal spiega, e comparte / Ogni poter celeste, ogni terreno, / Termine, stato, moto, sito, e seno, / Tempo, elementi, Ciel, natura, & Arte," (ibid.).

21. See Fonte, *The Worth of Women*, 134nn29–32. As Cox points out, Fonte clearly gives pride of place to Doglioni in this list, mentioning the much more renowned Magini only after praising her friend (134n29). The sixth person mentioned by Corinna, Lucio Scarano—to whom Marinella's *Nobility and Excellence of Women* is dedicated—does not appear to have authored any works on astronomy (134n30). As Pomata suggests, Scarano—a physician and member of the elite Accademia Veneziana—may well have initiated the publication of both Marinella's *Nobiltà delle donne* and Fonte's *Merito delle donne* ("Was There a *Querelle des Femmes* in Early Modern Medicine?," *Arenal* 20, no. 2 [2013]: 322).

22. Doglioni, *Compendio Historico Universale*, 3r: "deve la città abbracciar la scientia de' suoi cittadini." On the term *scientia*, see Pamela H. Smith, "Science on the Move: Recent Trends in the History of Early Modern Science," *Renaissance Quarterly* 62 (2009): 345n1.

23. See Ann Blair, *Too Much to Know: Managing Scholarly Information Before the Modern Age* (New Haven, Conn.: Yale University Press, 2010); for a mercantile context, see Harold Cook, *Matters of Exchange: Commerce, Medicine, and Science in the Dutch Golden Age* (New Haven, Conn.: Yale University Press, 2007).

24. On Venice's civic mythology, see, for example, Edward Muir, *Civic Ritual in Renaissance Venice* (Princeton, N.J.: Princeton University Press, 1986). On the implications of the civic mythology for women writers, see Margaret Rosenthal's discussion of another Venetian writer, Veronica Franco, in *The Honest Courtesan: Veronica Franco, Citizen and Writer in Sixteenth-Century Venice* (Chicago: University of Chicago Press, 1993).

25. Doglioni states that Fonte composed additional cantos for *Floridoro* that were not published (see Doglioni's *Vita*, in Fonte, *The Worth of Women*, 16). A letter from Fonte to Francesco de' Medici of Florence, dedicatee of *Floridoro* along with his consort Bianca Cappello, confirms Doglioni's assertion: Fonte writes that the poem is "completely planned out and will amount to more than fifty cantos" (*totalmente ordita e arriverà a meglio di cinquanta canti*) (see Eleonora Carinci, "Una lettera autografa inedita di Moderata Fonte al granduca di Toscana Francesco I," *Critica del testo* 5, no. 3 [2002]: 1–11). Emilio Zanette disapprovingly termed *Floridoro* "feminism's epic"

(*l'epopea del femminismo*) for its clear emphasis on independent female characters (cited in Fonte, *Tredici canti del Floridoro*, X).

26. In fact, Fonte warns her female readers to beware of Love: "Let ladies avoid even more than sin, / more than death, the undertaking of love, / for most men have ungrateful hearts" (Fonte, *Tredici canti del Floridoro*, V, 3: "Fuggan le donne pur più che 'l peccato, / Più che 'l morir l'officio dell'amare, / C'han la più parte il cor gli uomini ingrato").

27. See Virginia Cox, "Women as Readers and Writers of Chivalric Poetry in Early Modern Italy," in *Sguardi sull'Italia: Miscellanea dedicata a Franceso Villari*, ed. Gino Bedoni et al. (Leeds: Society for Italian Studies, 1997), 143; Fonte, *Floridoro: A Chivalric Romance*, 176n12; Stephen Kolsky, "Moderata Fonte's *Tredici canti del Floridoro*: Women in a Man's Genre," *Rivista di studi italiani* (1999): 170. As Cox notes, the *maga* is an "inherited character type so bereft of subjectivity and so intimately shaped by male fears and desires as to present a paradigmatic instance of the problems faced by women writers attempting to voice themselves through a male literary code" ("Women as Readers and Writers," 142).

28. "L'oro che sta ne le minere ascoso, / non manca d'esser or, benché sepolto, / E quando è tratto e se ne fa lavoro, / È così ricco e bel come l'altro oro" (*Floridoro*, IV, 1–4).

29. Indeed, as Stephen Kolsky notes, "It is almost inconceivable that Fonte was not attempting to rewrite the romance epic, while remaining within its structure, in order to highlight the unequal power relations between men and women" ("Moderata Fonte's *Tredici canti*," 181).

30. On Bianca Cappello [Cappella], see James Chater, "Bianca Cappella and Music," in *Renaissance Studies in Honor of Craig Smyth*, ed. Andrew Morrough, Fiorella Superbi Gioffredi, Piero Morselli, and Eve Borsook (Florence: Giunti Barbèra, 1985); on Fonte and Bianca Cappello, see Emilio Zanette, "Bianca Cappella e la sua poetessa," *Nuova antologia* 88 (1953): 455–468.

31. See Carinci, "Una lettera autografa," 669–670.

32. On these scientific interests, see Valentina Conticelli, "Lo studiolo di Francesco I e l'alchimia: nuovi contributi storici e iconologici, con un carteggio in appendice (1563–1581)," in *L'art de la Renaissance entre science et magie*, ed. Philippe Morel (Académie de France a Rome, Villa Medici: Somogy Editions d'Art, 2006), 207–268; Luciano Berti, *Il principe dello studiolo: Francesco I dei Medici e la fine del Rinascimento fiorentino* (Florence: Edam, 1967). Francesco was not the only Medici with an interest in alchemy: on alchemy at the Medici court under Cosimo I (Francesco's father), see Alfredo Perifano, *L'alchimie à la Cour de Côme Ier de Medicis: Savoir, culture et politique* (Paris: Honore Champion, 1997); and on father and son, Giulio Cesare Lensi Orlandi, *L'arte segreta: Cosimo e Francesco de' Medici alchimisti* (Florence: Convivio, 1978).

33. See *Orlando furioso*, VI.

34. Fonte, *Floridoro:* "grave e severa" (VII, 49); "cortese e gentilissima" (VII, 51); "sí giovene onesta e saggia a maraviglia" (VII, 52).

35. "M'insegna il ben ch'uscir può da quell'arte" (*Floridoro,* VIII, 25).

36. This paradigm, evident in Galileo and others, persisted well into the seventeenth century; see Findlen, *Possessing Nature,* 55–56.

37. In Kisacky, trans., *Floridoro,* VIII, 1; "Circe già in virtù d'erbe e di parole, / con alto studio oggi a nissuno espresso, / Poté oscurar l'illustre faccia al sole" (*Floridoro,* VIII, 1).

38. Eamon notes the contrast between della Porta's natural magic and Aristotelian natural philosophy: "In contrast to Aristotelian natural philosophy, whose aim was to explain the normal, everyday aspects of nature, natural magic explained the exceptional, the unusual, and the 'miraculous'" (William Eamon, *Science and the Secrets of Nature: Books of Secrets in Medieval and Early Modern Culture* [Princeton, N.J.: Princeton University Press, 1994], 210–211).

39. della Porta, *Magia naturalis,* bk. 1, chap. 2: "Nobis vero non nisi universae Naturae contemplationem esse videtur. Ex coelorum enim motus consideratione, stellarum, elementorum, eorumque transmutationibus, sic animalium, plantarum, mineralium, eroumque; ortus, & interitus occulta vestigantur arcana, ut tota scientia ex Naturae vultu dependere videatur, ut latius videbimus." The work was hugely popular and was soon translated into Italian (1560) as well as French (1565), Dutch (1566), and German (1612); I take the English translation from the 1658 edition printed by Thomas Young and Samuel Speed. On the *Magia naturalis,* see Laura Balbiani, "La ricezione della 'Magia Naturalis' di Giovan Battista Della Porta. Cultura e scienze dall'Italia all'Europa," *Bruniana e Campanelliana* 5 (1999): 277–303; G. Belloni, "Conoscenza magica e ricerca scientifica in Giovanni Battista Della Porta," in *Della Porta. Criptologia. Edizione, nota biografica, traduzione,* ed. G. Belloni (Rome, 1982), 43–101; William Eamon, "Science and Popular Culture in Sixteenth-Century Italy: The 'Professors of Secrets' and Their Books," *Sixteenth Century Journal* 16 (1985): 471–485"; Eamon, *Science and the Secrets of Nature,* 201–227; Luisa Muraro, *Giambattista della Porta mago e scienziato* (Milan: Feltrinelli, 1978), 15–19.

40. Fonte, *Floridoro: A Chivalric Romance,* VIII, 22–23.

41. Fonte, *The Worth of Women,* 129 (*Il merito delle donne,* 70, "Dal vento . . . il qual dovendo vagar per aria suo proprio luogo, viene a cacciarsi sotterra . . . e non trovando poi sì presto l'uscita; poiché naturalmente non può star chiuso, pone ogni suo sforzo per uscirvi, & con questa forza viene a scuoter, e crollar così forte la terra").

42. Fonte, *The Worth of Women,* 129n18. Della Porta, too, composed a treatise on meteorology, *De aeris trasmutationibus* (1610), which provided a variation on Aristotle's explanation in attributing earthquakes to the vapour produced by underground fires. On this subject, see Craig Martin, *Renaissance Meteorology: Pompanazzi to Descartes* (Baltimore: Johns Hopkins University Press, 2011); Paul Lettinck, *Aristotle's Meteorology and Its Reception in the Arab World* (Leiden: Brill, 1999).

43. "Quando alla nostra età gli uomini errando / Di lor medesmi son trasforma-tori" (Fonte, *Floridoro*, VIII, 3).

44. "E con tal facilità girsi mutando / Gli veggio senza oprar versi o liquor, / Che poco stima in ciò fo di quell'arte / Poi che 'l secol di noi n'ha tanta parte. / Ciascun dell'esser proprio è sì buon mago, / Che non ne seppe tanto ella in quel tempo / Quando spese in cangiar la nostra imago / Tant'erbe, tanto studio, e tanto tempo, / E d'uscir di sé stesso è così vago / Che di tornarvi poi non trova il tempo; / Di tutti no, ma ben del più ragion, / A cui piace parer quel che non sono" (ibid., VIII, 3–4).

45. "Ella ch'alle maniere ostar leggiadre / Non può con vari versi e varie ampolle, / Fece all'amante ogni scienza espresso / E per grader altrui nocque a se stessa" (ibid., VIII, 16).

46. "Non mancan sopra questi i propri segni, / Ariete, Toro, Gemini e i seguenti. / Par poi ch'ogni pianeta alberghi e regni / Sopra le case lor convenienti" (ibid., X, 18).

47. On early modern astrology, see Monica Azzolini, *The Duke and the Stars: Astrology and Politics in Renaissance Milan* (Cambridge, Mass.: Harvard University Press, 2013); see also William R. Newman and Anthony Grafton, eds., *Secrets of Nature: Astrology and Alchemy in Early Modern Europe* (Cambridge, Mass.: MIT Press, 2001).

48. "Nell'ultima facciata, che scolpita / Di dietro fu dove era poca luce, / Una giovane stavasi romita / E non ardia con gli altri uscir in luce, / Vergognandosi assai che troppo ardita / Aspirasse alla via ch'el ciel conduce" (ibid., Fonte, *Floridoro*, X, 36–37).

49. On these figures, see Fonte, *Floridoro*, 286nn11–12.

50. Ibid., 336n1.

51. See Findlen, *Possessing Nature*, 457.

52. Fonte, *The Worth of Women*, 161; *Il merito delle donne*, 107, "non vi giovarebbe quanta tiriaca fanno gli speciali . . . per riparaci a tanta malignità."

53. Suzanne Magninini notes, "Since for Fonte female emancipation is inextri-cably bound to female education, her quasi selva [e.g., the second day] functions as an instrument of liberation" ("Una selva luminosa," 283).

54. Fonte, *The Worth of Women*, 121–122 (*Il merito delle donne*, 65: "Procede anco l'amor di molte, sia di che qualità esser si voglia, appresso le ragioni predette, spesse volte dale influenze celesti, che ve le inclinano in tale maniera, che molte di loro cono-scendo assai bene la indegnità, & ingratitudine di questi huomini . . . si asteneriano volentieri dalla loro perverse pratica, ma la loro inclination, oltra le altre cause, è ca-gion potentissima, & la maggior di tutte per disponerle prima, & poi mantenerle nel loro errore"). On the role of the stars in influencing natural attraction, see, for ex-ample, Ebreo's *Dialoghi d'amore*.

55. Fonte, *The Worth of Women*, 122 (*Il merito delle donne*, 65, "Noi overo il cor nostro è disposto per la bontà del genio a ricever la forma del vero amore, ma gli huo-mini per natura, e per volontà disamorevoli non ponno ne anco esser molto inclinati a tal disposition, né quante stelle sono in cielo potriano fare, ch'essi ci amassero").

56. On the conceptualization of friendship in the early modern period, see Daniel T. Lochman, Maritere López, and Lorna Hurney, eds., *Discourses and Representations of Friendship in Early Modern Europe, 1500–1700* (Aldershot: Ashgate, 2010).

57. Fonte, *The Worth of Women,* 128 (*Il merito delle donne,* 69).

58. Fonte, *The Worth of Women,* 128 (*Il merito delle donne,* 69: "per l'amicitia et unione de gli elementi ne i nostri corpi si mantien la sanità; nell'aria i tempi chiari, nel mar la bonaccia, in terra le Città per la pace se costruggono, i Regni si accrescono; e tutte le creature si consolano").

59. As Corinna laments, discord is "the destroyer of everything! It is when men fall out of harmony with each other that wars begin"; leading to more discord: "provinces and families are exterminated, states overthrown, and whole peoples consumed" (Fonte, *The Worth of Women,* 128–129) ("Oh per questa discordia . . . ogni cosa va in desolatione; per la disunion de gli huomini, le guerre suscitano . . . le provincie, & le famiglie s'esterminano, gli stati si mutano, e i popoli si consumano. Nell'aria, le contrarietà cagionano tuoni e saette; le tempeste nel mare, e i terramoti nella terra" [*Il merito delle donne,* 70]).

60. See Martin *Renaissance Meteorology,* 65; for a discussion of these and other works on natural phenomena and disasters, see 60–79. On female audiences for such works see also Martin, "Meteorology for Courtiers and Ladies: Vernacular Aristotelianism in Renaissance Italy," *Philosophical Readings* 4, no. 2 (2012): 3–14.

61. Martin, "Meteorology for Courtiers," 12–13.

62. Fonte, *The Worth of Women,* 147.

63. Ibid., 132 (*Il merito delle donne,* 72, "Se volete dir di cosa instabile, qual cosa è più de gli huomini? Se di discorde il simile? . . . che tante astrologie? a noi non apartengono tali discorsi no, né sì fatti studij").

64. Fonte, *The Worth of Women,* 134 (*Il merito delle donne,* 72, "questo è un bellissimo studio e degno di elevatissimi intelletti").

65. Fonte, *The Worth of Women,* 132 (*Il merito delle donne,* 72, "Io non ho studiato . . . in alcuna scienza, e molto manco in questa.").

66. It is worth noting, as Cox points out, that Fonte—like Marinella after her—did not subscribe to the Copernican view of the earth, preferring the then still-orthodox Aristotelian/Ptolemaic position (Fonte, *The Worth of Women,* 131n21).

67. See Lesage, "Le savoir alimentaire," 225.

68. Fonte, *The Worth of Women,* 137 (*Il merito delle donne,* 75).

69. Fonte, *The Worth of Women,* 141–142 (*Il merito delle donne,* 78).

70. On this aspect, see Lesage, "Le savoir alimentaire." On early modern dietary regimes and advice, see Ken Albala, *Eating Right in the Renaissance* (Berkeley: University of California Press, 2002), and Sandra Cavallo and Tessa Storey, *Healthy Living in Late Renaissance* Italy (Oxford: Oxford University Press, 2013).

71. On Renaissance consumerism, see, for example, Lisa Jardine, *Worldly Goods: A New History of the Renaissance* (New York: Norton, 1996); Evelyn S. Welch, *Shopping in the Renaissance* (New Haven, Conn.: Yale University Press, 2005).

72. Fonte, *The Worth of Women*, 140 (*Il merito delle donne*, 78, "da ciò son buoni essi, cioè da uccellare, ingannare, e prendere, anzi questo è il lor proprio mestiero").

73. On early modern balneology, see, for example, D. S. Chambers, "Spas in the Italian Renaissance," in *Reconsidering the Renaissance*, ed. M. Di Cesare (Binghamton, N.Y.: Medieval and Renaissance Texts and Studies, 1992), 3–27; and Katharine Park, "Natural Particulars: Medical Epistemology, Practice, and the Literature of Healing Springs," in *Natural Particulars: Nature and the Disciplines in Renaissance Europe*, ed. Anthony Grafton and Nancy G. Siraisi (Cambridge, Mass.: MIT Press, 1999), 347–367.

74. Fonte likely also had in mind book 31 of Pliny's *Natural History* (see Fonte, *The Worth of Women*, 151n73); see also Bruce Moran, *Distilling Knowledge: Alchemy, Chemistry, and the Scientific Revolution* (Cambridge, Mass.: Harvard University Press, 2005), 11.

75. Fonte, *The Worth of Women*, 151 (*Il merito delle donne*, 98).

76. Fonte, *The Worth of Women*, 157 (*Il merito delle donne*, 103, "fate conto di veder gli uomini, che essendo inferiori a noi e perciò dovendo essi star bassi ed umili, vedete come s'inalzano, come ci soprastano contra ogni ragione, contra ogni giustizia, però non vi maravigliate se l'acqua, elemento basso, anch'ella presume d'ascendere all'altezza dei monti, ma pur ella torna ad abbassarsi di nuovo, dove gli uomini stanno sempre fermi nel lor rigore ed ostinazione").

77. Fonte, *The Worth of Women*, 169: " 'True balsalm,' said Corinna , '. . . is a divine liquor (as long as it isn't doctored in any way) and extremely good for our whole body, for a taste of it revives flagging spirits, restores strength to the body, and generally puts new life in you. Balsamic ointment applied to the living keeps the face looking fresh and young, while, applied to the dead bodies, it acts as a preservative against putrefaction and decay' " (Fonte, *Il merito delle donne*, 98, "Il vero balsamo, disse Corinna . . . è liquore divino, ottimo ne' corpi nostri, poiché gustandosi ricrea li spiriti smarriti, ritorna la virtù, e rinfranca la vita; mantien la sua ontion ne i vivi la freschezza della gioventù nella faccia, e preserva i morti dalla putretudine, & corruttione.").

78. Fonte, *The Worth of Women*, 169 (*Il merito delle donne*, 114, "Il balsamo . . . si dice pur che guarisse di tutte le infirmità.").

79. Fonte, *The Worth of Women*, 185 (*Il merito delle donne*, 128).

80. Fonte, *The Worth of Women*, 185 (*Il merito delle donne*, 110, "a questo che dite, non credo già, che vaglia alcuna di esse").

81. Fonte, *The Worth of Women*, 176 (*Il merito delle donne*, 101, "in vero è cosa mirabile a molte infirmità").

82. Fonte, *The Worth of Women*, 170 (*Il merito delle donne*, 99, "[coloquintida] . . . giova anco alla durezza della milza cioè la sua medulla con acqua di scolopendria; la sua decotion al dolor de i denti con aceto, il suo empiastro alli vermin, ma non è da adoprarsi sola, perchè è venenosa").

83. Fonte, *The Worth of Women*, 170: "In that case, it must resemble man . . . who is noxious when alone, and needs women's company as his antidote" (*Il merito delle*

donne, 99, "Debbe esser, disse Leonora, quest'herba come l'huomo, che solo è mortifero, ma la compagnia della donna è la sua teriaca").

84. Fonte, *The Worth of Women,* 160 (*Il merito delle donne,* 91, "egli è il più utile animale . . . nel dispensarci tante sorte di latticini utili . . . la pelle si sa di quanto utile sia in varij lavori; la carne a mangiare, le corna e l'unghie a diverse operationi.").

85. Fonte, *The Worth of Women,* 160 (*Il merito delle donne,* 91, "lisciar la pelle come di giovene di quindici anni").

86. Fonte, *The Worth of Women,* 161 (*Il merito delle donne,* 91, "non son tanto noti come bisogneria che ci sapressimo più schermire, che non sappiamo; e possiamo meglio intendere la proprietà de gli animali irragionevoli, ancor che ci dovria esser più occolta per esser tanto diversa dalla nostra ed anco perché non sanno essi parlare, che quella di questi falsi a noi simili di natura, ma diversi di qualità e volontà, che mai ci dicono il vero").

87. Fonte, *The Worth of Women,* 171, "After all, it would hardly have been in their interests, for wolf doesn't eat wolf, and men know very well which side of their bread is buttered: if we stopped loving them, they'd be in a fine state!" (*Il merito delle donne,* 116, "Non la scrive Galeno questa medicina, né altro auttore l'ha mai trovata . . . o se l'ha trovata non la lasciò scritta . . . perché lupo non mangia di lupo, troppo conoscono il lor danno gli uomini, se noi non gli amassimo guai a loro").

88. See Meredith K. Ray, *Writing Gender in Women's Letter Collections of the Italian Renaissance* (Toronto: University of Toronto Press, 2009), 45–80; see also Chapter 2 here.

89. Fonte, *The Worth of Women,* 186–187 (*Il merito delle donne,* 130). On medicinal, or "potable" gold, see David Gentilcore, *Medical Charlatanism in Early Modern Italy* (Oxford: Oxford University Press, 2006), 214–215.

90. On alchemical fraud see Tara Nummedal, *Alchemy and Authority in the Holy Roman Empire* (Chicago: University of Chicago Press, 2007), 147–176.

91. Fonte, *The Worth of Women,* 187 (*Il merito delle donne,* 130, "non so io che più bella alchimia per far oro ed argento si possa trovar quanto che l'uomo studi e s'affatichi per imparar virtù e che con le sue giuste fatiche sia solecito ad acquistarsi le facoltà e le richezze che questa è una alchimia che non falla mai.").

92. Fonte, *The Worth of Women,* 157 (*Il merito delle donne,* 89, "convertendosi co 'l tempo l'uno elemento nell'altro, trovasi che la terra a poco a poco si resolve in acqua, l'acqua in aria, e l'aria in foco. Et all'incontro il foco si tramuta in aria, l'aria in acqua, & l'acqua ritorna in terra.").

93. Fonte, *The Worth of Women,* 169–179.

94. "inspira il tuo intelletto, / Te picciol mondo in se stesso trasforma" (*Il merito delle donne,* 133).

95. Ibid., 133n28.

96. "The small world is the human being: he is also enclosed by a skin . . . so that his elements [do] not come into contact with the external substance. . . . That is why the human being is covered with skin, which defines the human being in order to sepa-

rate the two worlds from one another, the great one and the small one" (*On the Matrix*, H: 1:192, in Paracelsus, *Paracelsus Theophrastus Bombastus von Hohenheim, 1493–1541: Essential Theoretical Writings*, ed. and trans. Andrew Weeks [Leiden: Brill, 2008], 621).

97. "the physician is an inner *astronomus*, as well as an inner *philosophus*, born from the external astronomy and philosophy . . . [but] he must first become an alchemist. What is it that makes the pears ripe, what brings forth the grapes? Nothing other than natural alchemy" (*Paragranum*, Preface, H 2:13 in ibid., 83). On Paracelsus, see Walter Pagel, *Paracelsus: An Introduction to Philosophical Medicine in the Era of the Renaissance*, 2nd ed. (Basel: Karger, 1982); Allen Debus, *The Chemical Philosophy: Paracelsian Science and Medicine in the Sixteenth and Seventeenth Centuries* (2 vols., New York: Science History Publications, 1977); Andrew Weeks, *Paracelsus: Speculative Theory and the Crisis of the Early Reformation* (Albany: State University of New York Press, 1997); Charles Webster, *Paracelsus: Medicine, Magic and Mission at the End of Time* (New Haven, Conn.: Yale University Press, 2008).

98. On Giovanni Marinello and his work, see *Medicina per le donne nel Cinquecento. Testi di Giovanni Marinello e di Girolamo Mercurio* (Turin: Strenna UTET, 1992), 7–40 and 45–64. On the influence of Giovanni Marinello on his daughter, see Pomata, "Was There a *Querelle des Femmes* in Early Modern Medicine?"

99. On Marinella's husband, see Marinella, *The Nobility and Excellence of Women*, 6n11.

100. "peritissima nella filosofia morale e naturale" (Bronzini, *Della dignità e nobiltà*, 82 ; see also 112–113, where Marinella is described as "very well informed in philosophy" [*nella filosofia molto intendente*]). See also Luigi Scarano's praise of her knowledge of classical philosophy in his *Scenophylax* (1601; cf. Marinella, *The Nobility and Excellence of Women*, 5n10).

101. As Malpezzi Price and Ristaino point out, Fonte and Marinella are in fact the "chief defenders of women's reputation and worth in this episode of the multisecular *querelle des femmes*" (*Lucrezia Marinella*, 105–106).

102. See Marinella, *The Nobility of Women*, 55, 78–79.

103. While the *Nobility of Women*, with its strong protofeminist message and impressive display of the breadth of Marinella's learning, has received the most attention from scholars, Marinella was the author of numerous other works, most of them spiritual in nature, and also, at the end of her life, an *Essortationi* that some critics have taken, mistakenly, as a retraction of her earlier pro-woman arguments (see Malpezzi Price and Ristaino, *Lucrezia Marinella*, 120–155; Lynn Lara Westwater, "Lucrezia Marinella [1571–1653]," in *Encyclopedia of Women in the Renaissance: Italy, France, and England*, ed. Diana Maury Robin, Anne Larson, and Carol Levin (Santa Barbara: ABC-Clio, 2007), 234–237.

104. See Marinella, *The Nobility and Excellence of Women*, 4, for Scarano as the "vital link between [Marinella's] private studies and writing and the wider world of Venetian literary circles and publishing."

105. "Delle nobili attioni, & virtù delle donne, le quali quelle de gli huomini di gran lunga superano, come con ragioni, & essempi, si prova" (Marinella, *Nobiltà delle donne*, V, section 1).

106. Marinella, *Nobility and Excellence of Women*, 78–80. On the expanded 1601 edition, see Lynn Lara Westwater, "'Le false obiezioni de' nostri calunniatori.' Lucrezia Marinella responds to the Misogynist Tradition," *Bruniana e campanelliana* 12, no. 1 (2006): 1–15. See Laura Lazzari, *Poesia epica e scrittura femminile nel Seicento: L'Enrico di Lucrezia Marinella* (Leonforte: Insula, 2010), 178.

107. Marinella, *The Nobility and Excellence of Women*, 79 (*Nobiltà delle donne*, 52, "temendo di non perdere la signoria, et divenir servi delle donne, vietano a quelle ben spesso ancho il saper leggere, & scrivere").

108. "donne scientiate & di molte arti ornate" (*Nobiltà delle donne*, V, II). For an overview of the use of the "catalog" in texts pertaining to the debate over women, see Glenda McLeod, *Virtue and Venom: Catalogs of Women From Antiquity to the Renaissance* (Ann Arbor: University of Michigan Press, 1991).

109. "donne nelle scienze e nell'arti perite, & dotte" (*Nobiltà delle donne*, V, II; Marinella, *The Nobility and Excellence of Women*, 22).

110. Marinella, *The Nobility and Excellence of Women*, 92 (Marinella, *Nobiltà delle donne*, 43, "Ma le scienze, che tant'alto vanno, / E portan seco i sensi agri, e terrestri, / che poi rinchiusi nel corporeo velo / sappiamo come sta la terra, e il Cielo"). The citation is from Filippi, *Vita di Santa Caterina in ottava rima* (Venice, 1592).

111. Marinella, *The Nobility and Excellence of Women*, 92 (*Nobiltà delle donne*, 40, "Cassandra Fedele etiandio dottissima era, dispute publicamente in Padoa, & scrisse uno elegante libro dell'ordine delle scienze").

112. Marinella, *The Nobility and Excellence of Women*, 89 (*Nobiltà delle donne*, 41, "Isota Novarrolla veronese, la quale di filosofiche dottrine era adorna, faceva vita filosofica").

113. Marinella, *The Nobility and Excellence of Women*, 88 (*Nobiltà delle donne*, 40: "Ben voi, voi sola con l'eccelsa mente / A le cagion passando in ogni cosa, / Levate a la natura i suoi secreti. / E stando Apollo, e le sue muse intente / Al vostro dotto, stil, già gloriosa / Avanzate i filosofi, e i poeti"). See Giulio Camillo Dalminio, in *I fiori delle rime de' poeti illustri, nuovamente raccolti et ordinate da Girolamo Ruscelli* (Venice, Sessa, 1558), 109. Marinella repeats this linking of Apollo/poetry with scientific discourse in *Happy Arcadia*, in which an automaton-like statue of Apollo in Erato's kingdom serves, as Virginia Cox notes, "as an ingenious metaphor for the increasingly prestigious literary discourse of science, soon to be taken to new heights by Galileo" (*Prodigious Muse: Women's Writing in Counter-Reformation Italy* [Baltimore: Johns Hopkins University Press, 2011], 211).

114. See, for example, Virginia Cox's preface to Lazzari, *Poesia epica*, 12, in which she discusses Marinella's *Enrico* as part of a "microtradition" of epic poetry by women; see also Claire Lesage, "Femmes de lettres à Venise aux XIXe siècles: Moderata Fonte, Lucrezia Marinella, Arcangela Tarabotti," *Clio: Histoire, Femmes et Societe* (2001): 3.

115. On Marinella's referencing of Boccaccio's *De claris mulieribus,* along with Betussi's later edition of the work, see Marinella, *The Nobility and Excellence of Women,* 22.

116. Ravisius Textor, *Theatrum poeticarum et historicum, sive Officina Io. Ravisii Textoris* (Basileae: Leonhardi Ostenij, 1637), 558–563.

117. See Meredith K. Ray, "Un'officina di lettere: Le *Lettere di molte valorose donne* e la fonte della 'dottrina femminile,'" *Esperienze letterarie* 26, no. 3 (2001): 69–62. See also Walter J. Ong, "Commonplace Rhapsody: Ravisius Textor, Zwinger, and Shakespeare," in *Classical Influences on European Culture,* ed. Robert Ralph Bolgar (Cambridge: University of Cambridge Press, 1976), 91–126.

118. See Paolo Cherchi, *Polimatia di riuso: Mezzo secolo di plagio (1539–1589)* (Rome: Bulzoni, 1988).

119. Approximately 40 percent of Marinella's examples here are also found in Textor, as well as in other sources including Betussi, as cited in note 115. Other examples of particularly close textual correspondences include Sappho (Textor, *Officina,* 558; Marinella, *Nobilità delle donne,* 39); Praxilla (*Officina,* 558; *Nobilità delle donne,* 40); and Hypatia (Textor, *Officina,* 558; Marinella, *Nobilità delle donne,* 39).

120. Malpezzi Price and Ristaino, *Lucrezia Marinella,* 109.

121. Marinella, *The Nobility and Excellence of Women,* 135 (*Nobiltà delle donne,* 119).

122. At least, not in the *Nobiltà delle donne.* Marinella seemingly retracts many of her protofeminist views in her later *Essortazioni alle donne.* As Laura Benedetti, editor of the English translation of the *Essortazioni,* puts it, "Marinella's attitude toward Aristotle is complex"—in contrast to her consistent celebration of Plato, for example (Lucrezia Marinella, *Exhortations to Women and to Others If They Please,* ed. Laura Benedetti [Toronto: Centre for Reformation and Renaissance Studies, 2012], 12). On Paduan Aristotelianism, see John H. Randall, Jr., "Paduan Aristotelianism Reconsidered," in *Philosophy and Humanism: Renaissance Essays in Honor of Paul Oskar Kristeller,* ed. Edward Patrick Mahoney (Leiden: Brill, 1976), 275–282; see also Antonio Poppi, *Introduzione all'aristotelianismo padovano* (Padua: Antenore, 1970).

123. Marinella, *The Nobility and Excellence of Women,* 25.

124. See Malpezzi Price and Ristaino, *Lucrezia Marinella,* 37.

125. For a brief consideration of these figures in the context of early modern scientific culture, see Cox, *Prodigious Muse,* 209–212.

126. Like Fonte, Marinella also chooses a female dedicatee who belongs to a milieu in which scientific knowledge—especially the culture of experiments—was prized. On Vincenzo Gonzaga, see Valeria Finucci, *The Prince's Body: Vincenzo Gonzaga and Renaissance Medicine* (Cambridge, Mass.: Harvard University Press, 2014).

127. See Marinella, *Arcandia felice,* XVI, ed. Lavocat. Lavocat notes that while readers had little appetite for pastoral romance, there was still a market for pastoral drama; she speculates that Marinella may have hoped to adapt her *Arcadia* for theater. On pastoral drama, see Virginia Cox and Lisa Sampson, eds., *Flori: A Pastoral Drama* (Chicago: The University of Chicago Press, 2004); Meredith K. Ray, "La castità conquistata: The Function of the Satyr in Pastoral Drama," *Romance Languages*

Annual 9 (1996), 312–221; Lisa Sampson, *Pastoral Drama in Early Modern Italy: The Making of a New Genre* (London: Legenda, 2006).

128. Panizza describes this work as "out of character with [Marinella's] bold polemics" and highlights its call to women to embrace a life of seclusion focused on domestic tasks (Marinella, *The Nobility and Excellence of Women*, 15). Critical interpretations of Marinella's *Essortationi* are, however, beginning to piece together a nuanced view of this work that some see as a literary exercise (see Malpezzi Price and Ristaino on the *Essortationi* as an exercise in close reading and even *paignon,* a form of paradox [*Lucrezia Marinella,* 120–155]; Cox, *Women's Writing,* 223; Ross, *Birth of Feminism,* 296–299); others as a reaction to the exhausting literary polemics faced by women writers in Seicento Venice (see Lynn Lara Westwater, "The Disquieting Voice: *Women's Writing and Antifeminism in Seventeenth-Century Venice,* 2003, Ph.D. dissertation). Benedetti rejects the interpretation of Malpezzi Price and Ristaino in her modern edition and translation, which she herself characterizes as "hopelessly conservative" (Marinella, *Exhortations to Women,* 34).

129. Sanazzaro, *Arcadia* (Livorno: G. T. Masi, 1781), x.

130. "la varietà de' frutti sopra un sol ramo, la diversità de' fiori in un sol pianta, da altro non viene, che dallo inestare" (Marinella, *Arcadia felice,* 122).

131. See della Porta, *Magia naturalis,* III, 12. For a historical overview of grafting, see Mudge, "A History of Grafting," *Horticultural Reviews* 35 (2009): 438–490.

132. "Veggiamo eziandio talora diversi colori dipingerli il lampeggiante volto, e se la mia osservazione non è stata in vano, conobbi che quando di cerulee macchie cosperso si mostra denunciar abbondanti pioggie; quando acceso di fiammeggiante rossore vedrai senza dubbio i sonanti venti gittare a terra i non maturi pomi. Se mirerai nello splendido del suo aspetto alcune cerulee note remescolarsi ad un foscoso colore il tutto sarà dalla sonante grandine, da larghe piogge e da strepitosi venti rotto, stracciato, e molle ne oserà il misero navigante a fidarsi dell'infido mare" (*Arcadia felice,* 52). See 45–46 for Marinella's description of trees. As Lavocat notes, the disposition in which the nearly two dozen species are presented reflects that in Pliny's *Natural History* (Marinella, *Arcadia felice,* 45n219). Marinella may also have turned once again here to Textor's *Officina* ("De arborum speciebus," 1146).

133. On "sapere pastorale," see Marinella, *Arcadia felice,* 51n240.

134. "Sono, non 'l nego, rispose il dotto pastore, molte, anzi infinite cose, le quali sono cagione alle menti nostre di sommo stupore, che ben spesso leggieri, e di poca, anzi niuna considerazione sono" (ibid., 122). On curiosity and wonder in the early modern period, see, for example, Lorraine Daston and Katharine Park, eds., *Wonders and the Order of Nature, 1150–1750* (New York: Zone, 2001); R. J. W. Evans and Alexander Marr, eds., *Curiosity and Wonder From the Renaissance to the Enlightenment* (Aldershot: Ashgate, 2006).

135. Findlen, *Possessing Nature,* 1–2.

136. Ibid., 2–3.

137. "a cui non mancava altro che lo spirito" (Marinella, *Arcadia felice*, 117).

138. "non vero uomo, ma finta immagine e similitudine di uomo" (ibid.).

139. See Findlen, *Possessing Nature*, 3: "From the imaginary to the exotic to the ordinary, the [early modern] museum was designed to represent nature as a continuum."

140. See Bonardo, *La minera del mondo* (Venice, 1569), 68: *Radiano*, which Erimeno describes as deriving from the "capo di un antico gatto marino" [Marinella, *Arcadia felice*, 118]), is mentioned by Bonardo as deriving from the same source and possessing similar qualities: "La pietra radiano, è pietra nera tralucente trovasi nella testa di un gallo, alcuni dicono nel capo di un gatto marino . . . a chi la porta recca honore, e giova a commandare perche fa, che sia obedito." Probable sources also include Pliny ("Cheolonia . . . melle enim colutu ore impositam linguae futrum divinationem praestare promittunt," *Natural History*, XVII, 10); or Sanazzaro, who cites this passage in *Arcadia* (see Marinella, *Arcadia felice*, 119n509).

141. Marinella, *Arcadia felice*, 118–119: "Ecco l'Artemesia, che portata in lunghi viaggi con l'Elesifaco non lascia sentir la stanchezza, ed appesa sopra l'alte porte assicura dagli incanti." Compare Bonardo, *La minera del mondo*, 80: "Artemisia, e lo Elesiphaco portati adosso per viaggio non lasciano sentire stanchezza, l'Artemesia appiccicata sopre le porte delle case, fa sicuri gli habitatori dalle malie, e da gl'incanti."

142. "insegnavami sopra l'alto delle sommità degli inaccessibili monti le occulte virtù di molte erbe e pietre" (Marinella, *Arcadia felice*, 123).

143. "solamente gli immortali Dii l'avanzavano" (ibid., 123); "un carro tirato da quattro negri corsieri, sopra il quale in poche ore tutto il mondo scorreva" (ibid., 122).

144. "la gloriosa ninfa Erato, nel mezzo della bella camera venne con le chiome disciolte, infiammata negli occhi e nel viso, con moti orribili, la qual smaniando, ed a pena respirando potendo, con sonante e terribil voce disse: 'Diocleziano, e voi di lui compagni degni, chiedete ciò che più v'aggrada' " (ibid., 163).

145. "sovrana vergine"; "indovina vergine" (ibid., 170). On *Arcadia felice* and the cult of Diana, see Cox, *The Prodigious Muse*, 209–210.

146. See Adam Mosley, *Bearing the Heavens: Tycho Brahe and the Astronomical Community of the Late Sixteenth Century* (Cambridge: Cambridge University Press, 2007). For a list of Italian works describing the comet of 1577, see Dario Tessicini, "The Comet of 1577 in Italy: Astrological Prognostications and Cometary Theory at the End of the Sixteenth Century," in *Celestial Novelties on the Eve of the Scientific Revolution, 1540–1630*, ed. Patrick J. Boner and Dario Tessicini (Florence: Olschki, 2013), 57–84.

147. "chioma d'oro"; "rilucente coda" (Marinella, *Arcadia felice*, 170).

148. "noi da questa parte veggiamo il cerchio di latte, altro non essere, salvo che la riflessione de' raggi di una spessa moltitudine di picciole ed a pena visibili stelle, e non come piacque a quel Grande, la picciola fiamma di una arida e continua essalazione"). Compare Aristotle, *On Meteorology*, 1, 6 and 1, 8. For an overview of Aristotle's theory of exhalations, see Martin, *Renaissance Meteorology*, 21–37.

149. The Fourth Crusade was notable in that it did not reach the Holy Land but rather resulted in the conquest of Constantinople, a Christian (Eastern Orthodox) city. Marinella freely interprets the historical facts in *Enrico* to promote a "peculiarly Venetian and religious standpoint" (see Marinella, *L'Enrico*, 25). On the Fourth Crusade, through which Venice acquired a significant part of its territory in the Mediterranean, see Thomas Madden, *The Fourth Crusade: The Conquest of Constantinople* (Philadelphia: University of Pennsylvania Press, 1997); Madden, *The Fourth Crusade: Event, Aftermath;, and Perceptions* (Aldershot: Ashgate, 2008); and Madden, *Enrico Dandolo and the Rise of Venice* (Baltimore: Johns Hopkins University Press, 2003); see also John Godfrey, *The Unholy Crusade* (Oxford: Oxford University Press, 1980). As Rinaldina Russell points out, *L'Enrico* is also deeply influenced by Margherita Sarrocchi's *Scanderbeide* (1623), containing "characters, episodes, segments, details of battles, comparisons and descriptions that, notwithstanding the existence of previous models for both writers, show themselves to be closely patterned on similar ones in Sarrocchi's poem" (Sarrocchi, *Scanderbeide*, ed. Russell, 40–41).

150. See Lazzari, *Poesia epica*, 113. On Marinella's de-eroticization of this "maga" figure, see Cox, "Women as Readers and Writers," 143; Finucci, *Floridoro*, 25; Lazzari, *Poesia epica*, 102.

151. Marinella, *Enrico*, V, 62–63.

152. In sharp contrast to this example of the measured and just deployment of "marvelous" knowledge by a female character stands the figure of Esone, a *mago* who uses black magic to enchant and confuse the Venetian crusaders (see cantos XIII, 35–42; X, 1–78; XI, 14–16, 82–84; XIX, 57–70, 101, 111; and XX, 1–4, 7–9, and 14–16). On Marinella's use of Esone to illustrate the weakness of men, see Lazzari, *Poesia epica*, 128–131.

153. Marinella, *Enrico*, V, 81 (see *L'Enrico*, 141, "Col contemplar delle cagioni ascose / Gli alti principi e le mirabil opre, / Vien l'uomo divin, sue voglie gloriose, / Per cui l'ascoso e occulto appre et discopre").

154. Ibid., V, 89–90.

155. Ibid., V, 91 (see *L'Enrico*, 142, "Come la terra in acqua e 'n aria l'acqua / si cangi e fatta lieve in fuoco ascenda; / E come il fuoco in aria e l'aria in acqua, / poi l'acqua di vil terra forma prenda," V, 91).

156. "Come poi chiuso della terra in seno /vento cruddel gli alti edifici scuote," V, 92. Compare Fonte, *The Worth of Women*, 129 (*Il merito delle donne*, 70).

157. "Come di molte stelle il lume unito / e riflesse tra loro candida via / formi, qual latte sparso e disunito / Ch'al palagio di Giove i divi invia" (Marinella, *L'Enrico*, V, 93; compare *Arcadia felice*, 170–171).

158. On the role of optics and the "dilemmas, anxieties, and tensions" of the New Science as the human eye is gradually replaced by the lens, see Ofer Gal and Raz Chen-Morris, *Baroque Science* (Chicago: University of Chicago Press, 2013), 7.

159. See Barbara Fuchs, *Mimesis and Empire* (Cambridge: Cambridge University Press, 2004), 25–29.

160. "Tu lo semplice spirto e questa mente / Saggia, dotta e prudente a un tempo festi" (Marinella, *L'Enrico*, VI, 9).

161. Ibid., VI, 50.

162. "Gli aspetti osserva, i moti, i corsi e i giri / scherzo degli astri ne' stellanti giri" (ibid., VI, 51).

163. "Non dagli effetti alle cagioni occulte / Ma dalle cause a manifesti segni" (ibid., VI, 52). As Martin notes, Aristotle's *Meteorology* and Aristotelian thought in the early modern period approached the problem of signs and causes in varied ways. Emphasis on causes was "central to natural philosophy and a legacy of the influence of Aristotle's *libri naturalis* from the Middle Ages to the middle of the seventeenth century"; in some cases, however—as with the causes of earthquakes—natural signs provide evidence of probable causes (*Renaissance Meteorology*, 26).

164. "E di quanto sapea, ne partia meco / Il buono e 'l bello" (Marinella, *L'Enrico*, VI, 53).

165. "Divin poter serbollo, il trasse e scorse / Sicur per l'onde e 'l tolse a morte, a noia . . . Se non puoi contrastare al suo desio, / Sforzata al fin darai fedel consiglio: Buon non è ritener cuor che non sente / Dolce la pace e torbid'ha la mente" (Marinella, *L'Enrico*, VI, 3, 5).

166. See Marinella, *Enrico*, 13.

167. "Benché per l'aere puro il carro muova, / Non siamo in cielo o nella sfera ardente" (Marinella, *L'Enrico*, XI, 37).

168. See Marinella, *Enrico*, 60. Erina points out an area where the mythical phoenix was said to reside, also rich in myrrh and amomum (black cardamom), both of which had numerous medicinal uses with which Marinella may have been familiar (Marinella, *L'Enrico*, XI, 54).

169. "Meraviglioso, o mia celeste guida! / Chi a tante parti e a tanti luoghi affisse / il nome e dienne a noi certezza fida? / Chi fu a ritrovar contante gisse / i sconosciute terre in patria infida?" (ibid., XXI, 61).

170. "Alla più bella e più gioconda parte / Che 'l ciel vagheggi," (ibid., XXII, 1).

171. "Essa, che vede l'animo feroce / Tutto avvampare al marziale aspetto" (ibid., XXII, 63). Marinella's account of Venier's valiant return to battle is complicated by the fact that Venetians had been forbidden to launch an assault on Zara under pain of excommunication, a detail Marinella omits (see Marinella, *Enrico*, 20).

172. On familial networks and intellectual development in the cases of Marinella and Fonte, see Ross, *The Birth of Feminism*, 196–203.

173. See Lazzari, *Poesia epica*, 50: "celebra le imprese memorabili di una collettività dove l'eroe ne è l'attore simbolo." As Stampino reiterates, for Marinella this is an inherently nostalgic project: by 1635, "Venice's commercial empire was already past its peak," undermined by some of the very discoveries Erina points out to Venier (Marinella, *Enrico*, 11–12). On epic and imperialism see for example David Quint, *Epic and Empire* (Princeton, N.J.: Princeton University Press, 1993); Sergio Zatti, *Il modo epico*

(Rome: Laterza, 2000); Fuchs, *Mimesis and Empire*. See also Tasso's comments in *Discorsi del poema eroico.*

174. Among them: Giovanni Villifranco, *Colombo* (1601); Tommaso Stigliani, *Mondo Nuovo* (1617, 1628); Alessandro Tassoni, *Oceano* (1622). See Nathalie Hester, "Failed New World Epics in Baroque Italy," in *Poiesis and Modernity in the Old and New Worlds,* ed. Anthony J. Cascardi and Leah Middlebrook (Nashville: Vanderbilt University Press, 2012), 201–224. See also Fuchs, *Mimesis and Empire.*

175. Marinella's *Nobility of Women* "promoted an active debate about women that was to continue for at least fifty years after its publication, instrumental in forcing Passi to reevaluate his ideas and retract many of his negative statements about the female gender" (see Malpezzi Price and Ristaino, *Lucrezia Marinella,* 105).

176. Fonte, *The Worth of Women,* 180–181 (*Il Merito delle donne,* 125, "Anzi . . . è bene che noi ne impariamo per tenir da noi, acciò che non abbiamo bisogno dell'aiuto loro; e saria ben fatto che vi fussero anco delle donne addotrinate in questa materia, acciò essi non avessero questa gloria di valer in ciò piu di noi e che convenimo andar per le man loro").

177. Fonte, *The Worth of Women,* 184–185; *Il merito delle donne,* 128 "con tutto ciò che si è ragionato di stelle, di aria, di uccelli, di acque, di pesci e di tante qualità di animali, di erbe e di piante, non si ha già ritrovato cosa ancora di tal virtù, che potesse far mutar animo a questi uomini"

178. Fonte's seven characters are: Adriana, a widow and the eldest member of the group; Leonora, the hostess of the gathering, a young widow who does not wish to remarry; Cornelia, a married woman who looks upon her conjugal state with diffidence; Corinna, an unmarried woman who dedicates herself to study; Helena, a content young bride; Virginia, a naive young girl, as yet unmarried; and finally, Lucrezia, an older, married woman.

4. Scientific Circles in Italy and Abroad

1. As Virginia Cox notes, the "urge . . . to go beyond humanistic and literary pursuits and embrace 'the sciences' more generally seems to have been something of a trend of the period" (*The Prodigious Muse: Women's Writing in Counter-Reformation Italy* [Baltimore: Johns Hopkins University Press, 2011], 238). This marks a contrast from earlier decades in which access to education in natural philosophy for women was particularly contested: as Laura Giannetti notes, the obstacles faced by women wishing to study this subject formed the basis for various early modern comedies (*Lelia's Kiss: Imagining Gender, Sex, and Marriage in Italian Renaissance Comedy* [Toronto: University of Toronto Press, 2009], 62–67).

2. See Cox, *The Prodigious Muse;* and Eleonora Carinci, "Una 'speziala' padovana: *Lettere di philosophia naturale di Camilla Herculiana* (1584)," *Italian Studies* 2 (2013): 210–213.

3. "essendo che sola cagione che V. Signoria non habbia possuto alcune cose sapere, stimo io che sia l'esserle stata ascosa la lingua latina, colpa della mal usanza dei nostri tempi, la qual dapoi che le scienzie non sono nella lingua nostra, ne vieta ancora che le Donne non apprendin quella lingua in cui le si trovano e così ne impedisce che molte Donne non venghin ne gli studi de le lettere eccellentissime e rare" (Alessandro Piccolomini, *Della Sfera del mondo* [Venice: al segno del pozzo, 1552], aii-v). On Laudomia Forteguerra, see Diana Robin, *Publishing Women: Salons, The Presses, and the Counter-Reformation in Sixteenth-Century Italy* (Chicago: University of Chicago Press, 2007), 124–159.

4. "giovare a molti ch'io conosco d'intelletto buonissimo, e atto a filosofare; i quali non sapendo altra lingua che la italiana lor materna" (Piccolomini, *Instrumento della filosofia* [Venice: Giorgio de' Cavalli, 1565], A1r). Piccolomini further notes that women are not generally taught Latin, "Since it is not the custom in Italy to teach them any language other than that which they learn from their nurses" (*Non essendo costume in Italia di far loro apprender altra lingua, che quella che dalle nutrici imparano* [ibid., A4v]). On Piccolomini and women's learning, see also Carinci, "Una 'speziala' padovana," 210–211; on Piccolomini more generally, see Florindo Cerreta, *Alessandro Piccolomini: Letterato e filosofo Senese del cinquecento* (Siena: Accademia Senese degli Intronati, 1960).

5. Appended to Borro's work is a second dialogue in which a male character addresses six female interlocutors on the subject of the perfection of women (see Craig Martin, "Meteorology for Courtiers and Ladies: Vernacular Aristotelianism in Renaissance Italy," *Philosophical Readings* 4, no. 2 [2012]: 12). On Nicolò Vito di Gozze and Michiele Monaldi, see Simeone Gliubich, *Dizionario biografico degli uomini illustri della Dalmazia* (Vienna-Zara, 1836); on Ragusa in this period, see Francesco Maria Appendini, *Notizie istorico-critiche sulle antichità storia e letteratura de' Ragusei* (Ragusa: Antonio Martecchini, 1803), and Susan Moshard Stuard, *A State of Deference: Ragusa/Dubrovnik in the Medieval Centuries* (Philadelphia: University of Pennsylvania Press, 1992). Nicolò Vito di Gozze studied in Padua and wrote Neoplatonic dialogues as well as scientific treatises.

6. See Cox, *The Prodigious Muse*, 238 and Carinci, "Una 'speziala' padovana," 212. The work is mentioned briefly in Stuard, *A State of Deference*, 100. Gondola's letter is included in a volume forthcoming in the "Other Voice in Early Modern Europe" series: see *Renaissance Women's Writing between the Two Adriatic Shores*. I am grateful to the editors for sharing some preliminary pages of this edition with me. Born in Ragusa but raised in Italy and married to a Florentine nobleman, Bartolomeo Pescioni, in 1577, Fiore Zuzori (Cvijeta Zuzorić) was widely praised by Ragusan and Italian poets, including Tasso, for her beauty and intellect (see Torquato Tasso, *Rime*, ed. Bruno Basile [Torino, Einaudi 1994], 431; Josip Torbarina, *Tassovi soneti i madrigali u čast Cvijete Zuzorić Dubrovkinje, Hrvatsko kolo* no. 21 [Zagreb, 1940]). After returning to Ragusa, Zuzori and her husband settled in Ancona. On Zuzori, see Claudia Boccolini,

Flora Zuzzeri in Ancona (Ancona: Consiglio Regionale delle Marche, 2007), which, while speculative, contains transcriptions of some archival documents. On Maria Gondola, see Zdenka Janeković Römer, "Marija Gondola Gozze, *La querelle des femmes* u renesansnom Dubrovniku," in *Žene u Hrvatskoj: Ženska i kulturna povijest*, ed. Andrea Feldman (Zagreb: Ženska Infoteka, 2004), 105–123. I thank Velimir Jurdjevic for his assistance in translating this article.

7. "a lei indirizzando questa fatica del mio marito, acciò ch'ella sia saldo scudo contra quelli, che sono pronti per loro natural malignità morder & lacerare le più belle & virtuose cose, essendo lei fra le bellissime virtuoissima, & fra le virtuossisime bellissima; & avenga che molti poriano maravigliarsi della cagione, che mi mosse di far uscire questi presenti discorsi sotto la protettione, o difesa del sesso feminile, credendosi eglino forse, che sí come noi per natura non siamo habili all'essercitio dell'armi, cosí ancora naturalmente siamo prive della capacità delle scienze, e cognittione delle cose" (Nicolò Vito di Gozze, *Discorsi di M. Nicolò Vito di Gozze . . . Sopra le Metheore d'Aristotele* [Venice: Francesco Ziletti, 1584], unnumbered but 2r-v).

8. "lesse publicamente la filosofia naturale"; "Pithagora hebbe non solo la sorella Theoclea, dalla qual esso imparò tanta filosofia, ma ancor' hebbe una figliuola . . . che in Athene più dilettavano sentir essa parlare nella sua casa, che sentire Pithagora legger in Academia" (ibid., unnumbered but 4r-5r).

9. See Römer, "Marija Gondola Gozze," 106, 119. Francesca Maria Gabrielli posits that the letter was altered not because of its pro-woman content, but as a result of its attack on Ragusan society (see *Renaissance Women's Writing*). On Mara Gundulić see also Dunja Fališevac, "Women in Croatian Literary Culture, 16th to 18th centuries," in *A History of Central European Women's Writing*, ed. Celia Hawkesworth (New York: Palgrave Macmillan, 2001), 33–40.

10. "troveremo le donne essere più atte; che gli huomini, a' imparare ogni scienza." (Nicolò Vito di Gozze, *Dialogo della bellezza/Dijalog o Ljepoti. Dialogo d'Amore/ Dijalog o Ljubavi,* ed. and trans. Ljerka Schiffler [Zagreb: Matica Hrvatska, 2008], 53–54. Gondola echoes this comment in her letter to Zuzori: "women are quicker to learn . . . have sharper minds, and are more suited to learning than men" (*sono le donne più facili all'imparare . . . hanno intelletto più acuto, e più disposto alle discipline, che non hanno gli huomini,* Vito di Gozze, *Discorsi,* not numbered but 7r-v).

11. See Erculiani, *Letters,* d1-r: "Come dice Aristotele nel 4 della *Meteora.*" Di Gozze elaborates that while Aristotle considered the universal flood a natural phenomenon, "we Catholics believe Noah's flood was supernatural" (*noi cattolici crediamo che quel diluvio noetico fu soprannaturale, Discorsi,* 56r). Di Gozze spent time in Padua as a student; it is possible there could even have been a personal connection if the two had occasion to cross paths. Four extant copies of Erculiani's *Lettere* have been indentified, located at the Biblioteca Alessandrina (Rome); the Biblioteca Civica (Padua); the Houghton Library (Cambridge, Massachusetts); and the Biblioteca PAN (Kórnik, Poland). The author is identified in the title as *Camilla Herculiana.* The

spelling of her name varies in sources; for consistency, I have chosen to use "Erculiani."

12. Garnier, called Garnero in Erculiani's text, was the author of a treatise on plague, *Liber de peste, quae grassata est Venetiis del 1576*. On Garnero, see Carinci, "Una 'speziala' padovana," 209 and n35.

13. Although Erculiani's name is occasionally noted in bibliographic catalogs of works by women, only recently has she begun to attract scholarly attention: see Virginia Cox, *Women's Writing in Italy, 1400–1650* (Baltimore: Johns Hopkins University Press, 2008), 162, and Cox, *The Prodigious Muse*, 237–238; Carinci, "Una 'speziala' padovana"; see also Sandra Plastina, " 'Considerar la mutatione dei tempi e delli stati e degli uomini': Le *Lettere di philosophia naturale di Camilla Erculiani*," *Bruniana e Campanelliania* 20, no. 1 (2014): 145–156; and, for a brief mention of her work, Axel Erdmann, ed., *My Gracious Silence: Women in the Mirror of 16th Century Printing in Western Europe* (Luzern, Switzerland: Gilhofer und Ranschburg, 1999), 214.

14. The brief discussion of Camilla Erculiani's biography here is based on Carinci, "Una 'speziala' padovana," which provides valuable archival details regarding her historical identity.

15. "è vero che si potranno molto maravigliar, ch'io senza veder libri m'habbia posto a dar' fuori queste quattro mal composte righe" (Erculiani, "A lettori," in *Lettere di philosophia naturale*).

16. "e questo lo conosco naturalmente, senza guardare Galeno né Aristotele," (ibid., 2a-r).

17. See Carinci, "Una 'speziala' padovana," 209. On pharmacies as a locus of sociability, see Filippo de Vivo, "Pharmacies as Centres of Communication in Early Modern Venice," *Renaissance Studies* 21 no. 4 (2007): 506; Franco Franceschi, "La bottega come spazio di sociabilità," in *La grande storia dell'artigianato*, ed. Franco Franceschi and Gloria Fossi (Florence: Giunti, 1999), 65–84. For a detailed case study of an early modern apothecary in Florence, see James Shaw and Evelyn Welch, *Making and Marketing Medicine in Renaissance Florence*, Clio Medica 89/The Wellcome Series in the History of Medicine (Amsterdam: Rodopi), 2011.

18. On links between Padua and Poland, see Paolo Marangon, "Schede per una reinterpretazione dei rapporti culturali tra Padova e la Polonia nei secoli XIII–XVI" in *Italia Venezia e Polonia tra Medio Evo e Età moderna*, ed. Vittore Branaca and Sante Graciotti (Florence: Olschki, 1980), 177. On intellectual exchanges and influences more broadly between Italy and Poland, see also Jan Slaski, "La letteratura italiana nella Polonia fra il Rinascimento e il Barrocco," in *Cultura e nazione in Italia e Polonia dal Rinascimento all'Illuminismo*, ed. Vittore Branca and Sante Graciotti (Florence: Olschki, 1986), 219–252; and Paulina Buchwald-Pelcowa, "Il libro italiano in Polonia nel periodo del Rinascimento," in *Il Rinascimento in Polonia: atti dei colloqui italo-polacchi 1989–1992*, ed. Jolanta Zurawaska (Naples: Bibliopolis, 1994), 427–453.

19. Januszowki initiated the Polish edition of a commentary on Hermes Trismegistus's alchemical treatise *Pymander* (see Buchwald-Pelcowa, "Il libro italiano," 428n2), dedicated to Francesco Gonzaga, Ferdinando and Francesco de Medici, and Stephen Báthory.

20. On Camilla Erculiani's Polish connections, see Buchwald-Pelcowa, "Il libro italiano," 427–428. The fact that the name of Erculiani's shop figures prominently in the work's title also suggests that the publication may have served to advertise the Tre Stelle abroad.

21. Ibid., 428.

22. "Laudarunt alii Hippolyten, & Amazona peltis / Decertantem Penthelisam, / Mentitamque virum Semiramin" (Erculiani, *Lettere*, aiv-v).

23. See Harold Segal, *Renaissance Culture in Poland: The Rise of Humanism: 1470–1543* (Ithaca, N.Y.: Cornell University Press, 1989), 4.

24. For a concise overview of the "Polish Renaissance" and of this period in Polish history, see Anita J. Prażmowska, *A History of Poland* (Hampshire, UK: Palgrave Macmillan, 2004), 42–129; see also Bronisław Bilinski, "The Italian Renaissance in Poland," *Italian Books and Periodicals* 25, no. 2 (1982): 138–145; Tadeusz Ulewicz, "Polish Humanism and its Italian Sources: Beginnings and Historical Development," in *The Polish Renaissance in its European Context*, ed. Samuel Fizman (Bloomington: Indiana University Press, 1988), 215–235.

25. See Prażmowska, *A History of Poland*, 77–85.

26. See Anna Brzezínska, "Accusations of Love Magic in the Renaissance Courtly Culture of the Polish-Lithuanian Commonwealth," *East Central Europe* 20–23, no. 1 (1993–1006): 117–118. Hostility to Bona Sforza was due in large part to the promonarchy policies she tried to set in place, eliciting the ire of the Polish noble class. On Bona Sforza's connections to Italian writers and intellectuals, see Giorgio Petrocchi, "Bona Sforza regina di Polonia e Pietro Aretino," *Italia Venezia e Polonia tra Medio Evo e età moderna*, ed. Vittore Branaca and Sante Graciotti (Florence: Olschki, 1980), 325–331.

27. Bilinksi, "The Polish Renaissance in Italy," 141.

28. See Maria Bogucka, *Women in Early Modern Polish Society, against the European Background* (Aldershot: Ashgate, 2004), 153. The first letter included in Ortensio Lando's celebration of women, *Lettere di molte valorose donne* (Venice, 1548), for example, is addressed to Bona Sforza (3r-4r). On Cassandra Fedele's oration in her honor, see Cassandra Fedele, *Cassandra Fedele: Letters and Orations*, ed. and trans. Diana Robin (Chicago: University of Chicago Press, 2000), 162–164. On Bona Sforza's continued cultural ties with Italy, see Petrocchi, "Bona Sforza regina di Polonia," 325–331.

29. Among the works dedicated to Anna Jagiellon are a poem by Andrzej Glogowcyzk praising wise women, and the first edition of Piotr Skarga's *Lives of the Saints* (1579) (see Bogucka, *Women in Early Modern Polish Society*, 153).

30. "li quali non si stancano mai di predicare in ogni luoco la virtù, pietà, & giustitia sua"; "essendo fatta certissima da molti delli suoi creati, che li seranno grate, per conoscerla virtuosissima, & amatrice delle scientie," (Erculiani, *Lettere*, aii-r; aii-v).

31. On Anna Vasa (Wazówna) and other Polish women writers, see Ursula Phillips, "Polish Women Authors," in *A History of Central European Women's Writing*, ed. Celia Hawkesworth (New York: Palgrave Macmillan, 2001), 15. In addition to her own studies in botany, Anna Vasa financed the publication of an *Herbarium* by Syrenius in 1613 (see Bogucka, *Women in Early Modern Polish Society*, 145, 153).

32. Bogucka, *Women in Early Modern Polish Society*, 132.

33. "ho voluto con gli studi far conoscere al mondo, che noi siamo atte a tutte le scientie, come gli huomini" (Erculiani, *Lettere*, aii-v). Erculiani's words echo those of Vito di Gozze's dedication to Nika Zuzori in *Dialogo della bellezza* (see note 10).

34. "Mentre donna, con sommo ingegno, & arte. . . . Fate che di Helicona / e fonti suoi, / Da gli Hesperi, a gli Eo / Lodi si sentan poi / Di Camilla che sian degne, e di voi" ("From the river Helicon / and her sources, / to the Hesperides, to the Aeolian islands / Let praise resound, / worthy of Camilla, and of you") (Erculiani, *Lettere*, ai-v]).

35. See Maria Bogucka, *Nicholas Copernicus: The Country and Times* (Wrocław: The Ossolinki State Publishing House, 1973), 143.

36. See Andrzej Glaber, *Tales about the Harmony of Human Limbs*, dedicated to a woman, Lady Jadwiga Kościelecka, cited in Bogucka, *Women in Early Modern Polish Society*, 90, 90n61.

37. See Andrzej Frycz Modrzweski's *De republica emendanda*, chap. 21: "Women should not meddle in public affairs" (cited in Bogucka, *Women in Early Modern Polish Society*, 92).

38. "a woman can know everything that man knows . . . as philosophy teaches us, those who have a subtler body must also have a sharper wit. And since women have this, it is obvious they are more capable to understand subtle sciences than men are" (Lukasz Górnicki, *The Polish Courtier*, cited in ibid.). At other points in the text, however, it is maintained that women are less learned than men. On Górnicki's adaptation of Castiglione's work, see Peter Burke, *The Fortunes of the Courtier: The European Reception of Castiglione's Cortegiano* (Cambridge, UK: Blackwell, 1995), 29–31.

39. "Mirthis non fu nominata gigantessa, se non per la grandezza dei suoi fatti, e non fu cognominata tra il numero delli sette famosi re di Lidia, se non per la sapientia, & eloquentia sua; né Nicostrata moglie di Evandro sarebbe stata tenuta in tanta stima, senon fosse stata più che dotta" (Erculiani, *Lettere*, aii-v).

40. "Poi che Cicerone non finse ancora lui di non conoscere la gran dottrina di Cornelia, quando dice nella sua rettorica, che se il nome di donna, non havesse sbatuta Cornelia, ella tra tutti i philosophi meritarebbe esser singulare" (ibid.).

41. "ma hora non so da che stella maligna siano mossi a non volere conoscere per grandi, se non le cose fatte da loro" (ibid.).

42. "to whom I might have proposed to dedicate this work; but knowing him to be occupied with matters of war, I did not wish to ask this of him, since I knew that Your Majesty is equally able to defend this work from detractors" (*del quale haveva proposto di dedicare quest'opera; ma conoscendo la occupata nelle guerre, non ho voluto*

dargli questo travaglio, poi ch'ho conosciuto V[ostra] M[aestà] essere ancora lei attissima a diffendere quest'opera da malivoli, ibid., aiii-r). Báthory had studied in Padua, creating another obvious link to Erculiani. Báthory had a reputation for tolerance with regard to heresy; given that Camilla's book would come under scrutiny by the Inquisition for its unorthodox perspectives on matters such as the causes of the flood and the problem of original sin, it is plausible that the kind of protection she sought was not just of a financial or intellectual nature. On Báthory, see Jerzy Besala, *Stefan Batory* (Warsaw, 1992). Two poetic anthologies were published in his honor in Venice, edited by Ippolito Zucconello, a physician (see Carinci, "Una 'speziala' padovana," 214n54).

43. Erculiani mentions Piccolomini's *Seconda parte della filosofia naturale* in her *Letters* (f2v-f3r). Her comments also echo Moderata Fonte's observations in *The Worth of Women* on custom versus law (see Fonte, *The Worth of Women*, ed. and trans. Virginia Cox [Chicago: University of Chicago Press, 1997], 61).

44. "Parrà senza dubio maraviglia ad alcuno, ch'io donna mi sia posta, a scrivere e dare alla stampa cose che non s'appartengono (secondo l'uso del tempo) a donna; ma . . . trovarà che non è la donna priva di quelle providenze e virtù che si sian gl'huomini (Erculiani, *Letters,* aiii-v).

45. Such claims were not necessarily just a deflection of agency: Arcangela Tarabotti (1604–1652), for example, detailed her concerns that a manuscript of her *Tirannia paterna* might be published in France under the name of the man to whom she had entrusted it (see Arcangela Tarabotti, *Lettere familiari e di complimento,* ed. and trans. Meredith Ray and Lynn Westwater [Turin: Rosenberg e Sellier, 2005], 207–208 and Tarabotti, *Letters Familiar and Formal,* ed. and trans. Meredith Ray and Lynn Westwater [Toronto: Centre for Reformation and Renaissance Studies, 2012], 198–199).

46. "il buon animo delle Donne de nostri tempi, cosa invero da me molto desiderata" (Erculiani, *Lettere,* aiii-v).

47. "voglio, piacendo a Dio, nell'altre mie dirvi che cosa, e dove, e quando, et, in qual virtù si generi l'anima nostra" (Erculiani, "A lettori," s.n.). She will only publish this other composition once she has seen that the present work (the *Lettere*) has been well received: "Parrà senza dubio difficile il provar questo ad alcuni, ma a gli intelligenti non parerà cosa fuori della verità, e questo si darà in luce poco dappoi queste, secondo mi parrà, che queste poche righe siano tentute, & accettate con quel buon'animo che io le do in luce hora a voi" (ibid.).

48. "Né il far questo mi dà noia ancor ch'io habbia il travaglio d'allevar figliuoli, il peso del governo della casa, e l'obedienza al marito, e la mia complessione non troppo sana, quanto mi dà noia il conoscere che da molti velati da spirito maligno saranno queste mie fatiche, o scritti biasimate, e tanto più saranno tenute vane e di poca stima, per esser tenute tali le Donne de nostri tempi" (ibid., aiii-v).

49. "Ma con tutto ciò non voglio restar d'affaticarmi per ricuperar in parte l'honor delle spensierate, e sarò forsi una causa e svegliamento a gl'intelletti loro" (ibid., aiv-r).

50. "E son sicura che s'attendessero a questo, non havriano ardire i cavallieri es-terni di venir' in questa inclita città di Padova e volere con spada, e lanza provar' a tassar noi d'imperfectione; Oltre che son sicura che molti savi & intelligenti lettori di quest'opera non si faranno beffe dell'inventione di quella, & ammiraranno la volontà mia, insieme con il desiderio de i miei pensieri" (ibid., aiv-r).

51. "As to the work on the sun, I have written it, but have not had time to copy it; nor do I wish to send it [to] you as I do [this letter], without making a copy for myself to correct if necessary. You will do me a great favor by sending this one back to me, or bringing it when you come to Padua, so that I may keep a record of what I have written" (*Quanto alla cosa del sole, io l'ho scritta, ma non ho tempo di copiarla, né vorrei mandarla come faccio questa senza tenerne copia e correggerla dove facesse bisogno, et ancomi farete piacere se mi rimanderete questa in dietro, overo portarla quando ver-rete a Padoa, acciochè possa tener memoria di quello ch'io ho scritto*," ibid., f1v).

52. See Erculiani, *Letters,* br. I have found no further information about this figure; however, Montagnana is a province of Padua, so this *medico* was likely a resident of the area.

53. On the connections between epistolary writing and dialogue, especially in the context of early modern women writers, see Janet Levarie Smarr, *Joining the Con-versation: Dialogues by Renaissance Women* (Ann Arbor: University of Michigan Press, 2005), esp. 130–153.

54. See Virginia Cox, *Renaissance Dialogue: Literary Dialogue in its Social and Political Contexts, Castiglione to Galileo* (Cambridge: Cambridge University Press, 2008), xii.

55. For a reading of Castiglione's courtly dialogue as a "drama of doubt," rife with unease regarding "conventional didactic modes," see ibid., 47–60. Galileo's scientific *Dialogo sopra i due massimi sistemi del mondo,* without explicitly endorsing either hypothesis, famously attributed the Copernican theory to the author's own stand-in, Salviati, and the Ptolemaic view to the aptly named Simplicio.

56. See Craig Martin, *Renaissance Meteorology: Pompanazzi to Descartes* (Balti-more: Johns Hopkins University Press, 2011), 66–67.

57. "Eccellentissimo Sig[nor] Giorgio, non posso far di non scrivere un ragion-amento, ch'io ho fatto l'altro giorno con un Eccel[entissimo] huomo, il qual diceva, che se non era il peccato commesse Adamo, l'huomo viveva sempre" (Erculiani, *Lettere,* b1r). Elsewhere in her *Letters,* Erculiani hints at other work she is composing in re-sponse to other debates with this same gentleman: a discussion of the movement of the sun, and a consideration of the nature of generation. As she tells Garnero, she is "reserving many things for another letter, in which I will tell you the answer that the most excellent doctor Montagnana, gave me regarding the movement of the sun. He says it receives its heat from the movement of the earth, and I disagree. I also asked him whether nature can produces a living creature without generation, and other things, which I will keep for another letter" (*riservandomi molte cose per un altra mia,*

*nella qual'ho da dire la risposta, che mi diede l'Eccel. Montagnana, medico eccelentis-
simo, sopra il moto del Sole, che lui dice, che riceve il calore dalla terra per il moto che fa
in lei: & io lo nego. L'ho anco dimandato, se la natura può produrre animal vivente senza
generatione, & altre cose, ch'io mi riserbo a dire in un altra mia,* ibid., cii-v).

58. Janet Altman defines the "internal" reader as a "specific character represented
within the world of the narrative, whose reading of the letters can influence the writing
of the letters," whereas the "external" reader indicates members of the general public,
"who read the work as a finished product and have no effect on the writing of individual
letters" (Altman, *Epistolarity: Approaches to a Form* [Columbus: Ohio State University,
1982], 112). Classical theorists of epistolary writing including Demetrius and Cicero likened the letter to conversation (see Meredith K. Ray, *Writing Gender in Women's Letter
Collections of the Italian Renaissance* [Toronto: University of Toronto Press, 2009], 6).

59. On early modern epistolary collections and the question of revision and invention, see Ray, *Writing Gender,* 3–4.

60. Carinci suggests that Garnero met Erculiani while he was a student at the
University of Padua, from which he received a degree in 1576 ("Una 'speziala' padovana,'" 209). Although I have not found archival evidence of correspondence between
Erculiani and Garnero, such documentation, if it exists, would shed light on Erculiani's decision to address Garnero in her *Letters.*

61. "per haver così a cuore le scienze e virtù, reputando lei tutte l'altre cose baie
vane, e di nissun valore, fuor che saper le cose naturali" (Erculiani, *Lettere,* ciii-v).

62. "Quant'al primo, che dice quell'Ecc., ciò è, che se l'huomo non havesse peccato, havrebbe vissuto in eterno, per me non so come si debba intendere questo passo,
essendo che egli è tutto fuori della philosophia, e non essendo io theologo mi per-
donarete, se in questo son dubioso e di ciò ignorante" (ibid., civ-r).

63. "if man does not die, he cannot be rewarded: that is, he will not receive the
fruit of the works he did while his soul was still one with his body, and thus Saint
Paul rightly said, *Cupio dissolvi, et esse cum Christo,* because in fact he wished—indeed,
he longed—to die" (*l'huomo, se non morirà, non sarà fruttuoso, ciò è non riceverà il
frutto de gli suoi lavori che egl'ha fatti mentre era unita l'anima col corpo, e per questo
ben diceva S. Paolo: Cupio dissolvi, et esse cum Christo, perché in fatto desiderava, anzi
bramava morire,* ibid., ciii-v).

64. Ibid., bı-r.

65. On this episode, see Paola Zambelli, "Fine del mondo o inizio della propaganda?" in *Scienze, credenze occulte, livelli di cultura* (Olschki, Firenze, 1982), 291–368.
See also Ottavia Niccoli, *Prophecy and People in Renaissance Italy,* trans. Lydia G. Cochrane (Princeton, N.J.: Princeton University Press, 1990), 140–167; Martin, *Renaissance Meteorology,* 70–74.

66. Niccoli, *Prophecy and People,* 143.

67. Ibid., 126, 143. Given Erculiani's Polish connections and the publication of
her treatise in Kraków, it is perhaps worth noting that Ploniscus (Jan z Plonsk) was a

Polish astrologer whose work, composed in Latin, was also printed in Kraków. As Niccoli points out, the subject of the flood also inspired satire, as in Francesco Berni's *Capitolo del diluvio* (in *Rime,* ed. Giorgio Barberi Squarotti [Turin: Einaudi, 1969], 9–12).

68. See Ann Blair, "The *Problemata* as a Natural Philosophical Genre," in *Natural Particulars: Nature and the Disciplines in Renaissance Europe,* ed. Anthony Grafton and Nancy Siraisi (Cambridge, Mass.: MIT Press, 1999), 171.

69. Ibid., 173.

70. "[il diluvio] venne per esser cresciuti gl'huomini tanto sopra la terra, in numero, e grandezza di corpo, e longhezza di vivere, ch'havea appresso il peccato molto sminuito l'elemento della terra, come quella che dava la maggior parte di lei in quelli corpi così grandi: ne gl'era in tanti centinaia d'anni restituito, sí che si trovò tanto sminuita, che gli fu forza esser ingiottita da l'acque, come quelle ch'havevano poco contribuito della parte sua in quelli corpi" (Erculiani, *Lettere,* bi-v).

71. On this problem, see Martin, *Renaissance Meteorology,* 72.

72. "l'acqua, la quale putrefa, anichila, e consuma tutte quelle cose che con essa si compongono, o dentro si gli ripongono: e questo fa per la fredda et humida natura sua, essendoché con la freddezza, leva la virtù vegetativa, e con l'humidità putrefa e consuma, reducendo quasi tutte le cose nella natura sua; e questo fa in poco tempo, e più in una materia che nell'altra" (Erculiani, *Lettere,* bi-v).

73. "and I know this instinctively, without looking at Galen or Aristotle" (*e questo lo conosco naturalmente, senza guardare Galeno né Aristotele,* Erculiani, *Lettere,* 2a-r).

74. "non è fatta di terra semplice . . . ma gli concorrono anco tutti gl'altri elementi, per la contrarietà de quali, non si può fare, ch'in spatio di tempo l'uno non superi l'altro, e disciolga questo corpo, che mondo picciolo si chiama, e così ogn'uno de gl'elementi torni nella proprietà e regione sua: e che se questa non si facesse, si annihilarebbe questa macchina, che mondo si chiama" [ibid., bi]). See Paracelsus, *Selected Writings,* ed. Jolande Jacobi, trans. Norbert Guterman (New York: Pantheon Bollingen Series 28, 1958) 21–25, 152–154. See also Phillip Ball, *The Devil's Doctor: Paracelsus and the World of Renaissance Magic and Science* (New York: Farrar, Straus and Giroux, 2006), 144–145.

75. See Ball, *The Devil's Doctor,* 243. There is a large bibliography on Paracelus: see, among others: Allen Debus, *The Chemical Philosophy: Paracelsian Science and Medicine in the Sixteenth and Seventeenth Centuries* (New York: Science History Publications, 1977); William R. Newman, "The Homunculus and His Forebears: Wonders of Art and Nature," in *Natural Particulars,* 321–345; Walter Pagel, *Paracelsus: An Introduction to Philosophical Medicine in the Era of the Renaissance,* 2nd ed. (Basle: Karger, 1982); Charles Webster, *From Paracelsus to Newton: Magic and the Making of Modern Science* (Cambridge: Cambridge University Press, 1982); and Webster, *Paracelsus: Medicine, Magic and Mission at the End of Time* (New Haven, Conn.: Yale University Press, 2008).

76. On Paracelsus and theriac, see Ball, *The Devil's Doctor,* 174–175; William R. Newman, *Promethean Ambitions: Alchemy and the Quest To Perfect Nature* (Chicago: The University of Chicago Press, 2004), 48.

77. See, for example, William Eamon, *The Professor of Secrets: Mystery, Medicine and Alchemy in Renaissance Italy* (Washington, D.C.: National Geographic Society, 2010); 163–165; Paula Findlen, *Possessing Nature: Museums, Collecting, and Scientific Culture in Early Modern Italy* (Berkeley: University of California Press, 1994), 241–245.

78. "M. Giacomo le manda un vasetto di Teriaca, & è di quella istessa ch'habbiamo fatta quest'anno" (Erculiani, *Lettere,* ciiv-ciiir); "Hora m'affatico con il nostro Galeno, perch'io scrivo la natura, proprietà, e qualità de gl'ingredienti ch'entrano nella Teriaca, et con quali proprietà siano loro giovevoli contro i velleni" (ciii-r).

79. "Volevano . . . essi Stoici, che non solamente gl'homini si dessero a philoso-phare e contemplare le cose naturali, ma etiamdio volevano, che tutte le donne, fac-essero il simile: cosa in vero lodatissima da tutti i nostri maggiori, et hora da non so che raggione leggierissima è cosi odiata, e quasi dal mondo sbandita, non solamente dalle donne, ma che è piu, da gl'huomini istessi è cosí sprezzata, tal che par cosa vilis-sima il philosophare e sapere le cose naturale [*sic*]" (ibid., ciii-r-v).

80. "charissima come madre sempre osservandissima" (ibid., ciii-v); "tanto sono più lodati quei pochi, e pochissimi Donne, le quali con grand'ingegno dedicandosi del tutto alle dottrine e scienze, affaticandosi . . . con ogni suo potere, e sprezzando tutte le sue attioni e publiche e private solo per varcare nel studio della verità delle cose, credendo essi essere cosa assai più illustre investigare e saper la ragione delle cose humane e divine, che d'acquistar richezze, honori, et altre cose simili corrottibili e vane. Tra i quali meritevolmente V.S. deve esser annumerata, per haver cosí a cuore le scienze e virtù reputando lei tutte l'altre cose baie vane, e di nissun valore, fuore che saper le cose naturali" (ibid., ciii-v). Garnero's comments echo those made in pref-aces to many books of secrets, which often single out man's unique compulsion to seek explanations for natural phenomena (see Chapter 2).

81. On this figure, see Endre Veress, *Berzeviczy Márton, 1538–1596* (Budapest: A Magyar Történelm. Társulat Kiadása, 1911); and John Donnelly, "Antonio Possevino, S.J. as Papal Mediator between Emperor Rudolf II and King Stephan Bathory" (*Archivium Historicum Societatis Iesu* 69 (2000): 55.

82. "Alla qual dimanda rispondo, e gli dico non havere apresso autore alcuno letto, né credo che sia cosa lodevole il scrivere l'opinione d'altri autori come sua propria" (Erculiani, *Lettere,* fii-r).

83. "non nego ch'io non legga diversi autori speculando le diffinitioni loro . . . dove maravigliata da gl'ingegni e varie opinioni loro, mi son posta anch'io a scrivere il parer mio" (ibid., fii-r).

84. See ibid., biv-r-cii-r. On Erculiani's rethinking of the position of Venus, see also Carinici, "Una 'speziala' padovana," 218. While Erculiani's thinking here dem-onstrates her innovative spirit and willingness to take on established scientific doc-

trine, the reasoning behind her suggestion is still based on a Ptolemaic and earth-centric worldview.

85. "È ben vero che dalli sacri dottori e dai divini teologhi, sono tenute altre cause, e raggioni; ma a me basta ch'Iddio et la istessa Natura non opera contra quelle, ma si serve di quella nelle opere sue" (Erculiani, *Lettere,* fiv-r-v).

86. Cesare Cremonini, for example, was prosecuted for the avveroist heresy of "double truth." In a letter to Giacomo Buoncompagno, he described his personal philosophy as "Intus ut libet, foris ut moris est" (see Cesare Cantù, *Gli eretici d'Italia* [Turin: Unione Tipografico, 1861]).

87. Giacomo Menochio, *Consiliorum sive Responsorum,* vol. 8, n. 766 (Frankfurt: Seyler, 1676); see Carinci, "Una 'speziala' padovana," 221–229. Carinci posits that Erculiani's trial may have taken place between 1584, the year in which her *Letters* were published, and 1588, the year Menochio left Padua for Milan (ibid., 223). The documents pertaining to Erculiani's actual Inquisition proceedings are lost, but according to Carinci it is likely to have taken place in Padua, rather than Rome, and Erculiani was probably absolved or dismissed without severe punishment (ibid., 222n79, 223).

88. On Menochio's account of Erculiani's trial, see Carinci, "Una 'speziala' padovana," 221–229.

89. "V[ostra] S[ignoria] si degna communicar meco le sue cose altissime e profundissime richezze, e doni dello Spirito Santo larghissimi, dico dello Spirito santo come christiano ch'io sono, e faccio professione, perch'ogni bene e virtù viene de Dio pieno d'ogni bene, essendo lui sommo bene, virtù, vita e verità; ma lasciando per hora da canto la Theologia ritorno alli suoi raggionamenti" (Erculiani, *Lettere,* ciii-v).

90. In her responses, Erculiani maintained, "these are common disputes in philosophy and it was my intention to speak as a philosopher" *(Io dico che queste sono disputi, che si sogliono fare in Philosophia & così intendo d'haver parlato),* but admitted the theological justification for death: "From a theological perspective—with reference to the holy scripture—I admit that the flood and death resulted from sin" *(In theologia, reportandomi sempre alle sacre scritture, io confesso, ch'il Diluvio & la morte sono venuti per il peccato)* (cited in Carinci, "Una 'speziala' padovana," 225).

91. See Carinci, "Una 'speziala' padovana," 224–229. On women and the Inquisition in Italy, see, for example, Jonathan Seitz, *Witchcraft and Inquisition in Early Modern Venice* (New York: Cambridge University Press, 2011); Anne Jacobson Schutte, *Aspiring Saints: Pretense of Holiness, Inquisition, and Gender in the Republic of Venice (1618–1750)* (Baltimore: Johns Hopkins University Press, 2001).

92. Menochio, referring to the ambiguity of Erculiani's statements, sums up, "Quae sententia multo magis locum habet in idiotis et mulieribus, qui facilius solent excusari" (cited in Carinci, "Una 'speziala' padovana," 224).

93. As de Vivo notes, "The archive of the Holy Office abounds with evidence of pharmacies functioning as centres of heterodox religious discussion," and Inquisitors

would in certain cases order "special observation" of particularly suspect locales ("Pharmacies," 507).

94. As Daniel Stone writes, "[a]lmost alone in early modern Europe, Poland-Lithuania accepted religious pluralism." (*The Polish-Lithuanian State, 1386–1795* [Seattle: University of Washington Press, 2001], 120). Although the Polish-Lithuanian state condemned the Reformation from the outset, with Zygmunt I banning Luther's works from Poland and issuing edicts against heretics in 1522–1525, he was not "personally committed to maintaining Catholicism at all costs" and did not follow through on such policies (42). Italian heretics such as Bernardino Ochino, for example, sought refuge in Poland—and found it, at least briefly; see Janusz Tazbir, *A State without Stakes: Polish Religious Toleration in the Sixteenth and Seventeeth Centuries* (New York: The Kosciuszko Foundation Twayne Publishers, 1973), 71–72, 133. The Henrician articles were originally accepted by Henri of Valois, to whom Anna Jagiellon was first betrothed before marrying Stephen Báthory (when Henri returned home to ascend the throne of France upon the death of Charles IX, provoking a constitutional crisis in Poland); see Stone, *The Polish-Lithuanian State,* 120.

95. "ho conosciuto V[ostra] M[aestà] essere ancora lei attissima a diffendere quest'opera da malivoli [Padua, February 25, 1584]" (Erculiani, *Lettere,* aiiir).

96. Galileo first came to the attention of the Inquisition in 1611, and traveled to Rome to defend his Copernican views there; in 1616, he was called to Rome again by Pope Paul V and Cardinal Bellarmino and instructed not to teach the Copernican theory. The bibliography on Galileo is vast: see, inter alia: Galileo Galilei, *Le Opere di Galileo Galilei,* ed. Antonio Favaro (Florence: Barbera, 1890–1909); Mario Biagioli, *Galileo Courtier: The Practice of Science in the Culture of Absolutism* (Chicago: The University of Chicago Press, 1993); Michele Camerota, *Galileo Galilei e la cultura scientifica nell'età della Controriforma* (Rome: Salerno Editrice, 2004); Maurice A. Finocchiaro, *Retrying Galileo, 1633–1992* (Berkeley: University of California Press, 2005); Thomas S. Kuhn, *The Copernican Revolution* (Cambridge, Mass.: Harvard University Press, 1957); Pietro Redondi, *Galileo eretico* (Torino: Einaudi, 1983); William Shea and Mariano Artigas, *Galileo in Rome: The Rise and Fall of a Troublesome Genius* (Oxford: Oxford University Press, 2003); David Wootton, *Galileo: Watcher of the Skies* (New Haven, Conn.: Yale University Press 2010).

97. In a letter to Achillini, Marino would later brush off Sarrocchi's own criticism of him, writing, "it doesn't sadden me to have felt myself shot through with the sharp points of the quills of those Scanderbeidians" (*non mi attrista l'avermi sentito trafigere con acute punture delle penne scheccheratrici delle Scanderbeidi* [Giambattista Marino, *Epistolario,* ed. Angelo Borzelli (Bari: Laterza, 1911–1912), 251]). Sarrocchi's distaste for Marino's baroque style fueled her decision to abandon the Accademia degli Umoristi for the new Accademia degli Ordinati. For the verse exchanged by Marino and Sarrocchi in praise of one another, see Nadia Verdile, "Contributi alla biografia di Margherita Sarrocchi" *Rendiconti dell'Accademia di Archeologia, Lettere e Belle Arti di Napoli* 61

(1989–1990) 62:165–206. Stigliani, too, though a fierce adversary of Marino, mocked Sarrocchi's *Scanderbeide* in verse (see Verdile, "Contributi alla biografia," 197–198).

98. Sarrocchi is also said to have composed commentaries on della Casa and Petrarch, a translation from the Greek of Musaeus's *Hero and Leander,* and a treatise on predestination (see Verdile, "Contributi alla biografia," 167–168). Although her translation is lost, a highly complimentary letter to Sarrocchi from the editor and publisher Aldo Manuzio in Venice provides a glimpse of this work: "Therefore, I await your Musaeus, so that I too may participate in this wonderful work" *(Starò dunque aspettando il suo Museo, per dover essere anch'io partecipe di sì fatto bene).* Three letters from Manuzio to Sarrocchi are published in Aldo Manuzio, *Lettere volgari* (Rome: Santi e Comp., 1592).

99. "una maga turca pur favolosa . . . post[a] solo per dilettare ad imitazione d'altri poeti, & non per significato" (Margherita Sarrocchi, *La Scanderbeide, poema heroico* [Rome: Lepido Facij, 1606], "Benigni lettori," not numbered). For Calidora, see ibid., III, stanzas 37–38. This early version had only eleven cantos, as opposed to the twenty-three of the 1623 edition.

100. "Et quanto bisogni all'Epico essere intendente d'ogni scientia, & d'ogni arte; l'hanno dimostrato gli Autori de buoni Poemi, huomini tutti di profonda dottrina. Questa non manca in ciascheduna scientia alla Signora Sarrocchi" (ibid., 2v). The letter, signed "L'Arrotato Academico Raffontato," is surely intended to add academic (masculine) credibility to Sarrocchi's endeavor.

101. "in materia di scientie, si tratta de Cieli, dell'intelligenze di astrologia, d'uno studio di cose naturali molto curioso poste, & esplicate tutte opportunamente et poeticamente" (ibid., unnumbered but 3r).

102. A sonnet penned by Sarrocchi for the birth of Odoardo Farnese of Parma in 1612, for example, links the seasons to their corresponding astrological signs (see Angelo Colombo, "Il principe celebrato: Autografi poetici di Tomaso Stigliani e Margherita Sarrocchi," *Philo-logica: Rassegna di analisi linguistica ed ironia culturale* 1, no. 1 [1992]: 29). Her sonnet celebrating Felice Orsini, published in Muzio Manfredi, *Per donne romane. Rime di diversi raccolte e dedicate al Signor Giacomo Buoncompagni* (Bologna: Alessandro Benaco, 1575), compares Felice to the Orsa constellation ("l'Orsa nostra della celestea assai più vaga").

103. Sarrocchi's tutors included Rinaldo Corso, author of books on law and theology as well as an Italian grammar text, and the mathematician Luca Valerio, who would become her longtime friend.

104. Sarrocchi's poem appeared in Muzio Manfredi's anthology *Per donne romane.* Much of Sarrocchi's biography comes to us through the accounts of Bartolomeo Chioccarelli (Biblioteca Nazionale di Napoli, *Bartolomai Chioccarelli Illustres scriptores Regni Napoletani*) and Angelo Borzelli, who bases himself on Chioccarelli (*Note intorno a Margarita Sarrochi ed al suo poema La Scanderbeide* [Naples: Artigianelli, 1935]). See also Giulio Cesare Capaccio, *Illustrium mulierum et illustrium litteris virorum elogia*

(Naples: apud F. Iacobum Carlinum et Constantinum Vitalem, 1608); Janus Nicius Erythraeus (Gian Vittorio Rossi), *Pinacotecha imaginum illustrium doctrinae vel ingenii laude virorum, qui auctore superstite diem suum obierunt* (Wolfenbüttel: Jo. Christoph Meisner, 1729); and Nicolò Toppi, *Biblioteca napoletana, et apparato a gli huomini illustri in lettere di Napoli . . . Dalle loro origini, per tutto l'anno 1678* (Naples: Antonio Bulifon, 1678). More recent discussions of Sarrocchi include Antonio Favaro, "Margherita Sarrocchi," in *Amici e corrispondenti di Galileo* (Venice: Officine Grafiche C. Ferrari, 1894), 6–31; Ugo Baldini and Pier Daniele Napolitani, "Per una biografia di Luca Valerio: Fonti edite e inedite per una riscostruzione della sua carriera scientifica" in *Bollettino di storia delle scienze matematiche/Unione matematica italiana* 11 no. 2 (1991): 3–157; Margherita Sarrocchi, *Scanderbeide*, ed. and trans. Rinaldina Russell (Chicago: University of Chicago Press, 2006), 1–44; and Verdile, "Contributi alla biografia."

105. Rossi, *Pinacotheca*, cited in Baldini and Napoletani, "Per una biografia di Luca Valerio," 60. Little is known of Sarrocchi's husband except that his last name was Birago and he seems to have come from Piedmont (see Borzelli, *Note intorno*, 13; Sarrocchi, *Scanderbeide*, ed. Russell, 10). Her friendship with Luca Valerio, by contrast, is well documented, by Valerio himself and by contemporaries of Valerio and Sarrocchi. Scholars have speculated as to the nature of the relationship, but there is no clear evidence that it was more than a deep friendship and an intellectual partnership. On Valerio, see Baldini and Napolitani, "Per una biografia di Luca Valerio"; G. Gabrieli, *Luca Valerio linceo, un episodio memorabile della vecchia Accademia Lincea* (Rome: Accademia Nazionale dei Lincei, 1934); D. Freedberg, *The Eye of the Lynx: Galileo, His Friends, and the Beginnings of Modern Natural History* (Chicago: The University of Chicago Press, 2002).

106. "Donna virtuosissima, e di gran nome, per l'universal cognizione, che possiede delle più gravi scienze, e delle principali lingue. . . . E non solo perita in queste lingue, e nella Poesia . . . ma nella filosofia, teologia, geometria, logica, astrologica, e in tante altre nobilissime scienze e belle lettere versatissima" (Cristofano Bronzini, *Della dignità e nobilità, delle Donne. Dialogo di Cristofano Bronzini d'Ancona. Diviso in Quattro Settimane, e ciascheduna in esse in Sei Giornate* [Florence: Zanobi Pignoni, 1625], 130).

107. "dirò con brevi parole . . . che questa sí virtuosa donna, si deve meritamente annoverare fra le Corinne, le Saffe, le Hippazie, e fra le più faccenti donne, che mai fossero al mondo" (ibid., 131). Compare *Scanderbeide* (1606), 2r-v.

108. "il Sign. Camillo Paleotti, Bolognese, uno de' gran letterati di questa nostra età, persona di molta autorità, ed in tutte le scienze versatissimo, a nome della sua famosa città di Bologna (vera madre de' studi . . .) le fece instanza grandissima, ch'ella volesse accettare nel suo principalissimo studio, con mille, e più scudi di provisione l'anno . . . la catedra principale della lettura publica di Geometria, e Logica, o di qual'altra lezzione di filosofia che più le fosse piaciuta." Camillo Paleotti (c. 1482–1517) was a Bolognese senator and ambassador to Rome. According to Bronzini, Sarrocchi

had already been offered a similar position in Palermo (Bronzini, *Della dignità e nobiltà delle donne*, 134; see also Verdile, "Contributi alla biografia,"189).

109. "Margarita Sarrochia perspicassimi ingenii poetria nostra tempestate in universa Italia celeberrima, imo non in poeticis tanummodo studiis erudite, sed philosophiae etiam, atque omnium fere scientiarum ac disciplinarum ornamentis illustris, quam aetas nostra admirata est, ac celebrat" (Chioccarelli, *De illustribis scriptoribus*, c. 67r-68v, in Baldini and Napolitani, "Per una biografia di Luca Valerio," 135).

110. The sonnets and letters addressed to Sarrocchi are listed in Verdile, "Contributi alla biografia." Bartolomo Sereno, for example, upon meeting her at the court of the Colonna family, wrote: "[she has] attained such a degree of knowledge in the sciences that neither can she be heard to speak, nor her works seen, without amazement" (*a tanto colmo di saper nelle scienze è arrivata, che non può essere udita, non possono le compositioni sue vedersi senza stupore* [*Trattato de l'uso della lancia a cavallo. Del combattere a piedi alla sbarra et alle inventioni cavalleresche* (Naples: Nucci, 1610), 117]; in Verdile, "Contributi alla biografia," 181).

111. "mostro del sesso femminile" (Giulio Cesare Capaccio, *Il Forastiero* [Naples: Rongagliolo, 1634]), cited in Verdile, "Contributi alla biografia," 184; "donna altero / e raro mostro / mostro per certo, ma . . . / nuovo mostro, e miracolo del Cielo" (Maurizio Cataneo, cited in ibid., 188). Rossi, characterizing Sarrocchi as transcending gender divisions ("nec sicut quidam qui maligne eam laudebant, solit erat dicere, fuit inter mulieres vir, et inter viros mulier"), suggests that her intellect was such that it rendered the learned men she encountered mere women ("sed multiplici variaque doctirna, et carminis elegantia, assequabatur, ut multi nec imperiti nec indocti viri, cum ipsa comparati, mulierum instar habere videntur," cited in Baldini and Napolitani, "Per una biografia di Luca Valerio," 143). For praise of Sarrocchi, see also Francesco Iacobilli da Foligno, *Le conditioni del Cavaliero* (Rome: Appresso Carlo Vullietti, 1606), 41, who writes that God has blessed her with all the graces (*a cui par che Dio habbia donato il cumolo delle gratie*).

112. "Margarita Sarochia Neap[olita]na clariss[im]a foemina: ingenio, sap[ienti]a, literarum omnium genere supra sexus conditionem evecta: Philosophicis, theologicis, Mathematicis disciplinis instructiss[im]a" (Atto di morte, AVR [Archivio del Vicariato di Roma], *Mortuorum liber Ecclesiae Sanctissimi Salvatoris in suburra* [transcribed in Baldini and Napolitani, "Per una biografia di Luca Valerio," 124]). Muzio Manfredi praises Sarrocchi's poetic talents, comparing her to Vittoria Colonna and Veronica Gambara (*Lettere brevissime* [Venice: Maglietti, 1606], 142).

113. "la sua honorata habitazione, continuamente piena dei più nobili, e virtuosi spiriti, che habitino, e capitino in Roma d'ogni tempo" (Bronzini, *Della dignità e nobiltà delle donne*, 130).

114. "pregandola a conservarmi la gratia del S. Luca et di quegl'altri SS. i literati che conobbi in casa [di] V[ostra] [S[ignoria]" (Galileo Galilei to Margherita Sarrocchi, January 21, 1612; Galilei, *Opere*, 11:647).

115. On these figures and their roles as cultural mediators, see (for Colonna), Robin, *Publishing Women*, esp. 41–78; and (for Gonzaga), Ray, *Writing Gender*, 81–122.

116. On the controversy over Galileo's telescope, see Eileen Reeves, *Galileo's Glassworks: The Telescope and the Mirror* (Cambridge, Mass.: Harvard University Press, 2008); Lawrence Lipking, *What Galileo Saw: Imagining the Scientific Revolution* (Ithaca, N.Y.: Cornell University Press, 2014), 28–40.

117. "essendo la casa sua ricorso et academia dei primi virtuosi di Roma" (Guido Bettioli to Margherita Sarrocchi, June 4, 1611; in Favaro, *Amici e corrispondenti*, 86).

118. Gabriele Gabrieli speculates that Cesi's willingness to entertain female membership in the Lincei stemmed directly from the case of Sarrocchi: "È probabile che . . . avesse in mente qualche candidata più o meno linceabile: forse quella Margherita Sarrocchi Birago" ("Contributi alla storia," 474). I thank Virginia Cox for sharing with me a draft of her essay on women and early modern academies, "Members, Muses, Mascots: Women and Italian Academies." On women and academic culture in the Renaissance, see also Elisabetta Graziosi, "Arcadia femminile: presenze e modelli," *Filologia e critica* 17, no. 3 (1992): 321–358; Conor Fahy, "Women and Italian Cinquecento Literary Academies," in *Women In Italian Renaissance Culture and Society*, ed. Letizia Panizza (Oxford: European Humanities Research Center, 2000), 438–452; Robin, *Publishing Women*. Rossi counters his own praise of Sarrocchi's intellectual gifts with a sharp digression on her character, complaining that she is overly sensitive and cannot tolerate criticism: "Much greater than her merits were her vanity and pride. She put herself before anyone else, she tolerated no one, she complained as if injured if any of her supporters praised her above all women but failed to also praise her above all men, living and dead. She harbored unending enmity for anyone who neglected to agree continuously with anything she said, or dared contradict her" (*Sed longe meritis maior illi inerat vanitas atque superbia; omnibus se anteponere, neminem ferre, iniuriam sibi factam queri, si a laudatoribus suis audiret, se mulieribus tantum, ac non viris etiam, quot sunt, quotoque fuissent, praestare; immortales cum eo inimicitias suscipere, qui in disputando, non continuo iis, quae ab ipsa dicerentur, assensionem suam praebuisset, aut eisdem adversari ausus esset* [in Baldini and Napolitani, "Per una biografia di Luca Valerio," 143]).

119. "Si sentì questa singolar donna trattare con tanta eccellenza, della natura, e moto de' Cieli, e di cose veramente profittevoli, e celesti, che come col moto della sua lingua rese quasi immobili le menti di quelli, che l'udivano, cosí fu cagione, che ciascheduno con ammirazione grandissima la incominciasse intentamente a riguardare, et altamente poi ad interrogare; fra i quali, il Galileo di Toscana, alle cui quistioni, non solo pronta, e prudentemente rispose la saggia Donna, con altissime, e ben fondate rissoluzioni, ma ella appresso mosse a' lui dubbi sí profondi, & alti, che 'l diede molto che fare per buona pezza" (Bronzini, *Della dignità e nobiltà*, 135).

120. Della Chiesa recalls, "A woman of much learning in all the sciences, she enjoyed conversing with men of letters, and debating with them, whence on many oc-

casions I saw and heard her reciting beautiful poetry in the public Academies of Rome" (F. A. della Chiesa, *Theatro delle donne letterate* [Modovi: Giflaudi e Rossi, 1620], 253–254). Similarly, Rossi writes, "I saw her recite, to the applause of all present, well-crafted epigrams, elegant and clever" (*ego eam vidi, elegantissia argutissimaque conclusa epigrammata, summa eorum qui aderant approbatione, recitare* [cited in Baldini and Napolitani, "Per una biografia di Luca Valerio," 143]).

121. See Verdile, "Contributi alla biografia," 199 and n140; 200 and n143; Sarrocchi, *Scanderbeide,* ed. Russell, 14–15 and n34; Michele Maylender, *Storia delle Accademie d'Italia* (Bologna: Forni, 1976), 4:140–141, 375–380; Francesco Saverio Quadrio, *Della storia e della ragione d'ogni poesia* (Bologna: Pisarri, 1739–1752), 1:98; Girolamo Tiraboschi, *Storia della letteratura italiana* (Florence: Molini Landi, 1805–1812), 8:48–50.

122. Cited in Cox, "Members, Muses, Mascots," 21.

123. Both Valerio and Galileo were members. See Peter Armour, "Galileo and the Crisis in Italian Literature of the Early Seicento," in *Collected Essays on Italian Language and Literature Presented to Kathleen Speight,* ed. Giovanni Aquilecchia, Stephen N. Cristea, and Sheila Ralphs (Manchester: Manchester University Press, 1971), 148.

124. On women's epic poems, see Virginia Cox, "Fiction, 1560–1650," in *A History of Women Writing in Italy,* ed. Letizia Panizza and Sharon Woods (Cambridge: Cambridge University Press, 2000), 57–61; Cox, "Women as Readers and Writers of Chivalric Poetry in Early Modern Italy," in *Sguardi sull'Italia: Miscellanea dedicata a Franceso Villari,* ed. Gino Bedoni et al. (Leeds: Society for Italian Studies, 1997), 134–145; and Valeria Finucci's introduction to Moderata Fonte *Tredici canti del Floridoro* (Modena: Mucchi, 1995), x–xii.

125. See Serena Pezzini, "Ideologia della conquista, ideologia dell'accoglienza: 'La Scanderbeide' di Margherita Sarrocchi (1623)," *MLN* 120 no. 1 (2005): 219. The war with Hungary ended with the peace of Zsitvatorok in 1606.

126. See Sarrocchi, *Scanderbeide,* ed. Russell, 10n18.

127. Pezzini, "Ideologia della conquista," 218. On the New World in seventeenth-century epic poetry, see Nathalie Hester, "Failed New World Epics in Baroque Italy," in *Poiesis and Modernity in the Old and New Worlds,* ed. Anthony J. Cascardi and Leah Middlebrook (Nashville: Vanderbilt University Press, 2012), 201–224.

128. Sarrocchi, *Scanderbeide,* ed. Russell, 40–41; see also Cox, "Women as Readers and Writers," 144. On Marinella's *L'Enrico,* see also Laura Lazzari, *Poesia epica e scrittura femminile nel Seicento: L'Enrico di Lucrezia Marinelli* (Leonforte: Insula, 2010); on Marinella and natural philosophy more generally, see Chapter 3 here.

129. For a consideration and comparison of the works of Sarrocchi, Marinella, and Fonte, see also Cox, *The Prodigious Muse,* 164–212.

130. For example, Bona Sforza (poisoned in Bari after her return to Italy from Poland); Francesco I de' Medici and his wife, Bianca Cappello (said to have been poisoned on the same day). While some have argued that the cause of death for Francesco and Bianca was arsenic poisoning, others find they died of malaria: see Francesco

Mari et al., "The Mysterious Death of Francesco I and Bianca Cappello: An Arsenic Murder?" *BMJ* (2006): 333–1229; Gino Fornaciari et al., "Malaria Was the Killer of Francesco I de' Medici (1531–1587)," *American Journal of Medicine* 123, no. 6 (2010): 568–569. Caterina Sforza (discussed in Chapter 1) was accused of attempting to poison Pope Alexander VI.

131. Sarrocchi, *Scanderbeide,* ed. Russell, 2 (Sarrocchi, *La Scanderbeide,* 5–6: "Il prende ei che 'l sapore non fà divieto, / Né toglie al senso il natural diletto").

132. Sarrocchi, *Scanderbeide,* ed. Russell, 8–9 (Sarrocchi, *La Scanderbeide,* canto I, 91, 4–6, "E sentì non sò che pungersi al core, / ch'in poco spatio, tanto acuto il morse, / Ch'un quanco non provò simil dolore," canto I, 91, 4–6).

133. Sarrocchi, *Scanderbeide,* ed. Russell, 21 (Sarrocchi, *La Scanderbeide,* canto II, 1, 4–8: "Cresce al buono Duce la doglia aspra, e fella, / Sparso homai tutto d'angoscioso gielo / Fosco gli appar quanto d'intorno vede, / E muove incerto e vacillante il piede").

134. "By now the lethal poison was making its way unobstructed down the hero's veins" (Sarrocchi, *Scanderbeide,* ed. Russell, 2, 12); (Sarrocchi, *La Scanderbeide,* canto II, 3, 4–5: *Ma già il venen mortal, con forza molta, / S'apre la via, serpendo entro le vene*). As Mari points out, "White arsenic (arsenic trioxide), also known as arsenious acid or commercial arsenic, was certainly the best known and most commonly used poison in the Medici era" ("Mysterious Death"). Its symptoms include vomiting, and—as in Sarrocchi's description—a lengthy, painful course, cold sweats, aggressive restlessness, and seeming improvement four to five days after onset, followed by a return of symptoms (see Mari, "Mysterious Death").

135. "The more they studied his symptoms, the more certain they became that he had ingested a poison" (Sarrocchi, *Scanderbeide,* ed. Russell, 2, 13; Sarrocchi, *La Scanderbeide,* canto II, 4, 1–2: *Se cerca i segni più, più s'assicura / Ciascuno, ch'ei succo velenoso ha preso*).

136. See, for example, Cortese, *I secreti,* which contains recipes for mixing arsenic compounds ("Olio d'arsinico") as well as recipes for antidotes ("Contra veleno"). On books of secrets, see Chapter 2 here.

137. Giulio Iasolino, *De remedii naturali che sono nell'isola di Pithecusa: hoggi detta Ischia. Libri due* (Naples: appresso Giuseppe Cacchik, 1588). This work is dedicated to a female patron, Signora D. Geronima Colonna, d'Aragona, duchessa di Monteleone. Sarrocchi's poem appears at the conclusion of the work.

138. Sarrocchi, *Scanderbeide,* ed. Russell, 2, 14 (Sarrocchi, *La Scanderbeide,* canto II, 4, 5–6: "E con più d'un licor grata l'arsura, / Va mitigando, onde ha 'l cor tanto offeso").

139. Sarrocchi, *Scanderbeide,* ed. Russell, 2, 28 (Sarrocchi, *La Scanderbeide,* canto II, 22, 7–8: "Il Duce ascolta, e infondersi del Cielo / sente nel cor, non più sentito zelo"; and canto II, 2, 24, 1–8: "Miracol fu, ch'al fin delle parole, / Che più co 'l core, che con la lingua espresse: / Qual dare salute in fresca età più sole, / Tornar le membra alle lor forze stesse: / Tal splendon più chiare i rai del Sole / Ch'intempestiva nube a

case oppresse: / Già sano sorge, e già devoto stende / Le mani al Cielo, e gratie a Dio ne rende").

140. The Calidora episode is restored in the 1701 edition (Naples: Antonio Bulifon). On Sarrocchi's concerns about the poem's narrative unity, organization, and overall length, see her letter to Galileo dated January 13, 1612, discussed in this chapter. On the excision of this and other episodes from the final version of the poem, see Sarrocchi, *Scanderbeide,* ed. Russell, 25–26; and Cox, "Fiction, 1560–1650," 61–62.

141. Sarrocchi, *La Scanderbeide,* canto III, 37, 7–8: "Così d'ogni arte liberal nomata, / E con maschio valor femina ornata."

142. Ibid., 38, 1–8 "Penetra con l'acuto alto intelletto / Da le cagion primiere a le seconde, / Che producon tra noi diverso effetto, / Con qualitati, e sterili, e feconde, / Che con vario del ciel corso, & aspetto, / Dan le fortune averse, e le seconde, / E così de le stelle il lume, e 'l moto, / E di natura l'operar l'è noto."

143. "Tu sommo Dio de regni, alti e tonanti, / Che per giovare altrui se' Giove detto; / Tu virtù inspiri a vaghi augei volanti, / Gli alti secreti del tuo sacro petto: / Gli humani avvenimenti, e varij, e tanti, / Rivelarne hor con voce, hor con aspetto" (ibid., III, 76).

144. See, for example, Michael Maier's book of alchemical emblems, *Atalanta fugiens.* Although this work was published after the first edition of Sarrocchi's *Scanderbeide,* it compiles and comments upon widely used alchemical imagery and symbolism. The marriage of Sun and Moon is addressed in Emblem XXX: "That is the finall intent: a durable Marriage between the Sun and Moon, and when that is accomplished all embassies, contracts, congresses, mistrusts shall have an End."

145. "Forma di cera poi candida, e pura, / Da cui pur anco il mel dolce ven fuora, / A suo poter simile ad huom figura, / Cui del sangue del cor pinge, e colora. . . . E giunta l'hora in solitaria loco, / Et ambra, e mirra, & altro eletto odore, / D'hebano sparge sorpa un vivo fuoco, / Che senza fumo alcun nutre l'ardore. . . . Discoglie al vento poi l'aurate chiome, / Traendo il cor da tre colombe vive / Co'l ferro stesso, e di Serrano il nome / In quella carta co'l lor sangue scrive" (Sarrocchi, *La Scanderbeide* [1606], canto VII, 74–76).

146. "inchinano i nostri corpi al bene & al male; alle quali agevolmente l'huomo può opporsi, e contradire" (Bronzini, *Della dignità e nobiltà delle donne,* 134–135).

147. Bronzini, *Della dignità e nobiltà,* 134–135.

148. "In questi canti della Scanderbeide trovarete alcuna volta Fato, & Fortuna, dalle quali voci non s'argumenta, che realmente sia la necessità fatale, ne meno che la Fortuna habbia alcuna sussitenza personale. . . . Troverete ancora un'osservatione d'auguri all'antica; & similmente una fattura d'una maga Turca pur favolosa, poste solo per dilettare ad imitatione d'altri poeti, & non per significato, & forza che possino havere nel mondo finto & favoleggiato dal poeta" (Sarrocchi, *La Scanderbeide* [1606], "A benigni lettori").

149. See Galileo Galilei, *Discorsi e dimostrazioni matematiche intorno a due nuove scienze attenenti alla meccanica et i movimenti locali . . . con un appendice del centro di*

gravità d'alcuni solidi (Leiden: Elsevier, 1638), 30. Valerio recalls meeting Galileo in Pisa in a letter dated April 4, 1609 (Luca Valerio to Galileo Galilei, BNCF Mss Gal 88, c. 93).

150. See Armour, "Galileo and the Crisis," 147. See also Sarrocchi, *Scanderbeide,* ed. Russell, 25n56. For Galileo's literary writings, see Galilei, *Scritti letterari,* ed. Alberto Chiari (Florence: Le Monnier, 1970), 362–486, 487–635; and idem, *Rime.* Galileo held poetic gatherings, or "simposi poetici," in his house in Florence; see Baldassar Nardi's reference in a letter to Galileo dated April 19, 1633: "spero ben presto, piacendo a Dio, ritrovarmi un'altra volta ad un simposio poetico, come poco avanti fui favorito in casa di V[ostra] S[ignoria]" (in Galilei, *Opere,* 15:96). On this aspect of Galileo, see Crystal Hall, *Galileo's Reading* (Cambridge: Cambridge University Press, 2014); J. L. Heilbron, *Galileo* (Oxford: Oxford University Press, 2010).

151. See Hall, *Galileo's Reading,* 1.

152. "havevo il poema di Tasso legato con intersposizione di carta in carta di fogli bianchi, dove havevo non solamente registrati i riscontri de i luoghi di concetti simili in quello d'Ariosto, ma ancora aggiuntovi discorsi, secondo che mi parevano questo o quelli dovere essere anteposti. Tal libro mi andò male, né so in qual modo" (Galileo Galilei to Francesco Rinunccini [Galilei, *Opere,* 18:120]). There is continued debate over the date on which Galileo composed the *Considerations on Tasso;* see Armour, "Galileo and the Crisis," 149–150; Tibor Wlassics, *Galilei critico letterario* (Ravenna: Longo, 1974), 15–32.

153. Crystal Hall, "Galileo, Poetry, and Patronage: Giulio Strozzi's *Venetia edificata* and the Place of Galileo in Seventeenth Century Poetry," *Renaissance Quarterly* 66 no. 4 (2013): 1331. See also Wlassics, *Galilei critico letterario,* 15–32.

154. Sarrocchi's letters to Galileo are in the Biblioteca Nazionale Centrale di Firenze, Manoscritti Galileiani, busta 23, carte 8, 10, 12, 14, 16, 18, and 20 (see Galilei, *Opere,* 11:563, 579, 593, 596, 636, 647, 696; and Verdile, "Contributi alla biografia," and Baldini and Napolitani, "Per una biografia di Luca Valerio"). In citing from Sarrocchi's letters to Galileo, I have modernized accents but I follow the letters' original orthography and wording; I also provide the corresponding reference in Favaro. The sole letter of Galileo to Sarrocchi was found in the Archivio di Stato di Mantova and is published in Gilberto Govi, *Tre lettere di Galileo Galilei* (Rome: Tipografia delle scienze matematiche e fisiche, 1870), 9–10. It is almost certain that there were other letters that are lost, destroyed, or remain to be discovered.

155. Galileo's naming of the Medicean stars, as Richard S. Westfall writes, raised him "with one inspired blow from the level of an obscure professor of mathematics at the University of Padua to the status of the most desirable client in Italy" ("Science and Patronage: Galileo and the Telescope," *Isis* 76 [1986]: 11–30, at 20–21). For a reevaluation of Galileo and the telescope, see Reeves, *Galileo's Glassworks.*

156. "I won't forget to send you [my works], with some of the revisions I've made over the past year, and which I continue to make to my published works, which you have been so kind as to read, and also the eleven cantos of Signora Margarita Sarro-

chi's *Scanderbeide" (non mancherò d'inviargliele, col sagio anco d'alcuni miglioramenti ch'io fei l'anno passato, et vo' tuttavia facendo, ne' miei libri publicati, che V[ostra] S[ignoria] si è degnata di leggere, et con gli undici canti della Scanderbeide della S[igno]ra Margarita Sarrochi)* (Luca Valerio to Galileo Galilei, May 23, 1609; in Galilei, *Opere*, 10:221).

157. "il favore che io prencipalmente desidero da lei è che rivegga il mio poema, con quella diligenza, che sia maggiore, e con occhio inimico, acciò che ella vi noti ogni picciolo errore, e creda ch'io lo dica davero e che tutto quel male che ella me ne dirà io la pigliarò a segno di gran bontà, e di grande affettione" (Margherita Sarrocchi to Galileo Galilei, July 29, 1611 [BNCF Mss Gal 23, c. 8]; Galilei, *Opere*, 11:563).

158. "Il poema si attende a porre in netto, et cosí credo di mandarlo presto a V[ostra] S[ignoria] per ricevere il favore, che ella mi vuol fare del suo purgatiss[i]mo giudicio" (Margherita Sarrocchi to Galileo Galilei, September 10, 1611 [BNCF Mss Gal 23, c. 10]; Galilei, *Opere*, 11:579).

159. "Today I gave my *Scanderbeide* to the messenger" *(Io ho dato hoggi la mia Scanderbeide al procaccio)* (Margherita Sarrocchi to Galileo Galilei, January 13, 1612 [BNCF Mss Gal 23, c. 18]; Galilei, *Opere*, 11:643).

160. "Il sig[no]r Spinello m'ha scritto la buona volontà, che ha V[ostra] S[ignoria] di favorirmi nella revisione del mio poema, del che mi sono sommamente rallegrata, ancora ch'io non ne fussi in dubbio" (Margherita Sarrocchi to Galileo Galilei, October 15, 1611 [BNCF Mss Gal 23, c. 14]). Spinello Benci (b. 1565) served as secretary in several Italian courts, including that of the Medici in Florence.

161. "E chi potrebbe dubitare della cortesia del mio Sig[no]r Galileo ornato di tante vertù e amatore così de' letterati?" (BNCF Mss Gal 23, c. 14; Galilei, *Opere*, 11:596). For Galileo's interactions with other writers, see Nunzio Vaccalluzzo, *Galileo Galilei nella poesia del suo secolo* (Milan: Remo Sandron, 1910), xiv; Armour, "Galileo and the Crisis," 150).

162. Marino was celebrated as the "new Tasso," but his linguistic stye was more avant-garde than Sarrocchi's own. Marino himself was a supporter of Galileo, famously praising his telescope and describing him as a Columbus of the heavens in the *Adone* (10:42–45). See Giovanni Aquilecchia, "Da Bruno a Marino: Postilla all'*Adone* X 45," *Studi secenteschi* 20 (1979): 89–95.

163. "Facciame ancora gratia di riveder la lingua, ed emendarla, perchè io vorrei che la fusse toscana più che fusse possible almeno nella frase, pur che non guasti la grandezza del dire, essendo che la toscana è molto dolce. Il perché dove ella suol levare gli *r* qualche volta io hoccioli lasciati, come sarebbe per esempio che dove toscanamente si suol dire *trincea* io ho detto *trincera*, et cose simili . . . vorrei che V[ostra] S[ignoria] la rivedesse ancora, quanto all'ortografia. Vi troverà ancora molte rimesse e molti versi mutati" (Margherita Sarrocchi to Galileo Galilei, January 13, 1612 [BNCF Mss Gal 23, c. 18]; Galilei, *Opere*, 11:643).

164. See Biagioli, *Galileo Courtier;* Westfall, "Galileo and the Telescope."

165. "E perchè del secondo ne mando stampe a loro Altezze con mia lettera dell'incluso tenore, faccio adesso grandissimo capitale di V[ostra] S[ignoria], la quale

prego, per quell'amore che mi portò un tempo, ad aiutarmi del suo favore et insinu-
atione opportuna, che m'assicuro potrà giovarmi notabilmente a farmi ricevere qualche
segno di gratitudine" (Lorenzo Ceccarelli to Galileo Galilei, October 23, 1637, in Galilei, *Opere*, 17:205; see also Vaccalluzzo, *Galileo Galilei*, xvi).

166. "Riveduto poi che l'haverà V[ostra] S[ignoria], se le parerà cosa conveniente
circa alla dedicatione potrà d'esso fare quello che più le piacerà che io me rimetto in
tutto, e per tutto al suo sano consiglio" (Margherita Sarrocchi to Galileo Galilei, July
29, 1611 [BNCF Mss Gal 23, c. 8]; Galilei, *Opere*, 11:563).

167. "È bene il vero che la rassegna degli Italiani che hanno da andare in aiuto di
Scanderebech, non l'ho fatta per non havere a pieno determinato tutti coloro che vi
vorrò mandare e ancora per lasciare alcun loco [da] lodare alcun prencipe, sì che se
V[ostra] S[ignoria] mi manderà alcun[o] de' suoi, io honorarò le mie carte del nome
della sua casa et ancora, con buona occasione farò mentione di V[ostra] S[ignoria]
come di cosa futura" (ibid.).

168. "Cotal rassegna non fa nulla l'haverla sospesa perciocché a persona tanto es-
sercitata in simil materia com'io sono sarà fatica de quindici, o venti giorni" (ibid.).

169. "The poem is complete, except for the list of those who came to Scanderbeg's
aid, which I left incomplete so that I could insert the names of my friends and patrons, as you will see by many of the names, which I chose randomly, and then changed
to the names of my friends; and the list will take me eight, or maybe ten, days" *(Il
poema è compito se non che ci manca la rassegna del soccorso di Scandarebech, la quale ho
lasciata per potervi poner dentro de' miei amici, e p[ad]roni, come V[ostra] S[ignoria] vedrà
in molti nomi e quali io havea posto a caso e poi hogli mutati in nome degli amici miei.
A me la rassegna sarà la fatica di 8 overo 10 di)* (Margherita Sarrocchi to Galileo Galilei, January 13, 1612 [BNCF Mss Gal 23, c. 18]; Galilei, *Opere*, 11:643).

170. See, for example, Russell's edition of the *Scanderbeide*, at canto 18, 141.

171. "travagli domestici e . . . le continove malattie" (Margherita Sarrocchi to Galileo Galilei, July 29, 1611 [BNCF Mss Gal 23, c. 8]; Galilei, *Opere*, 11:563). Compare
Erculiani, *Lettere* (aiii-v). More than a century earlier, the humanist author Laura Cereta
wrote eloquently in her *Letters* of the demands placed on her by marriage and family,
with the subsequent consequence that her literary work was done at night (Laura Cereta,
Collected Letters of a Renaissance Feminist, ed. Diana Robin [Chicago: University of
Chicago Press, 1997], 32). Valerio's comment after the death of Sarrocchi's husband
that Sarrocchi now had more time for her interests in "philosophy" echoes this dilemma
(Luca Valerio to Galileo Galilei, August 31, 1613: "La S[igno]ra Margherita Sarrocchi,
la quale per innanzi havrà libero spazio di filosofare, sendo rimasta vedova" [quoted
in Baldini and Napolitani, "Per una biografia di Luca Valerio, 66]).

172. "guerra puerile, che pur le fanno talora gli ormai rochi e sprezzati parlatori"
(Luca Valerio to Galileo Galilei, October 23, 1610; Galilei, *Opere*, 8:112).

173. "io ho molto fatica per haver a trovar chi lo scriva corretto" (Margherita Sarrocchi to Galileo Galilei, July 29, 1611 [BNCF Mss Gal 23, c. 8]; Galilei, *Opere*, 11:563).

The only surviving letter of Galileo to Sarrocchi confirms that *Scanderbeide* arrived to him in poor condition, and that he was in equally poor health: "Your poem did not reach me in good condition, but it found me in even worse condition" (*Il poema di V[ostra] S[ignoria] non mi è giunto ben condizionato ma ben ha trovato me in malissima condizione;* Galileo Galilei to Margherita Sarrocchi, January 21, 1612 [in Govi, *Tre lettere,* 9–10]).

174. "our Lord God blessed me so that I am not at all enamoured of my own compositions, and taught me that just as print can demonstrate one's erudition, so it can sometimes show one's poor judgment. For this reason, not wishing to commit such a mistake, *in propria causa advocatum quero* (*il nostro Sig[no]re Iddio mi ha fatto gratia che io non sono inamorata punto delle mie compositioni, et mi ha fatto conoscere che sì come la stampa mostra il saper de gli huomini, così alcuna volta mostra il poco giudicio; la onde io, che non vorrei incorrere in simile errore* in propria causa advocatum quero) (Margherita Sarrocchi to Galileo Galilei, July 29, 1611 [BNCF Mss Gal 23, c. 8]; Galilei, *Opere,* 11:563). Sarrocchi repeats a virtually identical sentiment after she actually dispatches the manuscript, writing "sapendo bene che sì come le stampe mostrano il saper del'huomo, così palesano altresì l'ignoranza" (Margherita Sarrocchi to Galileo Galilei, January 13, 1612 [BNCF Mss Gal 23, c. 18]; Galilei, *Opere,* 11:643).

175. "Dessiderarei ancora che V[ostra] S[ignoria] me favorisse de devidere questo poema, col suo giuditio in più canti, percioché questi mi paiono troppo longhi. Le dirò ancora che mi sono forzata di far questo poema secondo le regole di Aris[totele], di Falareo, di Herm[ogene], di Lugn.o et di Eustat[io], i quali convengano tutti in uno, e però mi sono forzata, col verso d'immitare le cose e cosí nelle cose di guerra ho cercato inalzarlo e nelle cose di amore addolcirlo" (ibid.).

176. "l'ha letto ancora la Signora Margarita Sarrocchi . . . et giudica del facitore l'istesso ch'io" (Luca Valerio to Galileo Galilei, April 4, 1609 [BNCF Mss Gal 88, c. 93]).

177. "Io e il Sig.r Luca lo leggeremo, con ogni affetto, et con ammiratione" (Margherita Sarrocchi to Galileo Galilei, June 9, 1612 [BNCF Mss Gal 23, c. 20]; Galilei, *Opere,* 11:696).

178. See Camerota, *Galileo Galilei,* 235. Maria Celeste's letters to her father show that she followed his discoveries and the ensuing events in his life closely and with great interest, on occasion requesting that he send her copies of his works. See Dava Sobel, *Galileo's Daughter: A Historical Memoir of Science, Faith, and Love* (New York: Walker, 1999); for Maria Celeste's letters, see Sobel, *Letters to Father: Suor Maria Celeste to Galileo, 1623–1633* (New York: Walker, 2001); and Maria Celeste Galilei, *Lettere al padre,* ed. Bruno Basile (Rome: Salerno, 2002).

179. "Prego V.S. a non lasciarsi tanto trar dalle stelle, ch'ella non seguiti l'opera dei vari motii terrestri; sì come ancor ne la prega la Sig[no]ra Margherita, fatta non men di me del valore di V[ostra] S[ignoria] predicatrice" (Luca Valerio to Galileo Galilei, May 29, 1610; in Galilei, *Opere,* 10:362–363). Valerio's reference is probably to Galileo's studies of motion rather than the revolutions of the earth.

180. Westfall, "Science and Patronage," 22.

181. Lodovio Cardi (il Cigoli) to Galileo, August 23, 1611 (in Galilei, *Opere*, 11:175; Westfall, "Science and Patronage," 23).

182. Federico Cesi to Galileo Galilei, August 20, 1611 (in Galilei, *Opere*, 11:572).

183. Valerio affirms his statement as a "filosofo per amore della verità . . . perché, vendendole occasione, ella possa, con questa mia scrittura di mia mano, assicurare alcuni di questi ritrosi, atti a sparger la fama, ch'io non sono di contrario parere a quell ch'io mi contento che, come mio, apparisca per iscrittura" (Luca Valerio to Marc'Antonio Baldi, May 30, 1611 [BNCF Mss Gal, Par. VI, T. XIV, c. 29]).

184. Ibid.

185. "Many reverend Fathers here are protesting against signor Galileo . . . I would like a response to the objection I am hearing raised, which seems quite persuasive to me, that is that the [scope] makes things appear to be, which are not" *(Qua tra questi Padri Reverendi è un gran rumore contro il signor Galileo. . . . Desidererei la risposta a una ragione quale sento, che mi pare assai concludente, cioé che l'occhiale faccia apparire quello che non è)* (Cosimo Sassetti to Monsignor Piero Dini, May 14, 1611). The complete letter, with Galileo's response of May 21, 1611, is found in Galilei, *Opere*, 6:163–175; Favaro pieces together the incident in *Amici e corrispondenti*, 15–18 and in *Galileo Galilei e lo Studio di Padova* (Florence: Le Monnier, 1883), 1:396.

186. "Li mirabili effetti che di continuo si odono del Cannone, o occhiale che dir volemo, del sig.r Galileo Galilei, di continuo dà da dire ad ogni uno l'openione sua, mi ha fatto essere prosuntuoso di pigliare la penna et far riverenza a V[ostra] S[ignoria] et pregarla a favorirmi dell'openion sua, essendo ella perettamente compita d'ogni scienza, spero perfetta notizia del vero, poiché anch'Ella vi haverà fatto mille prove" (Guido Bettoli to Margherita Sarrocchi, June 4, 1611; Galilei, *Opere*, 11:537). As is characteristic of most letters involving Sarrocchi, the letter closes with a greeting to Valerio as well: "Perché bacio le mani con una mia al Sign.r Luca Valerio, non farò di lui altra memoria, essendo al uno et l'altro divotissimo servitore et di nuovo facendogli riverenza gli bacio le mani" (ibid.).

187. "Con questa digressione ho voluto disingannare V[ostra] S[ignoria] se alle purgate orecchie di Lei, e di altri virtuosissimi, fusse capitata tal lettera o openione, che il Sig[no]r Galileo pretende di rispondere, della quale qua da questi Signori non se ne sa se non quanto dal Sign[o]r Galileo ne vien tocco, cosa che veramente ha dato non poco disturbo, né so come se la passeranno. So quanto Ella sia magnanima et virtuosissima et defenditrice de' virtuosi, et per questo non mi stenderò più in longo, solo starò spettando risposta et che mi facci degno de' suoi commandamenti" (ibid.).

188. "che tutto quello che se ne dice del ritrovamento delle stelle del Signor Galileo è vero, cioé, che con Giove son quattro stelle erranti con moto proprio, sempre ugualmente distante da Giove, ma non fra di loro; et io con li proprii occhi l'ho vedute mediante l'ochiale del Sig.r Galileo, et fattele vedere a diversi amici: il che tutto il

mondo sa. Con Saturno sono due stelle una da un lato et l'altra dall'altro, che quasi lo toccano. Venere, quando si congiunge al Sole, si vede illuminare et diventar, come la luna, corniculata, infine a tanto che la si vede poi tutta piena. . . . Molti matematici grandi, et in particolare il P[ad]re Claudio col P[ad]re Gambergere [Grienberger] negavano questo da principio, et di poi si sono disdetti, essendosene certificati, et ne hanno fatte publiche lettioni" (Margherita Sarrocchi to Guido Bettoli, August 27, 1611; Galilei, *Opere,* 11:574).

189. "Ho voluto che V[ostra] S[ignoria] veda tutto quello che passa" (Margherita Sarrocchi to Galileo Galilei, September 10, 1611, BNCF Mss Gal 23, c, car. 10; Galilei, *Opere,* 11:579).

190. "havendogli ancora scritto io la verità delle stelle, et lodato l'ingegno di V[ostra] S[ignoria] . . . egli me rispose una lettera la quale m'alterò molto et per ciò gli replicai, come pareva a me che convenisse. . . . Egli replicò, come potrà V[ostra] S[ignoria] vedere, perciochè le mando anche ambe due l'ultime sue lettere" (ibid.).

191. "È bene il vero che quel frate par che la vogli meco, et che mi voglia pigliare in parole volendo intender da me la significatione d'alcuni vocaboli mentre ch'io voleva applicar le stelle di nuovo trovate alla astrologia quasi che voglia dire che non sia vero il ritrovamento di queste stelle; ma io ho chiarito altra barba delle sue e così spero di far lui avegna che io sia donna, ed egli frate maestro" (Margherita Sarrocchi to Galileo Galilei, October 12, 1611 [BNCF, Mss Gal 23, c. 12]; Galilei, *Opere,* 11:593).

192. See Monica Azzolini, *The Duke and the Stars: Astrology and Politics in Renaissance Milan* (Cambridge, Mass.: Harvard University Press, 2013), 17. Despite a papal bull forbidding judicial astrology, it was widely practiced nonetheless, even among ecclesiasticals: see Ugo Baldini, "The Roman Inquisition's Condemnation of Astrology: Antecedents, Reasons, and Consequences," in *Church Censorship and Culture in Early Modern Italy,* ed. Gigliola Fragnito (Cambridge: Cambridge University Press, 2001), 107.

193. "io le feci il piacere della natività, [et] egli ne fece chiedere un'altra d'una fanciulla, alla quale era succeduto un accidente maraviglioso" (Margherita Sarrocchi to Galileo Galilei, September 10, 1611 [BNCF Mss Gal 23, c. 10]; Galilei, *Opere,* 11:579). The episode described by Sarrocchi refers to a girl in Lucca who was abandoned by her mother. Galileo, too, was known to cast natal charts, sketching one for Cosimo II in a manuscript of *Sidereus Nuncius* (see Guglielmo Righini, "L'oroscopo galileaiano di Cosimo II de' Medici," *Annali dell'Istituto e Museo di Storia della Scienza di Firenze* I [1976]: 29–36).

194. "ho conosciuto ch'ella è troppo affettionata al Sig.r Galileo, onde si lascia trasportare più dall'affettione che forsi dalla verità'" (Fra Innocenzio Perugino to Girolamo Perugino, in Favaro, *Amici e corrispondenti,* 28).

195. "Havrei prima d'ora dato risposta alla gratissima vostra, ma il dubbio di non far sdegnare la Sig[no]ra Margarita mi ha trattenuto" (Fra Innocenzio Perugino to Girolamo Perugino (in Favaro, *Amici e corrispondenti,* 30).

196. Erythraeus [Rossi], *Pinachoteca,* 259–261. Rossi's account, a lengthy and hyperbolic description of Sarrocchi's extreme self-involvement, must be taken with a healthy dose of skepticism.

197. See Baldini and Napolitani, "Per una biografia di Luca Valerio," 66.

198. "Audio famigeratam illam nostri Saeculi Musam ex humanis sublatam; quid de collega futurum credis? An ne violentis illius suggestionibus liber et exemptus?" (Federico Cesi to Ioannes Faber, Acquasparta, November 6, 1617 (BANL, Archivio di S. Maria in Aquiro, filza 423, cc. 102–104; transcribed in Baldini and Napolitani, "Per una biografia di Luca Valerio," 121).

199. Baldini and Napolitani, "Per una biografia di Luca Valerio," 65.

200. Ibid., 68: "L'altera poetessa non avrebbe mai perdonato un affronto del genere a Galileo, e avrebbe fatto di tutto per impedire che Valerio potesse appoggiarlo."

201. "Quanto al mio poema V[ostra] S[ignoria] . . . mi farà favore rimandarmelo, perché ci ho fatto molte mutationi, di modo che quello non è più buono. Io lo farò di nuovo copiare et lo mandarò a V[ostra] S[ignoria]" (Margherita Sarrocchi to Galileo Galilei, June 9, 1612 [BNCF Mss Gal 23, c. 20]; Galilei, *Opere,* 11:696).

202. "E sarà in miglior tempo, perciò che spero che ella all'hora starà con sanità" (ibid.).

203. "Però se io sarò breve in rispondere alla sua cortessisima lettera, et in rendergli le debite grazie del continuar ella con tanta benignità in conferirmi de' suoi favori scuserà l'impotenza mia, la quale non mi permette di affaticare il pensiero, non che la mano, senza grandissimo nocumento. Ma perchè lei non stesse con pensiero del buon ricapito del poema, li ho volute scriver queste poche righe, ricordandogli insieme la servitù mia . . . con ogn'affetto di cuore gli bacio le mani, et dal S[ignor] Dio gli prego felicità" (Galileo Galilei to Margherita Sarrocchi, January 21, 1612; in Galilei, *Opere,* 11:647).

204. Antonio Favaro, *La libereria di Galileo Galilei* (Rome: La tipografia delle scienze matematiche e fisiche, 1887): see entries 278 (Boccaccio), and 394 (Costa, *La flora feconda,* 1640) and 438 (Costa, *Li buffoni,* 1641).

205. The frontispiece to Costa's *Li buffoni,* for example, depicts a figure using a telescope; while a Florentine manuscript of her equestrian ballet, *Festa reale per balletto a cauallo* (later published in France), may include references to the Medicean stars (I thank Jessica Goethals and Sara Diaz for drawing my attention to these references). For a discussion of the uses of the Medicean stars in other literary works and court performances, see Biagioli, who, however, argues that after 1633 their "association with Galileo was on the wane" (*Galileo Courtier,* 149).

206. See Favaro, *La libreria di Galileo Galilei,* entries 124 (Piccolomini, *La sfera,* 1572), 125 (Piccolomini, *Delle stelle fisse,* 1570), 256 (Valerio, *De quadrature parabola per simplex falsum,* 1606), and 276 (Valerio, *De centro gravitatis solidorum,* 1604).

207. Margherita Sarrocchi to Galileo Galilei, June 9, 1612 (BNCF Mss Gal 23, c. 20; Galilei, *Opere,* 11:696).

208. "Nel tempo che le mandarò il mio poema, la pregarò a riveder le cose mie liriche. Intanto leggeremo il suo trattato, et scriverò più lungo poi a V[ostra] S[ignoria]" (ibid.).

209. See Vaccalluzzo, *Galileo Galilei,* i–lxx.

210. Ibid., 123.

Epilogue

1. On terminology, see Pamela H. Smith, "Science on the Move: Recent Trends in the History of Early Modern Science," *Renaissance Quarterly* 62 (Summer 2009): 358.

2. See Galileo Galilei, *Lettera a Cristina di Lorena,* ed. Franco Motta (Genoa: Marietti, 2000), 76–77. As Mario Biagioli notes, Galileo cultivated a patronage relationship with the Grand Duchess Cristina, "the most powerful" of the brokers surrounding her son Cosimo II (*Galileo Courtier: The Practice of Science in the Culture of Absolutism* [Chicago: University of Chicago Press, 1993], 22–23).

3. For a discussion of this aspect of the *Lettera,* see Jean Dietz Moss, "Galileo's *Letter to Christina:* Some Rhetorical Considerations," *Renaissance Quarterly* 36, no. 4 (Winter 1983): 547–576.

4. On Tarabotti, see Emilio Zanette, *Suor Arcangela monaca del Seicento* (Rome-Venice: Istituto per la collaborazione culturale, 1960); and the essays collected in *Arcangela Tarabotti: A Literary Nun in Baroque Venice,* ed. Elissa Weaver (Ravenna: Longo, 2006). In 1630 Tarabotti's father, Stefano Bernardino di Marc'Antonio, was involved in a dispute with Hacma Juda, a Levantine Jew, over the rights to market various "*sollimati* [sublimates], cinnabars, and other alchemical mixtures," including mercury and lead, to the Venetian Republic (see Mario Infelise, "Books and Politics in Arcangela Tarabotti's Venice," in *Arcangela Tarabotti,* 57n1).

5. See Lynn Lara Westwater, "A Rediscovered Friendship in the Republic of Letters: The Unpublished Correspondence of Arcangela Tarabotti and Ismaël Boulliau," *Renaissance Quarterly* 65 (2012): 67–134. On Tarabotti's ties to France and to the French diplomatic community in Venice, see Westwater, "A Cloistered Nun Abroad: Arcangela Tarabotti's International Writing Career," in *Women Writing Back/Writing Women Back: Transnational Perspectives From the Late Middle Ages to the Dawn of the Modern Era,* ed. Anke Gilleir, Alicia C. Montoya, and Suzan van Dijk (Leiden: Brill, 2010), 283–307; see also Meredith K. Ray, "Letters and Lace: Arcangela Tarabotti and Convent Culture in Seicento Venice," in *Early Modern Women and Transnational Communities of Letters,* ed. J. Campbell and A. Larsen (Aldershot: Ashgate, 2009), 45–72.

6. See Westwater, "A Rediscovered Friendship," 67.

7. Little is known about Marie Meurdrac, including her dates of birth and death. *La Chymie charitable et facile, en faveur des Dames* (Paris, 1666) is dedicated to Marguerite Louise Susanne de Béthune, Comtesse de Guiche (see Luisa Tosi, "Marie Meurdrac: Paracelsian Chemist and Feminist," *Ambix* 48, no. 2 [2002]: 69–82).

8. English translations are from Tosi, "Marie Meurdrac" (71–72). As Tosi points out, Poullain de la Barre's statement (following the Cartesian method to its conclusion) that "mind has no sex" came seven years after that of Meurdrac (72).

9. Ibid., 72.

10. Meurdrac, *Chymie,* preface; trans. in Tosi, "Marie Meurdrac," 72.

11. On Margaret Cavendish, see, for example, Patricia Demers, *Women's Writing in English* (Toronto: University of Toronto Press, 2005), 225–234; Lisa T. Sarasohn, *The Natural Philosophy of Margaret Cavendish: Reason and Fancy during the Scientific Revolution* (Baltimore: Johns Hopkins University Press, 2010).

12. See Londa Schiebinger, "Women of Natural Knowledge," in *Cambridge Histories Online.*

13. See Paula Findlen, "Science as a Career in Enlightenment Italy: The Strategies of Laura Bassi," *Isis* 84 (1993): 441, 443. On Laura Bassi, see also Findlen, "The Scientist's Body: The Nature of a Woman Philosopher in Enlightenment Italy," in *The Faces of Nature in Enlightenment Europe,* ed. Gianna Pomata and Lorraine Daston (Berlin: Berliner Wissenschafs-Verlag, 2003), 211–236; Gabriella Berti Logan, "The Desire to Contribute: An Eighteenth-Century Italian Woman of Science," *American Historical Review* 99 (1994): 785–812; Marta Cavazza, "Laura Bassi e il suo gabinetto di fisica sperimentale: Realtà e mito," *Nuncius* 10 (1995): 715–753 and Cavazza, "Between Modesty and Spectacle: Women and Science in Eighteenth-Century Italy," in *Italy's Eighteenth Century: Gender and Culture in the Age of the Grand Tour,* ed. Paula Findlen (Redwood City, Calif.: Stanford University Press, 2009), 275–302; Beate Ceranski, *"Und sie fürchtet sich vor niemandem." Die Physikerin Laura Bassi (1711–1778)* (Frankfurt: Campus Verlag, 1996).

14. On Maria Gaetana Agnesi (1718–1799), see Giovanna Tilche, *Maria Gaetana Agnesi: La scienziata santa del '700* (Milan: Rizzoli, 1984); Massimo Mazzotti, "Maria Gaetana Agnesi: The Unusual Life and Mathematical Work of an Eighteenth-Century Woman," *Isis* 92 (2001): 657–83 and Mazzotti, *The World of Maria Gaetana Agnesi, Mathematician of God* (Baltimore: Johns Hopkins University Press, 2007). Paula Findlen speculates that Agnesi may have known and been influenced by the work of Lucrezia Marinella: see Findlen, *The Contest for Knowledge: Debates Over Women's Learning in Eighteenth-Century Italy,* ed. and trans. Rebecca Messbarger and Paula Findlen (Chicago: University of Chicago Press, 2005), n16. On Anna Morandi Manzolini (1714–1774), see Rebecca Messbarger, *The Lady Anatomist: The Life and Work of Anna Morandi Manzolini* (Chicago: University of Chicago Press, 2010).

Bibliography

Primary Sources

MANUSCRIPT

Archivio di Stato di Firenze, *Mediceo avanti il Principato,* f. 70, c. 87; f. 77, cc. 11, 85r-v, 85bis r-v, 129; f. 78, cc. 63, 181r-v, 181bis r-v, 302, 306; f. 125, cc. 10, 19, 71, 202, 243, 258; f. 128, c. 228.

Biblioteca Nazionale Centrale di Firenze, *Fondo Magliabechiano,* XI.

Biblioteca Nazionale Centrale di Firenze, *Fondo Galileiano,* b. 23 cc. 8–20.

Biblioteca Nazionale di Napoli. *Bartolomai Chioccarelli Illustres scriptores Regni Napoletani,* MS XIV A.28.

"Experimenti de la Ex[ellentissi]ma S[igno]ra Caterina da Furlj Matre de lo inllux[trissi]mo S[ignor] Giovanni de Medici." Private collection.

PRINT

Agrippa, Henricus Cornelius. *Declamation on the Nobility and Preeminence of the Female Sex.* Edited and translated by Albert J. Rabil Jr. Chicago: University of Chicago Press, 1996.

Alberti, Leon Battista. *The Family in Renaissance Florence.* Edited and translated by Renée Neu Watkins. Prospect Heights, Ill.: Waveland Press, 1994.

Aretino, Pietro. *Ragionamenti dialogo.* Edited by Giorgio Barberi Squarotti and
   Carlo Forni. Milan: Rizzoli, 1998.
Ariosto, Lodovico. *Herbolato di M. Lodovico Ariosto, nel quale figura Mastro Antonio
   Faentino, che parla della nobiltà dell'huomo, et dell'arte della Medicina cosa non
   meno utile che dilettevole, con eloquente stanze del medesimo novamente stampate.*
   Venice: Fratelli da Sabio, 1545.
————. *Satire.* Edited by Cesare Segre. Turin: Einaudi, 1987.
Bairo, Pietro. *Secreti medicinali.* Venice: Sansovino, 1561.
Barbarino, Francesco da. *Reggimento e costumi delle donne.* Edited by Giuseppe E.
   Sansone. Rome: Zauli, 1995.
Bernardi (Novacula), Andrea. *Cronache forlivesi di A.B. (Novacula) dal 1476-al 1517.*
   Edited by Giuseppe Mazzatinti. 2 vols. Bologna: R. Deputazione Storia Patria,
   1895; 1897.
Berni, Francesco. *Rime.* Edited by Giorgio Barberi Squarotti. Turin: Einaudi, 1969.
Biringoccio, Vanuccio. *Pirotechnia. Li diece libri della pirotechnia nelli quali si tratta
   non solo la diuersità delle minere, ma ancho quanto si ricerca alla prattica di esse: e
   di quanto s'appartiene all'arte della fusione ouer getto de metalli, e d'ogni altra cosa
   a questa somigliante.* Venice: Curtio Troiano Navò, 1558.
Boccaccio. *Famous Women.* Edited and translated by Virginia Brown. Cambridge:
   Harvard University Press, 2001.
Bonardo, Pietro. *La minera del mondo.* Venice, 1569.
Borro, Girolamo. *Dialogo sul flusso e riflusso del mare.* Lucca: Busdraghi, 1571.
Bronzini, Cristofano. *Della dignità e nobiltà delle Donne. Dialogo di Cristofano
   Bronzini d'Ancona. Diviso in Quattro Settimane, e ciascheduna in esse in Sei
   Giornate.* Florence: Zanobi Pignoni, 1625.
Buoni, Giacomo. *Terremoto dialogo.* Modena: Galdaldini, ca. 1571.
Campiglia, Maddelena. *Flori: A Pastoral Drama.* Edited by Virginia Cox and Lisa
   Sampson. Chicago: University of Chicago Press, 2004.
Capaccio, Giulio Cesare. *Illustrium mulierum et illustrium litteris virorum elogia.*
   Naples: apud F. Iacobum Carlinum et Constantinum Vitalem, 1608.
Celebrino, Eustachio. *Opera nova intitolata dificio delle ricette.* Venice: Giovanni
   Antonio et fratelli da Sabbio, 1528.
————. *Opera nuova piacevole . . . per far bella ciaschuna donna.* Venice: Bindoni,
   1551.
Cereta, Laura. *Collected Letters of a Renaissance Feminist.* Edited and translated by
   Diana Robin. Chicago: University of Chicago Press, 1997.
Cobelli, Leone. *Cronache Forlivesi di L.C. dalla fondazione della città sino all'anno
   1498.* Edited by G. Carducci and E. Frati, with notes by F. Guarini. Bologna,
   1874.
[Cortese, Isabella]. *I secreti della signora Isabella Cortese, ne' quali si contengono cose
   minerali, medicinali, arteficiose, & alchimiche, et molte de l'arte profumatoria,
   appartenenti a ogni gran Signora.* Venice: Bariletto, 1561.

———. *I secreti della signora Isabella Cortese, ne' quali si contengono cose minerali, medicinali, arteficiose, & alchimiche, et molte de l'arte profumatoria, appartenenti a ogni gran Signora.* Edited by Chicca Gagliardo. Milan: La vita felice, 1995.

Costa, Margherita. *Li buffoni, commedia ridicola.* Florence: Amadore Massi & Lorenzo Landi, 1641.

———. *Festa reale per balletto a cauallo, opera di Margherita Costa romana . . .* Paris: per Sebastiano Cramoisy, 1647.

Da Foligno, Francesco Iacobili. *Le conditioni del Cavaliero.* Rome: appresso Carlo Vullietti, 1606.

Del giardino de' poeti in lode del Serenissimo Re di Polonia . . . libro secondo. Venice: Fratelli Guerra, 1583. In *Viridiarium poetarum . . . in laudes Serenissimi atque potentissimi D.D. Stephani Regis Poloniae . . . in duos libros divisum,* edited by Ippolito Zucconello. Venice: ad Signum Hyppographi, 1583.

Delicado, Francisco. *Portrait of Lozana, the Lusty Andalusian Woman.* Translated by Bruno M. Damiani. Potomac, Md.: Scripta Humanistica, 1987.

———. *Ragionamento del Zoppino.* Edited by Mario Ciognani. Milan: Longanesi, 1969.

———. *Retrato de la Lozana Andaluza.* Venice, 1528.

della Chiesa, F. A. *Theatro delle donne letterate.* Modovi: Giflaudi e Rossi, 1620.

della Porta, Giambattista. *Magia naturalis.* Frankfurt: Andreae Wecheli heredes, 1591.

de Pizan, Christine. *The Book of the City of Ladies.* Translated by Earl Jeffrey Richards. New York: Persea Books, 1982.

Doglioni, Giovanni Niccolò. *Compendio Historico Universale di tutte le cose notabili già successe nel Mondo, dal principio della sua creatione sin'hora.* Venice, 1605.

———. *Vita della Sig.ra Modesta Pozzo de' Zorzi.* 1593. In *The Worth of Women,* edited by Virginia Cox. Chicago: University of Chicago Press, 1997.

Dolce, Lodovico. *Dialogo della institutione delle donne.* Venice: Giolito, 1545.

Ebreo, Leone. *Dialoghi d'amore.* Venice: Aldo Manuzio, 1541.

Erculiani, Camilla. *Lettere di philosophia naturale.* Kraków: Lazaro, 1584.

Erythraeus, Janus Nicius [Gian Vittorio Rossi]. *Pinacotecha imaginum illustrium doctrinae vel ingenii laude virorum, qui auctore superstite diem suum obierunt.* Wolfenbüttel: Jo. Christoph Meisner, 1729.

Falloppio [pseud.]. *Secreti diversi.* Venice: Zaltieri, 1588.

Fedele, Cassandra. *Cassandra Fedele: Letters and Orations.* Edited and translated by Diana Robin. Chicago: University of Chicago Press, 2000.

Ficino, Marsilio. *Three Books on Life: A Critical Edition and Translation with Introduction and Notes.* Edited by Carol V. Kaske and John R. Clark. Binghamton, N.Y.: The Renaissance Society of America, 1989.

Fioravanti, Leonardo. *Compendio dei secreti rationali.* Venice, 1564.

Firenzuola, Agnolo. *On the Beauty of Women.* Edited and translated by Konrad Eisenbichler and Jacqueline Murray. Philadelphia: University of Pennsylvania Press, 1992.

Fonte, Moderata [Modesta Pozzo]. *Floridoro. A Chivalric Romance.* Edited by Valeria Finucci. Translated by Julia Kisacky. Chicago: University of Chicago Press, 2007.

———. *La passione di Cristo descritta in ottava rima da Moderata Fonte.* Venice: D. and G.B. Guerra, 1582.

———. *La resurrettione di Gesù Christo nostro Signore . . . descritta in ottava rima.* Venice: G.D. Imberti, 1592.

———. *Le feste: Rappresentazione avanti al Serenissimo Prencipe di Venetia Nicolò da Ponte il giorno di S. Stefano, 1581.* Venice: D. and G. B. Guerra, 1582.

———. *Tredici canti del Floridoro.* Edited by Valeria Finucci. Modena: Mucchi, 1995.

———. *Il merito delle donne.* Venice, 1600.

———. *Il merito delle donne.* Edited by Adriana Chemello. Venice: Eidos, 1988.

———. *The Worth of Women.* Edited and translated by Virginia Cox. Chicago: University of Chicago Press, 1997.

Foresti da Bergamo, J. F. *De plurimus claris selectisque mulieribus.* Ferrara: Lorenzo Rosso da Valenza, 1497.

Galilei, Galileo. *Discorsi e dimostrazioni matematiche intorno a due nuove scienze attinenti alla meccanica et i movimenti locali . . . con un appendice del centro di gravità d'alcuni solidi.* Leiden: Elsevier, 1638.

———. *Lettera a Cristina di Lorena.* Edited by Franco Motta. Genova: Marietti, 2000.

———. *Le Opere di Galileo Galilei.* Edizione Nazionale. 20 vols. Edited by Antonio Favaro. Florence: Barbera, 1890–1909; repr. 1929–1939 and 1964–1966.

———. *Rime.* Edited by Antonio Marzo. Rome: Salerno, 2001.

———. *Scritti letterari.* Edited by Alberto Chiari. Florence: Le Monnier, 1970.

Galilei, [suora] Maria Celeste. *Lettere al padre.* Edited by Bruno Basile. Rome: Salerno, 2002.

Garzoni, Tomaso. *La piazza universale di tutte le professioni del mondo.* Edited by Giovanni Battista Bronzini, with Pina de Meo and Luciano Carcereri. Florence: Olschki, 1996.

Gozze, Nicolò Vito di (Gučetić, Nikola Vitov). *Dialogo della bellezza/Dijalog o Ljepoti. Dialogo d'Amore/Dijalog o Ljubavi.* Edited and translated by Ljerka Schiffler. Zagreb: Matica Hrvatska, 2008.

———. *Discorsi di M. Nicolò Vito di Gozze . . . Sopra le Metheore d'Aristotele.* Venice: Francesco Ziletti, 1584.

Guasco, Annibal. *Discourse to Lady Lavinia His Daughter.* Edited and translated by Peggy Osborn. Chicago: University of Chicago Press, 2003.

Iasolino, Giulio. *De remedii naturali che sono nell'isola di Pithecusa: hoggi detta Ischia.* Libri due. Naples: appresso Gioseppe Cacchik, 1588.

Lando, Ortensio. *Lettere di molte valorose donne.* Venice: Giolito, 1548.

Machiavelli, Niccolò. *Arte della guerra e scritti politici minori.* Edited by S. Bertelli. Milan, 1961.

———. *Discorsi sopra la prima deca di Tito Livio.* Ed. Francesco Bausi, Edizione Nazionale delle Opere di Niccolò Machiavelli. 2 vols. Rome: Salerno Editrice, 2001.

———. *Le istorie fiorentine.* Edited by L. Passerini and G. Milanesi. Florence and Rome: Tipografia Cenniniana, 1874.

———. *Il principe.* Edited by Luigi Firpo. Turin: Einaudi, 1984.

Maier, Michael. *L'Atalanta fugiens.* Oppenheim: Johann Theodori de Bry, 1617.

Manfredi, Muzio. *Lettere brevissime.* Venice: Maglietti, 1606.

———. *Per donne romane. Rime di diversi raccolte e dedicate al Signor Giacomo Buoncompagni.* Bologna: Alessandro Benaco, 1575.

Manuzio, Aldo. *Lettere volgari.* Rome: Santi e Comp., 1592.

Marinella, Lucrezia. *Arcadia felice.* Edited by Françoise Lavocat. Florence: Olschki, 1998.

———. *Della Nobiltà et eccellenza delle donne co' difetti e mancamenti de gli huomini.* Venice: G.B. Ciotti, 1600.

———. *Enrico; or, Byzantium Conquered. A Heroic Poem.* Edited and translated by Maria Galli Stampino. Chicago: University of Chicago Press, 2009.

———. *Exhortations to Women and to Others If They Please.* Edited by Laura Benedetti. Toronto: Centre for Reformation and Renaissance Studies, 2012.

———. *L'Enrico, overo Bizanzio acquistato.* Venice: Ghirardo Imberti, 1635.

———. *La nobiltà et eccellenza delle donne, co' diffetti et mancamenti degli huomini.* Venice: Ciotti, 1601.

———. *The Nobility and Excellence of Women and the Defects and Vices of Men.* Edited by Letizia Panizza. Translated by Anne Dunhill. Chicago: University of Chicago Press, 1999.

Marinello, Giovanni. *De gli ornamenti delle donne.* Venice: Valgrisio, 1574.

———. *Delle medicine partenenti alle infermità delle donne.* Venice: Valgrisio, 1574.

Marino, Giambattista. *Epistolario.* Edited by Angelo Borzelli. Bari: Laterza, 1911–1912.

———. *L'Adone.* Edited by G. Pozzi. Milan, 1988.

Maylender, Michele. *Storia delle Accademie d'Italia.* 5 vols. Bologna: Forni, 1976.

Medicina per le donne nel Cinquecento. Testi di Giovanni Marinello e di Girolamo Mercurio. Edited by M. L. Altieri Biagi et al. Turin: Strenna UTET, 1992.

Menochio, Giacomo. *Consiliorum sive Responsorum.* 14 vols. Frankfurt: Seyler, 1676.

Mercurio, Girolamo. *La Commare.* Venice, 1596.

Meurdrac, Marie. *La chymie charitable et facile, en faveur des dames.* Paris, 1666.

Niccolini, Sister Giustina. *The Chronicle of Le Murate.* Edited and translated by Saundra Weddle. Toronto: Iter Inc., Centre for Reformation and Renaissance Studies, 2011.

Nifo, Agostino. *Philosopho contra il falso giudicio che debba vegnir il diluvio . . .* Venice: Da Sabio, 1521.

Palmieri, Matteo. *La vita civile*. Edited by Gino Belloni. Florence: Sansoni, 1982.

Paracelsus. *Paracelsus Theophrastus Bombastus von Hohenheim, 1493–1541: Essential Theoretical Writings*. Edited and translated by Andrew Weeks. Leiden: Brill, 2008.

———. *Selected Writings*. Edited by Jolande Jacobi. Translated by Norbert Guterman. New York: Pantheon Bollingen Series 28, 1958.

Passi, Giuseppe. *I donneschi difetti*. Venice, 1599.

Petrarca, Francesco. *Secretum. Il mio segreto*. Edited by Enrico Fenzi. Milan: Mursia, 1992.

Piccolomini, Alessandro. *Della sfera del mondo*. Venice: al segno del pozzo, 1552.

———. *Instrumento della filosofia*. Venice: Giorgio de' Cavalli, 1565.

———. *La economica di Xenofonte, tradotta da lingua greca in lingua Toscana . . .* Venice: Comin da Trino, 1540.

———. *La Raffaella, dialogo della bella creanza delle donne*. Edited by Mario Cicognani. Milan: Longanesi, 1969.

———. *Raffaella of Master Alessandro Piccolomini*. Translated by John Nevison. Glasgow: Robert MacLehose. 1968.

Pliny the Elder. *Naturalis Historia*. 10 vols. Cambridge, Mass.: Harvard University Press, 1949–1962.

Quadrio, Francesco Saverio. *Della storia e della ragione d'ogni poesia*. 7 vols. Bologna: Pisarri, 1739–1752.

Quattrami, Evangelista. *La vera dichiaratione di tutte le metafore, similitudini, & enimmi de gl'antichi filosofi alchimisti*. Rome: Appresso Vincentio Accolti, 1587.

Ricettario Bardi. Cosmesi e tecnica artistica nella Firenze medicea. Edited by Antonio P. Torresi. Intro. Franco Cardini. Ferrara: Liberty House di Lucio Scardino, 1994.

Ricettario di bellezza di Caterina Sforza. Edited by Luigi Pescasio. Castiglione delle Stiviere, 1971.

Ricettario galante. Del principio del secolo XVI. Edited by Olindo Guerrini. Bologna: Gaetano Romagnoli, 1883.

Rosselli, Stefano. *Mes secrets: A Florence au temps des Medicis, 1593: patisserie, parfumerie, medecine*. Edited by Roderigo de Zayas. Paris: J.-M. Place, 1996.

Rossello, Timoteo. *Della summa de' secreti universali in ogni materia*. Venice: Bariletto, 1561.

Rosetti, Giovanventura. *Notandissimi secreti de l'arte profumatoria*. Venice: Neri Pozza, 1973.

Rossi, Gian Vittorio. See Erythraeus, Janus Nicius.

Ruscelli, Girolamo. *Secreti nuovi*. Venice: Heredi di M. Sessa, 1567.

[Ruscelli, Girolamo?]. *La prima parte de' secreti del reverendo donno Alessio Piemontese*. Pesaro: Appresso gli Heredi di Bartolomeo Cesano, 1562.

Sanazzaro. *Arcadia*. Livorno: G.T. Masi, 1781.

Sarrocchi, Margherita. *La Scanderbeide, poema heroico*. Rome: Lepido Facij, 1606.

———. *Scanderbeide.* Edited and translated by Rinaldina Russell. Chicago: University of Chicago Press, 2006.

Sereno, Bartolomo. *Trattato de l'uso della lancia a cavallo. Del combattere a piedi alla sbarra et alle inventioni cavalleresche.* Naples: Nucci, 1610.

[Sforza, Caterina]. *Experimenti de la Ex.ma S.ra Caterina da Furlj Matre, matre de lo inllux.mo signor Giovanni de Medici, copiati dagli autografi di lei dal Conte Lucantonio Cuppano; pubblicati da Pier Desiderio Pasolini.* Imola: Ignazio Galeati e Figlio, 1894.

Sforza, Isabella [Ortensio Lando]. *Della vera tranquillità dell'animo.* Venice: Manuzio, 1544.

Tarabotti, Arcangela. *Lettere familiari e di complimenti.* Edited and translated by Meredith Ray and Lynn Westwater. Turin: Rosenberg e Sellier, 2005.

———. *Letters Familiar and Formal.* Edited and translated by Meredith K. Ray and Lynn Lara Westwater. Toronto: Centre for Reformation and Renaissance Studies, 2012.

Tasso, Torquato. *Discorsi del poema eroico.* In *Prose.* Edited by Ettore Mazzali. Milan: Ricciardi, 1995.

———. *Rime.* Edited by Bruno Basile. Torino: Einaudi 1994.

Textor, Ravisius. *Theatrum poeticarum et historicum, sive Officina Io. Ravisii Textoris.* Basileae: Leonhardi Ostenij, 1637.

Tiraboschi, Girolamo. *Storia della letteratura italiana.* 9 vols. Florence: Molini Landi, 1805–1812.

Toppi, Nicolò, *Biblioteca napoletana, et apparato a gli huomini illustri in lettere di Napoli . . . Dalle loro origini, per tutto l'anno 1678.* Naples: Antonio Bulifon, 1678.

Zapata, Giovanni Battista. *Maravigliosi secreti di medicina e chirurgia.* Turin: appresso gli heredi del Bevilacqua, 1586.

Secondary Sources

Agrimi, Jole, and Chiara Crisciani. "Per una ricerca su '*experimentum-experimenta*': Riflessione epistemologica e tradizione medica (secoli XIII–XIV)." In *Presenza del lessico greco e latino nelle lingue contemporanee: Ciclo di lezioni tenute all'Universita di Macerata nell'a.a. 1987/88,* edited by Pietro Janni and Innocenzo Mazzini, 9–49. Macerata: Universita degli studi, 1990.

Ait, Ivana. *Tra scienza e mercato: gli speziali a Roma nel tardo medioevo.* Rome: Istituto Nazionale di Studi Romani, 1996.

Albala, Ken. *Eating Right in the Renaissance.* Berkeley: University of California Press, 2002.

Allen, Sally G., and Joanna Hubbs. "Outrunning Atalanta: Feminine Destiny in Alchemical Transmutation." *Signs: Journal of Women in Culture and Society* 6, no. 1 (1980): 210–219.

Altman, Janet. *Epistolarity: Approaches to a Form.* Columbus: Ohio State University, 1982.

Appendini, Franceso Maria. *Notizie istorico-critiche sulle antichità storia e letteratura de' Ragusei.* Ragusa: Antonio Martecchini, 1803.

Aquilecchia, Giovanni. "Da Bruno a Marino: Postilla all'*Adone* X, 45." *Studi secenteschi* 20 1979 (1980): 89–95.

Archer, Jayne Elizabeth. "Women and Chemistry in Early Modern England: The Manuscript Receipt Book (c. 1616) of Sarah Wigges." In *Gender and Scientific Discourse in Early Modern Culture,* edited by Kathleen P. Long, 191–216. Aldershot: Ashgate, 2010.

Armour, Peter. "Galileo and the Crisis in Italian Literature of the Early Seicento." In *Collected Essays on Italian Language and Literature Presented to Kathleen Speight,* edited by Giovanni Acquilecchia, Stephen N. Cristea, and Sheila Ralphs, 144–169. Manchester: Manchester University Press, 1971.

Aromatico, Andrea. *Alchemy: The Great Secret.* New York: Harry N. Abrams, 2000.

Arrizabalaga, John, John Henderson, and Roger French, eds. *The Great Pox: The French Disease in Renaissance Europe.* New Haven, Conn.: Yale University Press, 1997.

Azzolini, Monica. *The Duke and the Stars: Astrology and Politics in Renaissance Milan.* Cambridge, Mass.: Harvard University Press, 2013.

Balbiani, Laura. "La ricezione della 'Magia Naturalis' di Giovan Battista Della Porta. Cultura e scienze dall'Italia all'Europa." *Bruniana e Campanelliana* 5 (1999): 277–303.

Baldini, Ugo. "The Roman Inquisition's Condemnation of Astrology: Antecedents, Reasons, and Consequences." In *Church Censorship and Culture in Early Modern Italy,* edited by Gigiola Fragnito, 79–110. Cambridge: Cambridge University Press, 2001.

Baldini, Ugo, and Pier Daniele Napolitani. "Per una biografia di Luca Valerio: Fonti edite e inedite per una riscostruzione della sua carriera scientifica." *Bollettino di storia delle scienze matematiche/Unione matematica italiana* 11, no. 2 (1991): 3–157.

Ball, Phillip. *The Devil's Doctor: Paracelsus and the World of Renaissance Magic and Science.* New York: Farrar, Straus and Giroux, 2006.

Baschet, Armand, and Felix Feuillet de Conches. *Les femmes blondes selon les peintres de L'école de Venise.* Paris: Aubry, 1885.

Bausi, Francesco. "Machiavelli and Caterina Sforza." *Archivio storico italiano* (1991), disp. IV: 887–892.

Bayer, Penny. "From Kitchen Hearth to Learned Paracelsianism: Women's Alchemical Activities in the Renaissance." In *"Mystical Metal of Gold": Essays on Alchemy and Renaissance Culture,* edited by Stanton Linden, 365–386. Brooklyn: AMS Press, 2007.

———. "Madame de la Martinville, Quercitan's Daughter and the Philosopher's Stone: Manuscript Representations of Women Alchemists." In *Women and Scientific Discourse in Early Modern Europe,* edited by Kathleen Long, 165–182. Aldershot: Ashgate, 2010.

Bell, Rudolph. *How to Do It: Guides to Good Living for Renaissance Italians.* Chicago: University of Chicago Press, 2000.

Belloni, G. "Conoscenza magica e ricerca scientifica in Giovanni Battista Della Porta." In *Della Porta. Criptologia. Edizione, nota biografica, traduzione,* edited by G. Belloni, 43–101. Rome: 1982.

Bellucci, Novella. "Lettere di molte valorose donne . . . e di alcune pettegolette, ovvero: di un libro di lettere di Ortensio Lando." In *Le carte messaggieri: Retorica e modelli di communicazione epistolare: per un indice dei libri di lettere del Cinquecento,* edited by Amedeo Quondam, 255–276. Rome: Bulzoni, 1981.

Benson, Pamela. *The Invention of the Renaissance Woman.* University Park: Pennsylvania State University Press, 1992.

Benton, John F. "Trotula, Women's Problems and the Professionalization of Medicine in the Middle Ages." *Bulletin of the History of Medicine* 59 (1985): 30–53.

Berti, Luciano. *Il principe dello studiolo: Francesco I dei Medici e la fine del Rinascimento fiorentino.* Florence: Edam, 1967.

Besala, Jerzy. *Stefan Batory.* Warsaw, 1992.

Biagioli, Mario. *Galileo Courtier: The Practice of Science in the Culture of Absolutism.* Chicago: University of Chicago Press, 1993.

Bilinski, Bronislaw. "The Italian Renaissance in Poland." *Italian Books and Periodicals* 25, no. 2 (1982): 138–145.

Blair, Ann. "The *Problemata* as a Natural Philosophical Genre." In Grafton and Siraisi, *Natural Particulars,* 171–204.

———. *Too Much to Know: Managing Scholarly Information Before the Modern Age.* New Haven, Conn.: Yale University Press, 2010.

Boccolini, Claudia. *Flora Zuzzeri in Ancona.* Ancona: Consiglio Regionale delle Marche, 2007.

Bogucka, Maria. *Nicholas Copernicus: The Country and Times.* Wrocław: The Ossolinski State Publishing House, 1973.

———. *Women in Early Modern Polish Society, against the European Background.* Aldershot: Ashgate, 2004.

Borrelli, Arianna. "Giovan Battista Della Porta's Neapolitan Magic and His Humanistic Meteorology." In *Variantology 5: Neapolitan Affairs. On Deep Time Relations of Arts, Sciences and Technologies,* edited by Siegfried Zielinski and Eckhard Furlus in cooperation with Daniel Irrgang, 103–130. Cologne: Verlag Der Buchhandlung Walter König, 2011.

Borzelli, Angelo. *Note intorno a Margarita Sarrocchi ed al suo poema La Scanderbeide.* Naples: Artigianelli, 1935.

Braschi, Angelo. *Caterina Sforza*. Rocca San Casciano, 1965.

Breisach, Ernest. *Caterina Sforza. A Renaissance Virago*. Chicago: University of Chicago Press, 1967.

Brennan, Michael G. "The Medicean Dukes of Florence and Friar Lawrence's 'Distilling Liquor' (*Romeo and Juliet,* IV.i.94)." *Notes and Queries* 38, no. 4 (1991): 473–476.

Broomhall, Susan. *Women's Medical Work in Early Modern France*. Manchester: Manchester University Press, 2004.

Brzezínksa, Anna. "Accusations of Love Magic in the Renaissance Courtly Culture of the Polish-Lithuanian Commonwealth." *East Central Europe* 20–23, no. 1 (1993–1994): 117–140.

Buchwald-Pelcowa, Paulina. "Il libro italiano in Polonia nel periodo del Rinascimento." In *Il Rinascimento in Polonia: atti dei colloqui italo-polacchi 1989–1992*, edited by Jolanta Zurawaska, 427–453. Naples: Bibliopolis, 1994.

Burke, Peter. *The Fortunes of the Courtier: The European Reception of Castiglione's Cortegiano*. Cambridge, UK: Blackwell, 1995.

———. "The Uses of Literacy in Early Modern Italy." In *The Historical Anthropology of Early Modern Italy: Essays on Perception and Communication,* edited by Peter Burke. Cambridge: Cambridge University Press, 1987.

Burrièl, Antonio. *Vita di Caterina Sforza Riario Contessa d'Imola, e Signora di Forlí, descritta in tre libri dall'Abate A.B. sacerdote spagnolo e dedicata all'Illustrissimo Senato di Forlí*. 3 vols. Bologna, 1795.

Butters, Suzanne. *The Triumph of Vulcan: Sculptors' Tools, Porphyry, and the Prince in Ducal Florence*. Florence: Leo S. Olschki, 1996.

Camerota, Michele. *Galileo Galilei e la cultura scientifica nell'età della Controriforma*. Rome: Salerno Editrice: 2004.

Camillo, Elena. "Ancora su Donno Alessio Piemontese: Il libro di segreti tra popolarità e accademia." *Giornale storico della letteratura italiana* 162 [1985]: 539–553.

Camporesi, Piero. "Speziali e ciarlatani." In *Cultura popolare dell'Emilia Romagna*, 139–159. Milan: Silvana editoriale, 1981.

Cantù, Cesare. *Gli eretici d'Italia*. Turin: Unione Tipografico, 1861.

Carbonelli, Giovanni. *Sulle fonti storiche della chimica e dell'alchimia in Italia*. Rome: Istituto Nazionale Medico-Farmacologico, 1925.

Carinci, Eleonora. "Una lettera autografa inedita di Moderata Fonte al granduca di Toscana Francesco I." *Critica del testo* 5, no. 3 (2002): 667–671.

———. "Una 'speziala' padovana: *Lettere di philosophia naturale di Camilla Erculiana* (1584)." *Italian Studies* 2 (2013): 202–229.

Carra, G. "Speziali e spezierie nella Mantova dei Gonzaga." *Civiltà mantovana* 12 (1978): 245–278.

Carson, Rachel. *Silent Spring*. Boston: Houghton Mifflin, 1962.

Castagnola, Raffaela. "Alchimia fra scienza e gioco." In *Passare il Tempo: La letteratura del gioco e dell'intrattenimento dal XII al XVI secolo,* edited by E. Malato, 511–527. Rome: Salerno, 1993.

Cavallo, Sandra, and Tessa Storey. *Healthy Living in Late Renaissance Italy.* Oxford: Oxford University Press, 2013.

Cavazza, Marta. "Between Modesty and Spectacle: Women and Science in Eighteenth-Century Italy." In *Italy's Eighteenth Century: Gender and Culture in the Age of the Grand Tour,* edited by Paula Findlen, 275–302. Redwood City, Calif.: Stanford University Press, 2009.

———. "Laura Bassi e il suo gabinetto di fisica sperimentale: Realtà e mito." *Nuncius* 10 (1995): 715–753.

Ceranski, Beate. *"Und sie fürchtet sich vor niemandem." Die Physikierin Laura Bassi (1711–1778).* Frankfurt: Campus Verlag, 1996.

Cerreta, Florindo. *Alessandro Piccolomini: Letterato e filosofo Senese del cinquecento.* Siena: Accademia Senese degli Intronati, 1960.

Chambers, D. S. "Spas in the Italian Renaissance." In *Reconsidering the Renaissance,* edited by M. Di Cesare, 3–27. Binghamton, N.Y.: Medieval and Renaissance Texts and Studies, 1992.

Chater, James. "Bianca Cappella and Music." In *Renaissance Studies in Honor of Craig Smyth,* vol. 1, edited by Andrew Morrough, Fiora Superbi Gioffredi, Piero Morselli, and Eve Borsook, 569–572. Florence: Giunti Barbèra, 1985.

Chemello, Adriana. "La donna, il modello, l'immaginario: Moderata Fonte e Lucrezia Marinella." In *Nel cerchio della luna: Figura di donna in alcuni testi del XVI secolo,* edited by Marina Zancan, 95–170. Venice: Marsilio, 1983.

Cherchi, Paolo. *Enciclopedismo e politica della riscrittura: Tommaso Garzoni.* Pisa: Pacini, 1981.

———. *Polimatia di riuso: Mezzo secolo di plagio (1539–1589).* Rome: Bulzoni, 1988.

Cherchi, Paolo, and Beatrice Collina. "Invito alla lettura della 'Piazza.'" In *Tomaso Garzoni. La piazza universale di tutte le professioni del mondo,* XI–LXVI, edited by Paolo Cherchi and Beatrice Collina. Turin: Einaudi, 1996.

Ciasca, Raffaele. *L'arte dei medici e speziali nella storia e nel commercio fiorentino dal secolo XII al XV.* Florence: Olschki, 1927.

Clough, Cecil H. "The Sources for the Biography of Caterina Sforza and for the History of Her State during Her Rule, with Some Hitherto Unpublished Letters Illustrative of Her Chancery Archives." *Atti e Memorie, Deputazione di storia patria per le provincie di Romagna* 15–16 (1963): 57–143.

Clubb, Louise. *Giambattista Della Porta, Dramatist.* Princeton, N.J.: Princeton University Press, 1965.

Cockram, Sarah. "Epistolary Masks: Self-Presentation and Dissimulation in the Letters of Isabella d'Este." *Italian Studies* 64, no. 1 (2009): 20–37.

Cohen, Mark R., ed. *Autobiography of a Seventeenth-Century Venetian Rabbi: Leon Modena's Life of Judah.* Princeton, N.J.: Princeton University Press, 1988.

Collina, Beatrice. "Moderata Fonte e il *Merito delle donne.*" *Annali di italianistica* (1989): 142–164.

Colombo, Angelo. "Il principe celebrato: Autografi poetici di Tomaso Stigliani e Margherita Sarrocchi." *Philo-logica: Rassegna di analisi linguistica ed ironia culturale* 1, no. 1 (1992): 7–29.

Comelli, Giovanni. *Ricettario di bellezza di Eustachio Celebrino, medico e incisore del Cinquecento.* Florence, 1960.

Conticelli, Valentina. *Guardaroba di cose rare et preziose: lo studiolo de Francesco I de' Medici: arte, storia, e significati.* Lugano: Lumieres Internationales, 2007.

———. *L'alchimia e le arti: la fonderia degli Uffizzi da laboratorio a stanze delle meraviglie.* Livorno: Sillabe, 2012.

———. "Lo studiolo di Francesco I e l'alchimia: nuovi contribute storici e iconologici, con un carteggio in appendice (1563–1581)." In *L'art de la Renaissance entre science et magie,* edited by Philippe Morel, 207–268. Académie de France a Rome, Villa Medici: Somogy Editions d'Art, 2006.

Cook, Harold. *Matters of Exchange: Commerce, Medicine, and Science in the Dutch Golden Age.* New Haven, Conn.: Yale University Press, 2007.

Corsi, Dinora, Lada Hordynsky-Caillat, and Odile Redon. "Les secrés des dames, tradition, traductions." *Médiévales* 14 (1988): 47–57.

Coudert, Allison. *Alchemy: The Philosopher's Stone.* London: Wildwood House, 1980.

Covoni, P. F. *Don Antonio de' Medici al Casino de San Marco.* Florence: Tipografia cooperativa, 1893.

Cox, Virginia. "Fiction, 1560–1650." In *A History of Women Writing in Italy,* edited by Letizia Panizza and Sharon Woods, 57–61. Cambridge: Cambridge University Press, 2000.

———. "Fonte, Moderata." Available at www.lib.uchicago.edu.proxy.uchicago.edu /efts/IWW/BIOS/ A0016.html.

———. "Members, Muses, Mascots: Women and Italian Academies." (Forthcoming).

———. *The Prodigious Muse: Women's Writing in Counter-Reformation Italy.* Baltimore: Johns Hopkins University Press, 2011.

———. *Renaissance Dialogue: Literary Dialogue in its Social and Political Contexts, Castiglione to Galileo.* Cambridge: Cambridge University Press, 2008.

———. "The Single Self: Feminist Thought and the Marriage Market in Early Modern Venice." *Renaissance Quarterly* 48, no. 3 (Autumn 1995): 513–581.

———. "Women as Readers and Writers of Chivalric Poetry in Early Modern Italy." In *Sguardi sull'Italia: Miscellanea dedicata a Franceso Villari,* edited by Gino Bedoni et al., 134–145. Leeds: Society for Italian Studies, 1997.

———. *Women's Writing in Italy, 1400–1650*. Baltimore: Johns Hopkins University Press, 2008.

Cox, Virginia, and Lisa Sampson, eds. *Flori: A Pastoral Drama*. Chicago: University of Chicago Press, 2004.

Crisciani, Chiara. "From the Laboratory to the Library: Alchemy according to Guglielmo Fabri," in Grafton and Siraisi, *Natural Particulars*, 295–319.

Crisciani, Chiara, and Michela Pereira. *L'arte del sole e della luna. Alchimia e filosofia nel medioevo*. Spoleto: Centro italiano di studi sull'alto medioevo, 1996.

Daenens, Francine. "Donne valorose, eretiche, finte sante: note sull'antologia giolitina del 1548." In *Per lettera: La scrittura epistolare femminile*, edited by Gabriella Zarri, 181–207. Rome: Viella, 1999.

———. "Superiore perchè inferiore: il paradosso della superiorità delle donne in alcuni trattati italiani del Cinquecento." In *Transgressione tragica e norma domestica: esemplari di tipologie femminili dalla letteratura europea*, edited by Vanna Gentili, 11–50. Rome: Edizioni di storia e letteratura, 1983.

da Gorzano, Carlo Coppi. "Il Conte Lucantonio Coppi detto Cuppano, ultimo condottiere delle Bande Nere e dimenticato Governatore Generale di Piombino (1507–1557). *Rivista araldica* 3 (1960): 87–105.

Daston, Lorraine, and Katharine Park. *Wonders and the Order of Nature, 1150–1750*. New York: Zone Books, 2001.

Debus, Allen. *The Chemical Philosophy: Paracelsian Science and Medicine in the Sixteenth and Seventeenth Centuries*. 2 vols. New York: Science History Publications, 1977.

De Jong, H. M. E. *Michael Maier's Atalanta fugiens. Sources of an Alchemical Book of Emblems*. Leiden: Brill, 1969.

Demers, Patricia. *Women's Writing in English*. Toronto: University of Toronto Press, 2005.

de Roover, Raymond. *The Rise and Decline of the Medici Bank, 1397–1494*. Washington, D.C.: Beard Books, 1999.

De Vivo, Filippo. "Pharmacies as Centres of Communication in Early Modern Venice." *Renaissance Studies* 21, no. 4 (2007): 505–521.

de Vries, Joyce. "Casting Her Widowhood: The Contemporary and Posthumous Portraits of Caterina Sforza." In *Widowhood and Visual Culture in Early Modern Europe*, edited by Allison Levy. Aldershot: Ashgate, 2003.

———. *Caterina Sforza and the Art of Appearances: Gender, Art and Culture in Early Modern Italy*. Aldershot: Ashgate, 2010.

———. "Caterina Sforza's Portrait Medals: Power, Gender, and Representation in the Italian Renaissance Court." *Women's Art Journal* 24, no. 1 (2003): 23–28.

DeVun, Leah. *Prophecy, Alchemy and the End of Time: John of Rupecissa in the Late Middle Ages*. New York: Columbia University Press, 2009

Di Leonardo, Micaela, ed. *Gender at the Crossroads of Knowledge: Feminist Anthropology in the Postmodern Era*. Berkeley: University of California Press, 1991.

di Meo, Michelle, and Sarah Pennell, eds. *Reading and Writing Recipe Books, 1500–1800*. Manchester, UK: Manchester University Press, 2013.

Dionisotti, Carlo. "Letteratura italiana all'epoca del Concilio di Trento." In *Geografia e storia della letteratura italiana*. Turin: Einaudi, 1967.

Dixon, Laurinda S. *Perilous Chastity: Women and Illness in Pre-Enlightenment Art and Medicine*. Ithaca, N.Y. and London: Cornell University Press, 1995.

Donnelly, John. "Antonio Possevino, S.J. as Papal Mediator between Emperor Rudolf II and King Stephan Bathory." *Archivium Historicum Societatis Iesu* 69 (2000): 3–60.

Eamon, William. "Alchemy in Popular Culture: Leonardo Fioravanti and the Search for the Philosopher's Stone." *Early Science and Medicine* 5, no. 2 (2000): 196–213.

———. "Arcana Disclosed: The Advent of Printing, the Books of Secrets Tradition, and the Development of Experimental Science in the Sixteenth Century." *History of Science* 22 (1984): 111–150.

———. "Court, Academy, and Printing House: Patronage and Scientific Careers in Late Renaissance Italy." In *Patronage and Institutions: Science, Technology and Medicine at the European Court*, edited by Bruce T. Moran, 25–50. Rochester, N.Y.: Boydell Press, 1991.

———. "How to Read a Book of Secrets." In *Secrets and Knowledge in Medicine and Science 1500–1800*, edited by Elaine Leong and Alisha Rankin, 23–46. Aldershot: Ashgate, 2011.

———. "Masters of Fire: Italian Alchemists in the Court of Philip II of Spain." In *Chymia: Science and Nature in Early Modern Europe (1450–1750)*, edited by Miguel Lopez Perez and Didier Kahn, 138–156. Cambridge: Cambridge Scholars, 2010.

———. *The Professor of Secrets: Mystery, Medicine, and Alchemy in Renaissance Italy*. Washington, D.C.: National Geographic, 2010.

———. "Science and Popular Culture in Sixteenth-Century Italy: The 'Professors of Secrets' and Their Books." *Sixteenth Century Journal* 16 (1985): 471–485.

———. *Science and the Secrets of Nature: Books of Secrets in Medieval and Early Modern Culture*. Princeton, N.J.: Princeton University Press, 1994.

———. "The *Secreti* of Alexis of Piedmont." *Res Publica Litterarum* 2 (1979): 43–55.

———. "Technology as Magic in the Late Middle Ages and the Renaissance." *Janus* 70, nos. 3–4 (1983): 171–203.

Eamon, William, and Françoise Peheau, "The Accademia Segreta of Girolamo Ruscelli: A Sixteenth-Century Scientific Society." *Isis* 75 (1984): 327–342.

Eliade, Mircea. *The Forge and the Crucible*. New York: Harper and Row, 1971.

Erdmann, Axel, ed. *My Gracious Silence: Women in the Mirror of 16th Century Printing in Western Europe.* Lucerne: Gilhofer und Ranschburg, 1999.

Erspammer, E. "I Secreti di Alessio Piemontese e la rivoluzione scientifica." In *Du Pô à la Garonne: Recherches sur les échanges entre Italie e la France à la Renaissance,* edited by A. Fiorato, 376–377. Agen: Centre Matteo Bandello d'Agen, 1990.

Evans, Robert J. W. *Rudolf II and His World: A Study in Intellectual History, 1576–1612.* London: Thames and Hudson, 1997.

Evans, R. J. W., and Alexander Marr, eds. *Curiosity and Wonder from the Renaissance to the Enlightenment.* Aldershot: Ashgate, 2006.

Fahy, Conor. "Three Early Renaissance Treatises on Women." *Italian Studies* (1956): 30–55.

———. "Women and Italian Cinquecento Literary Academies." In *Women In Italian Renaissance Culture and Society,* edited by Letizia Panizza, 438–452. Oxford: European Humanities Research Center, 2000.

Fališevac, Dunja. "Women in Croatian Literary Culture, 16th to 18th Centuries." In *A History of Central European Women's Writing,* edited by Celia Hawkesworth, 33–40. New York: Palgrave Macmillan, 2001.

Favaro, Antonio. *Amici e corrispondenti di Galileo.* Venice: Officine Grafiche C. Ferrari, 1894.

———. *Galileo Galilei e lo Studio di Padova.* Florence: Le Monnier, 1883.

———. *La libreria di Galileo Galilei.* Rome: La tipografia delle scienze matematiche e fisiche, 1887.

Ferguson, John. *Bibliographic Notes on Histories of Inventions and Books of Secrets.* London: Holland Press, 1959.

Ferguson, Margaret W. *Dido's Daughters: Literacy, Gender, and Empire in Early Modern England and France.* Chicago: University of Chicago Press, 2003.

Findlen, Paula. *Possessing Nature: Museums, Collecting, and Scientific Culture in Early Modern Italy.* Berkeley: University of California Press, 1994.

———. "Science as a Career in Enlightenment Italy: The Strategies of Laura Bassi." *Isis* 84 (1993): 441–469.

———. "The Scientist's Body: The Nature of a Woman Philosopher in Enlightenment Italy." In *The Faces of Nature in Enlightenment Europe,* edited by Gianna Pomata and Lorraine Daston, 211–236. Berlin: Berliner Wissenschafs-Verlag, 2003.

Findlen, Paula, and Rebecca Messbarger, eds. and trans. *The Contest for Knowledge: Debates over Women's Learning in Eighteenth-Century Italy.* Chicago: University of Chicago Press, 2005.

Finocchiaro, Maurice A. *Retrying Galileo, 1633–1992.* Berkeley: University of California Press, 2005.

Finucci, Valeria. *The Prince's Body: Vincenzo Gonzaga and Renaissance Medicine.* Cambridge, Mass.: Harvard University Press, 2014.

Fissell, Mary. *Vernacular Bodies: The Politics of Reproduction in Early Modern England*. Oxford: Oxford University Press, 2004.

Fiszman, Samuel. *The Polish Renaissance in Its European Context*. Bloomington: Indiana University Press, 1988.

Fornaciari, Gino, Valentina Giuffra, Ezio Perroglio, and Raffaella Bianucci. "Malaria Was the Killer of Francesco I de' Medici (1531–1587)." *American Journal of Medicine* 123, no. 6 (2010): 568–569.

Forshaw, Peter J. "Marsilio Ficino and the Chemical Art." In *Laus Platonici Philosophi: Marsilio Ficino and His Influence*, edited by Stephen Clucas, Peter J. Forshaw, and Valery Rees, 198–249. Leiden: Brill, 2011.

Foschi, Umberto. "Fantasia e superstizione delle ricette di Caterina Sforza." *Bollettino economico. Organo ufficile della C.C.I.A.A. di Ravenna* 43, no. 2 (1988): 31–36.

Franceschi, Franco. "La bottega come spazio di sociabilità." In *La grande storia dell'artigianato*, edited by Franco Franceschi and Gloria Fossi, 65–84. Florence: Giunti, 1999.

Freccero, John. "Medusa and the Madonna of Forlì: Political Sexuality in Machiavelli." In *Machiavelli and the Discourse of Literature*, edited by Albert Russell Ascoli and Victoria Kahn, 161–178. Ithaca, N.Y.: Cornell University Press, 1993.

Freedberg, David. *The Eye of the Lynx: Galileo, His Friends, and the Beginnings of Modern Natural History*. Chicago: University of Chicago Press, 2002.

Fuchs, Barbara. *Mimesis and Empire: The New World, Islam, and European Identities*. Cambridge: Cambridge University Press, 2004.

Gabrieli, G. "Contributi alla storia dell'accademia dei Lincei." Vol. 1. Rome: Accademia Nazionale dei Lincei, 1989.

———. *Luca Valerio linceo, un episodio memorabile della vecchia Accademia Lincea*. Rome: Accademia Nazionale dei Lincei, 1934.

Gabrielli, Francesca Maria, ed., and Shannon McHugh, Melissa Swain, and Francesca Maria Gabrielli, trans. *Renaissance Women's Writing Between the Two Adriatic Shores*. Toronto: Centre for Reformation and Renaissance Studies. Forthcoming.

Gal, Ofer, and Raz Chen-Morris. *Baroque Science*. Chicago: University of Chicago Press, 2013.

Galassi, Adriano and Romano Sarzi. *Alla Syrena: Spezeria del '600 in Mantova*. Mantua: Editoriale Sometti, 2000.

Garzanti, G. "Un banco ebreo in Forlí." *Romagna* 5 (1908): 266–279.

Gentilcore, David. *Healers and Healing in Early Modern Italy*. Manchester: Manchester University Press, 1998.

———. *Medical Charlatanism in Early Modern Italy*. Oxford: Oxford University Press, 2006.

Gentile, Sebastiano, and C. Gilly, eds. *Marsilio Ficino e il ritorno di Ermete Trismegisto/Marsilio Ficino and the Return of Hermes Trimegistus*. Florence: Olschki, 1999.

Giannetti, Laura. *Lelia's Kiss: Imagining Gender, Sex, and Marriage in Italian Renaissance Comedy.* Toronto: University of Toronto Press, 2009.

Giannetto, Raffaella Fabiano. *Medici Gardens: From Making to Design.* Philadelphia: University of Pennsylvania Press, 2008.

Gliubich, Simeone. *Dizionario biografico degli uomini illustri della Dalmazia.* Vienna-Zara, 1836.

Godfrey, John. *The Unholy Crusade.* Oxford: Oxford University Press, 1980.

Gordon, Robin L. *Searching for the Soror Mystica: The Lives and Science of Women Alchemists.* Lanham, Md.: University Press of America, 2013.

Govi, Gilberto. *Tre Lettere di Galileo Galilei.* Rome: Tipografia delle scienze matematiche e fisiche, 1870.

Grafton, Anthony, and Nancy G. Siraisi, eds. *Natural Particulars: Nature and the Disciplines in Early Modern Europe.* Cambridge, Mass.: MIT Press, 1999.

Grafton, Anthony, Glenn W. Most, and Salvatore Settis, eds. "Automata." In *The Classical Tradition,* 109–110. Cambridge, Mass.: Harvard University Press, 2010.

Graziani, Natale, and Gabriella Venturelli. *Caterina Sforza.* Milan: Dall'Oglio, 1987.

Graziosi, Elisabetta. "Arcadia femminile: presenze e modelli." *Filologia e critica* 17, no. 3 (1992): 321–358.

Green, Monica H. "From Diseases of Women to 'Secrets of Women': The Transformation of Gynecological Literature in the Later Middle Ages." *Journal of Medieval and Early Modern Studies* 30 (2000): 5–39.

———. "In Search of an 'Authentic' Women's Medicine: The Strange Fates of Trota of Salerno and Hildegard of Bingen." *Dynamis* 19 (1999): 25–52.

———. "Reconstructing the *Oeuvre* of Trota of Salerno." In *La scuola medica salernitana: Gli autori e i testi,* Edizione Nazionale La Scuola media Salernitana, edited by Danielle Jacquart and Agostino Paravicini Bagliani, 183–233. Florence: SISMEL/Edizioni del Galluzzo, 2007.

———. *Women's Healthcare in the Medieval West.* Aldershot: Ashgate, 2000.

———, ed. *The Trotula: A Medieval Compendium of Women's Medicine.* Philadelphia: University of Pennsylvania Press, 2001.

Grendler, Paul. *Critics of the Italian World.* Madison: University of Wisconsin Press, 1969.

———. *Schooling in Renaissance Italy: Literacy and Learning, 1300–1600.* Baltimore: Johns Hopkins University Press, 1989.

Guccini, Anna Maria. "L'arte dei Semplici: Alchimie e medicina naturalistica tra conoscenza e credenza all'epoca di Caterina." In Novielli, *Caterina Sforza,* 131–138.

Hairston, Julia. "Out of the Archive: Four Newly-Identified Figures in Tullia d'Aragona's *Rime della Signora Tullia di Aragona et di diversi a lei* (1547)." *MLN* 118 (2003): 257–263.

———. "Skirting the Issue: Machiavelli's Caterina Sforza," *Renaissance Quarterly* 53, no. 3 (2000): 687–712.

Hall, Bert S. *Weapons and Warfare in Renaissance Europe: Gunpowder, Technology, and Tactics.* Baltimore: Johns Hopkins University Press, 1997.

Hall, Crystal. "Galileo, Poetry, and Patronage: Giulio Strozzi's *Venetia edificata* and the Place of Galileo in Seventeenth Century Poetry." *Renaissance Quarterly* 66, no. 4 (2013): 1296–1331.

———. *Galileo's Reading.* Cambridge: Cambridge University Press, 2014.

Heilbron, J. L. *Galileo.* Oxford: Oxford University Press, 2010.

Hester, Nathalie. "Failed New World Epics in Baroque Italy." In *Poiesis and Modernity in the Old and New Worlds,* edited by Anthony J. Cascardi and Leah Middlebrook, 201–224. Nashville: Vanderbilt University Press, 2012.

Hill, John. *A History of the Materia Medica.* London: Longman, Hitch and Hawes, 1751.

Hughes, Jonathan. "Alchemy and the Exploration of Late Medieval Sexuality." In *Medieval Virginities,* edited by Anke Bernau, Ruth Evans, and Sarah Salih, 140–166. Toronto: University of Toronto Press, 2003.

Hunter, Lynette. *Men, Women, and the Birthing of Science.* DeKalb: Northern Illinois University Press, 2005.

———. "Women and Domestic Medicine: Lady Experimenters 1570–1620." In Hunter and Hutton, *Women, Science and Medicine,* 89–107.

Hunter, Lynette, and Sarah Hutton, eds. *Women, Science and Medicine 1500–1700: Mothers and Sisters of the Royal Society.* Stroud: Sutton, 1997.

Infelise, Mario. "Books and Politics in Arcangela Tarabotti's Venice." In *Arcangela Tarabotti: A Literary Nun in Baroque Venice,* edited by Elissa B. Weaver, 57–72. Ravenna: Longo, 2006.

Jacquart, Danielle, and Agostino Paravicini Bagliani, eds. *La scuola medica salernitana: Gli autori e i testi,* Edizione Nazionale La Scuola media Salernitana. Florence: SISMEL/Edizioni del Galluzzo, 2007.

Jardine, Lisa. *Worldly Goods: A New History of the Renaissance.* New York: Norton, 1996.

Jordan, Constance. "Bocaccio's In-Famous Women: Gender and Civic Virtue in the *De Claris Mulieribus.*" In *Ambiguous Realities: Women in the Middle Ages and Renaissance,* edited by Carole Levine and Jeannie Watson, 25–47. Detroit: Wayne State University Press, 1987.

———. *Renaissance Feminism: Literary Texts and Political Models.* Ithaca, N.Y.: Cornell University Press, 1990.

Jütte, Daniel. "Trading in Secrets: Jews and the Early Modern Quest for Clandestine Knowledge." *Isis* 103, no. 4 (December 2012): 668–686.

Kahn, Didier. "Alchimia." In *Storia della scienza. Vol. 5: La rivoluzione scientifica.* Treccani, 2002, 389–398. Available at www.treccani.it/enciclopedia/la-rivoluzione-scientifica-i-domini-della-conoscenza-alchimia_(Storia-della-Scienza)/.

Kang, J. M. "Wonders of the Mathematical Magic: Lists of Automata in the Transition From Magic to Science." *Comitatus* 33 (2002): 113–139.

Katritzky, M. A. *Women, Medicine and Theatre 1500–1700: Literary Mountebanks and Performing Quacks.* Aldershot: Ashgate, 2007.

Kavey, Allison. *Books of Secrets: Natural Philosophy in England 1550–1600.* Urbana: University of Illinois Press, 2007.

Kelly, Joan. "Early Feminist Theory and the *Querelle des Femmes,* 1400–1789." *Signs* 8, no. 1 (Autumn 1982): 4–28.

Kelso, Ruth. *Doctrine for the Lady of the Renaissance.* Urbana: University of Illinois Press, 1956.

Kirsop, Wallace. "L'éxegèse alchimique des texts litteraires à la fin du XVIe siècle." *XVIIe Siècle* 120 (1978): 145–156.

Kolsky, Stephen. "The Literary Career of Lucrezia Marinella (1571–1653): The Constraints of Gender and the Writing Woman." In *Religion and Culture in the Italian Renaissance. Essays in Honour of Ian Robertson,* edited by F. W. Kent and Charles Zika, 325–342. Turnhout: Brepols, 2005.

———. "Moderata Fonte, Lucrezia Marinella, Giuseppe Passi: An Early Seventeenth-Century Feminist Controversy." *Modern Language Review* 96 (2001): 973–989.

———. "Moderata Fonte's *Tredici Canti del Floridoro*: Women in a Man's Genre." *Rivista di studi italiani* (1999): 165–184.

———. "Per la carriera poetica di Moderata Fonte: Alcuni documenti poco conosciuti." *Experienze letterarie* (1999): 3–17.

———. "Wells of Knowledge: Moderata Fonte's *Il Merito delle donne.*" *Italianist* 13 (1993): 57–96.

Kuhn, Thomas S. *The Copernican Revolution.* Cambridge, Mass.: Harvard University Press, 1957.

———. *The Structure of Scientific Revolutions.* Chicago: University of Chicago Press, 1962.

LaBalme, Patricia. "Venetian Women on Women: Three Early Modern Feminists." *Archivio Veneto* 5, no. 117 (1981): 81–109.

———. "Women's Roles in Early Modern Venice: An Exceptional Case." In *Beyond Their Sex: Learned Women of the European Past,* edited by Patricia H. LaBalme, 129–152. New York: New York University Press, 1980.

Laghi, Anna. *Cristoforo de Brugora speziale della duchessa Bona e della corte sforzesca.* In *Atti del nono convegno culturale e professionale dei farmacisti dell'alta Italia,* 141–147. Pavia, 1957.

Larner, John. *The Lords of Romagna: Romagnol Society and the Origins of the Signorie.* Ithaca, N.Y.: Cornell University Press, 1965.

Laroche, Rebecca. *Medical Authority and Englishwomen's Herbal Texts, 1550–1650.* Aldershot: Ashgate, 2009.

Lazzari, Laura. *Poesia epica e scrittura femminile nel seicento: L'Enrico di Lucrezia Marinelli.* Leonforte: Insula, 2010.

Le May, Helen Rodnite. *Women's Secrets: A Translation of Pseudo-Albertus Magnus' De Secretis Mulierum with Commentaries.* Albany: State University of New York Press, 1992.

Lensi, Giulio Cesare Orlandi. *L'arte segreta: Cosimo e Francesco de' Medici alchimisti.* Florence: Convivio, 1978.

Leong, Elaine. "Collecting Knowledge for the Family: Recipes, Gender and Practical Knowledge in the Early Modern English Household." *Centaurus* 55, no. 2 (2013): 81–103.

———. "Making Medicines in the Early Modern Household." *Bulletin of the History of Medicine* 82 (2008): 145–168.

Leong, Elaine, and Alisha Rankin, eds. *Secrets and Knowledge in Medicine and Science, 1500–1800.* Aldershot: Ashgate, 2011.

Leong, Elaine, and Sara Pennell. "Recipe Collections and the Currency of Knowledge in the Early Modern 'Medical Marketplace.'" In *Medicine and the Market in England and Its Colonies, c. 1450–c. 1850,* edited by Mark S. R. Jenner and Patrick Wallis, 133–152. New York: Palgrave, 2007.

Lesage, Claire. "Femmes de lettres à Venise aux XVIe et XVIIe siècles: Moderata Fonte, Lucrezia Marinella, Arcangela Tarabotti." *Clio: Histoire, Femmes et Societe* (2001): 135–144.

———. "La littérature des 'secrets' et *I secreti di Isabella Cortese.*" *Chroniques italiennes* 36 (1993): 145–178.

———. "Le savoir alimentaire dans *Il Merito delle donne* de Moderata Fonte." In *La table et ses dessous: culture, alimentation et convivialité en Italie, XIV–XVIe siècles,* edited by Adelin Charles Fiorato and Anna Fontes Baratto, 223–234. Paris: Presses de la Sorbonne Nouvelle, 1999.

Lettinck, Paul. *Aristotle's* Meteorology *and Its Reception in the Arab World.* Leiden: Brill, 1999.

Lev, Elizabeth. *The Tigress of Forlí: Renaissance Italy's Most Courageous and Notorious Countess, Caterina Riario Sforza de' Medici.* New York: Houghton Mifflin, 2012.

Levack, Brian P. *The Witch-Hunt in Early Modern Europe.* London: Longman, 1994.

Lipking, Lawrence. *What Galileo Saw: Imagining the Scientific Revolution.* Ithaca, N.Y.: Cornell University Press, 2014.

Lochman, Daniel T., Maritere López, and Lorna Hurney, eds. *Discourses and Representations of Friendship in Early Modern Europe, 1500–1700.* Aldershot: Ashgate, 2010.

Lochrie, Karma. *Covert Operations: The Medieval Uses of Secrecy.* Philadelphia: University of Pennsylvania Press, 1999.

Logan, Gabriella Berti. "The Desire to Contribute: An Eighteenth-Century Italian Woman of Science." *American Historical Review* 99 (1994): 785–812.

Long, Kathleen P. "Odd Bodies: Reviewing Corporeal Difference in Early Modern Alchemy." In *Gender and Scientific Discourse in Early Modern Culture,* edited by Kathleen P. Long, 63–86. Aldershot: Ashgate, 2010.

Long, Pamela O. *Artisan/Practitioners and the Rise of the New Science, 1400–1600.* Corvallis: Oregon State University Press, 2011.

————. *Openness, Secrecy, Authorship: Technical Arts and the Culture of Knowledge from Antiquity to the Renaissance.* Baltimore: Johns Hopkins University Press, 2001.

Loreti Biondi, M. "Le ricette di Caterina Sforza." *Romagna medica* XI.2 (1969): 246–253.

Lowe, Kate. *Nuns' Chronicles and Convent Culture in Renaissance and Counter-Reformation Italy.* Cambridge: Cambridge University Press, 2003.

Luzio, Alessandro, and Rodolfo Renier. "Buffoni, nani e schiavi dei Gonzaga ai tempi di Isabella d'Este." *Nuova antologia* 34 (1891): 618–650; 25 (1891): 112–146.

————. "Il lusso di Isabella d'Este, Marchesa di Mantova: accessori e segreti della toilette." *Nuova antologia di scienze, lettere ed arti* (1896): 666–688.

Madden, Thomas. *Enrico Dandolo and the Rise of Venice.* Baltimore: Johns Hopkins University Press, 2003.

————. *The Fourth Crusade: The Conquest of Constantinople.* Philadelphia: University of Pennsylvania Press, 1997.

————. *The Fourth Crusade: Event, Aftermath, and Perceptions.* Aldershot: Ashgate, 2008.

Magnanini, Suzanne. "Una selva luminosa: The Second Day of Moderata Fonte's *Il Merito delle donne.*" *Modern Philology* 101, no. 2 (2003): 278–296.

Malpezzi Price, Paola. "Lucrezia Marinella." In *Italian Women Writers: A Bio-Bibliographical Sourcebook,* edited by Rinaldina Russell, 234–242. Westport, Conn.: Greenwood Press, 1994.

————. "Moderata Fonte, Lucrezia Marinella and Their 'Feminist' Work." *Italian Culture* 12 (1994): 201–214.

————. *Moderata Fonte: Women and Life in Sixteenth-Century Venice.* Teaneck, N.J.: Fairleigh Dickinson University Press, 2003.

————. "Venezia Figurata and Women in Sixteenth-Century Venice: Moderata Fonte's Writings." In *Italian Women and the City. Essays,* edited by Janet Levarie Smarr and Daria Valentini, 18–34. Madison, N.J.: Fairleigh Dickinson University Press, 2003.

————. "A Woman's Discourse in the Italian Renaissance: Moderata Fonte's *Il Merito delle donne.*" *Annali d'Italianistica* 7 (1989): 165–181.

Malpezzi Price, Paola, and Christine M. Ristaino. *Lucrezia Marinella and the "Querelle des Femmes" in Seventeenth-Century Italy.* Teaneck, N.J.: Fairleigh Dickinson University Press, 2008.

Marangon, Paolo. "Schede per una reinterpretazione dei rapporti culturali tra Padova e la Polonia nei secoli XIII–XVI." In *Italia Venezia e Polonia tra Medio Evo e Età moderna,* edited by Vittore Branaca and Sante Graciotti, 165–180. Florence: Olschki, 1980.

Marciani, Corrado. "Troiano Navò di Brescia e suo figlio Curzio librai-editori del secolo XVI." *La Bibliofilia* 73 (1971): disp.1, 49–60.

Mari, Francesco et al. "The Mysterious Death of Francesco I and Bianca Cappello: An Arsenic Murder?" *BMJ* (2006): 1299–1301. Available at http://www.bmj.com /content/333/7582/1299.

Marr, Alexander. "*Gentille curiosité:* Wonder-working and the Culture of Automata in the Late Renaissance." In *Curiosity and Wonder from the Renaissance to the Enlightenment,* edited by R. J. W. Evans and Alexander Marr, 149–170. Aldershot: Ashgate, 2006.

———. "Understanding Automata in the Late Renaissance." *Journal de la Renaissance* (2004): 205–222.

Marra, Massimo. *Il pulcinello filosofo chimico: uomini e idee dell'alchimia a Napoli nel periodo del viceregno.* Milan: Mimesis, 2000.

Marshall, Peter. *The Magic Circle of Rudolf II: Alchemy and Astrology in Renaissance Prague.* New York: Walker, 2006.

Martin, Craig. "Meteorology for Courtiers and Ladies: Vernacular Aristotelianism in Renaissance Italy." *Philosophical Readings* 4, no. 2 (2012): 3–14.

———. *Renaissance Meteorology: Pompanazzi to Descartes.* Baltimore: Johns Hopkins University Press, 2011.

Martin, Ruth. *Witchcraft and the Inquisition in Venice 1550–1650.* Oxford: Blackwell, 1989.

Martinón-Torres, Marcos. "The Tools of the Chymist: Archeological and Scientific Analyses of Early Modern Laboratories." In *Chymists and Chymistry: Studies in the History of Early Modern Chemistry,* edited by Lawrence M. Principe, 149–164. Sagamore Beach, Mass.: Watson Publishing International, 2007.

Massey, Lyle. "The Alchemical Womb: Johann Remmelin's *Captotrum Microcosmicum.* In McCall, Roberts, and Fiorenza, *Visual Cultures of Secrecy,* 208–228.

Mayr, O. *Authority, Liberty, and Automatic Machinery in Early Modern Europe.* Baltimore: Johns Hopkins University Press, 1986.

Mazzotti, Massimo. "Maria Gaetana Agnesi: The Unusual Life and Mathematical Work of an Eighteenth-Century Woman." *Isis* 92 (2001): 657–683.

———. *The World of Maria Gaetana Agnesi, Mathematician of God.* Baltimore: Johns Hopkins University Press, 2007.

Mazzuchelli, Giammaria. *Gli scrittori d'Italia,* v. 1. Brescia: G. Bossini, 1753.

McCall, Timothy, Sean E. Roberts, and Giancarlo Fiorenza, eds. *Visual Cultures of Secrecy in Early Modern Europe.* Kirksville, Mo.: Truman State University Press, 2013.

McClure, George. *The Culture of Profession in Late Renaissance Italy.* Toronto: University of Toronto Press, 2004.

McGough, Laura. *Gender, Sexuality and Syphilis in Early Modern Venice: The Disease that Came to Stay.* New York: Palgrave Macmillan, 2011.

McLeod, Glenda. *Virtue and Venom: Catalogs of Women From Antiquity to the Renaissance.* Ann Arbor: University of Michigan Press, 1991.

Merchant, Carolyn. *The Death of Nature: Women, Nature and the Scientific Revolution*. San Francisco: Harper and Row, 1980.

Messbarger, Rebecca. *The Lady Anatomist: The Life and Work of Anna Morandi Manzolini*. Chicago: University of Chicago Press, 2010.

Milanesi, Carlo. "Lettere di Giovanni de' Medici, detto delle Bande Nere." *Archivio storico italiano*. Nuova serie, v. IX, part 2. Florence: Vieusseux, 1859.

Mola, Luca. *The Silk Industry of Renaissance Venice*. Baltimore: Johns Hopkins University Press, 2000.

Mombello, G. *Sur les traces d'Alexis Jure de Chieri. Le problème des francisants piémontais au XVI siècle*. Geneva: Slatkine, 1984.

Moran, Bruce. "Courts and Academies." In *The Cambridge History of Science. Vol. 3: Early Modern Science*, edited by Lorraine Daston and Katharine Park, 251–271. Cambridge: Cambridge University Press, 2006.

———. *Distilling Knowledge: Alchemy, Chemistry, and the Scientific Revolution*. Cambridge, Mass.: Harvard University Press, 2005.

———. "Introduction." *Isis* 102, no. 2 (2011): 300–304.

Moran, Bruce, ed. *Patronage and Institutions. Science, Technology, and Medicine at the European Court, 1500–1750*. Rochester, N.Y.: Boydell Press, 1991.

Morison, S. *Eustachio Celebrino da Udine Calligrapher, Engraver, and Writer for the Venetian Printing Press*. Paris: Pegasus Press, 1929.

Mosley, Adam. *Bearing the Heavens: Tycho Brahe and the Astronomical Community of the Late Sixteenth Century*. Cambridge: Cambridge University Press, 2007.

Moss, Jean Dietz. "Galileo's *Letter to Christina*: Some Rhetorical Considerations." *Renaissance Quarterly* 36, no. 4 (Winter 1983): 547–576.

Mudge, Ken. "A History of Grafting." *Horticultural Reviews* 35 (2009): 438–490.

Muir, Edward. *Civic Ritual in Renaissance Venice*. Princeton, N.J.: Princeton University Press, 1986.

Muraro, Luisa. *Giambattista Della Porta mago e scienziato*. Milan: Feltrinelli, 1978.

Navarrini, Roberto. "La guerra chimica di Vincenzo Gonzaga." *Civiltà mantovana* 4 (1969): 43–47.

Newman, William R. "The Homunculus and His Forebears: Wonders of Art and Nature." In Grafton and Siraisi, *Natural Particulars*, 321–345.

———. *Promethean Ambitions: Alchemy and the Quest to Perfect Nature*. Chicago: University of Chicago Press, 2004.

———. "What Have We Learned from the Historiography of Alchemy?" *Isis* 102, no. 2 (2011): 313–321.

Newman, William R., and Anthony Grafton, eds. *Secrets of Nature: Astrology and Alchemy in Early Modern Europe*. Cambridge, Mass.: MIT Press, 2001.

Niccoli, Ottavia. *Prophecy and People in Renaissance Italy*. Translated by Lydia G. Cochrane. Princeton, N.J.: Princeton University Press, 1990.

Nigrisoli, V. "Cenni sul ricettario di Caterina Sforza." *La Piê* 9 (1928): 178–182.

———. "Spigolature dal ricettario di Caterina Sforza." *La Piê* 9 (1929): 34–36, 38–39.

Novielli, Valeria, ed. *Caterina Sforza: Una donna del Cinquecento.* Imola: Mandragora, 2000.

Nummedal, Tara. "Alchemical Reproduction and the Career of Anna Maria Zieglerin." *Ambix* 48 (2001): 56–68.

———. *Alchemy and Authority in the Holy Roman Empire.* Chicago: University of Chicago Press, 2007.

———. "Anna Zieglerin's Alchemical Revelations." In *Secrets and Knowledge in Medicine and Science, 1500–1800,* edited by Elaine Leong and Alisha Rankin, 125–142. Aldershot: Ashgate, 2011.

———. "Words and Works in the History of Alchemy." *Isis* 102, no. 2 (2011): 330–337.

Odorisio, Ginevra Conti. *Donna e società nel Seicento.* Rome: Bulzoni, 1979.

Ogrinc, Will H.L. "Western Society and Alchemy, 1200–1500." *Journal of Medieval History* 6 (1980): 127–137.

Oliva, Fabio. *Vita di Caterina Sforza Signora di Forlì scritta da F.O. Forlivese.* Forlì, 1821.

Ong, Walter J. "Commonplace Rhapsody: Ravisius Textor, Zwinger, and Shakespeare." In *Classical Influences on European Culture,* edited by Robert Ralph Bolgar, 91–126. Cambridge: Cambridge University Press, 1976.

Osler, Margaret J. "The Gender of Nature and the Nature of Gender in Early Modern Natural Philosophy." In *Men, Women, and the Birthing of Modern Science,* edited by Judith P. Zinsser, 71–85. DeKalb: Northern Illinois University Press, 2005.

Osler, Margaret J., ed. *Rethinking the Scientific Revolution.* Cambridge: Cambridge University Press, 2000.

Outram, Dorinda. "Gender." In *Cambridge History of Science,* vol. 3., edited by Lorraine Daston and Katherine Park, 797–817. Cambridge: Cambridge University Press, 2006.

Pagel, Walter. *Paracelsus: An Introduction to Philosophical Medicine in the Era of the Renaissance,* 2nd ed. Basel: Karger, 1982.

Pal, Carol. *Republic of Women: Rethinking the Republic of Letters in the Seventeenth Century.* Cambridge: Cambridge University Press, 2012.

Palmero, Giuseppe. "La distillazione: I suoi prodotti ed i suoi usi nell'ambito della "letteratura dei segreti" tra Quattrocento e Cinquecento." In *Grappa & Alchimia. Un percorso nella millenaria storia della distillazione.* Atti del convegno "Dall'Alchimia alla grappa: un percorso millenario nella distillazione," Greve in Chianti, 11 settembre 1999, edited by Allen J. Grieco, 17–32. Rome: Agra Editrice, 1999.

Panofsky, Erwin. *Galileo as a Critic of the Arts.* The Hague: Nijhoff, 1954.

Park, Katharine. "Dissecting the Female Body: From Women's Secrets to the
Secrets of Nature." In *Crossing Boundaries,* edited by Jane Donawerth and Adele
Seef, 29–47. Newark, Del.: University of Delaware Press, 2000.

———. *Doctors and Medicine in Renaissance Florence.* Princeton, N.J.: Princeton
University Press, 1985.

———. "Medicine and Magic: The Healing Arts." In *Gender and Society in
Renaissance Italy,* edited by Judith C. Brown and Robert C. Davis, 129–149. New
York: Longman, 1998.

———. "Natural Particulars: Medical Epistemology, Practice, and the Literature
of the Healing Springs." In Grafton and Siraisi, *Natural Particulars,* 347–367.

———. *Secrets of Women: Gender, Generation, and the Origin of Human Dissection.*
New York: Zone Books, 2006, repr. 2010.

Park, Katharine and Lorraine Daston. "Introduction: The Age of the New." In
The Cambridge History of Science, vol. 3, edited by Katharine Park and Lorraine
Daston, 1–18. Cambridge: Cambridge University Press, 2006.

Pasolini, Pier Desiderio. *Caterina Sforza.* 3 vols. Rome: Loescher, 1893.

———. "Nuovi documenti su Caterina Sforza." *Atti e Memorie: Deputazione di
Storia per la Romagna* 15 (1897): 72–206.

Patai, Raphael. *The Jewish Alchemists.* Princeton, N.J.: Princeton University Press,
1995.

Pennell, Sara. "Perfecting Practice? Women's Manuscript Recipes in Early Modern
England." In *Early Modern Women's Manuscript Writing,* edited by V. Burke and
J. Gibson, 237–258. Aldershot: Ashgate, 2004.

Perifano, Alfredo. *L'alchimie à la Cour de Côme Ier de Médicis: Savoirs, culture et
politique.* Paris: Honore Champion, 1997.

Peterson, Mark A. *Galileo's Muse: Renaissance Mathematics and the Arts.* Cambridge,
Mass.: Harvard University Press, 2011.

Petrocchi, Giorgio. "Bona Sforza, regina di Polonia, e Pietro Aretino." In *Italia
Venezia e Polonia tra Medio Evo e Età moderna,* edited by Vittore Branaca and
Sante Graciotti, 325–331. Florence: Olschki, 1980.

Pezzini, Serena. "Ideologia della conquista, ideologia dell'accoglienza: 'La Scanderbeide' di Margherita Sarrocchi (1623)." *MLN* 120, no. 1 (2005): 190–222.

Phillips, Ursula. "Polish Women Authors." In *A History of Central European Women's
Writing.* Edited by Celia Hawkesworth. New York: Palgrave Macmillan, 2001.

Pizzagalli, Daniela. *La signora del rinascimento. Vita e splendori di Isabella d'Este alla
corte di Mantova.* Milan: Rizzoli, 2001.

Plastina, Sandra. "'Considerar la mutatione dei tempi e delli stati e degli uomini':
Le *Lettere di philosophia naturale di Camilla Erculiani." Bruniana e Campelliania*
20, no. 1 (2014): 145–156.

Pollock, Linda. *With Faith and Physic: The Life of a Tudor Gentlewoman, Lady Grace
Mildmay, 1552–1620.* New York: St. Martin's Press, 1993.

Pomata, Gianna. "Medicina delle monache. Pratiche terapeutiche nei monasteri femminili di Bologna in età moderna." In *I monasteri femminili come centri di cultura fra Rinascimento e Barocco,* edited by Gianna Pomata and Gabriella Zarri, 331–363. Rome: Edizioni di storia e letterature, 2005.

———. "Was There a *Querelle des Femmes* in Early Modern Medicine?" *Arenal* 20, no. 2 (2013): 313–341.

Pomata, Gianna, and Nancy G. Siraisi, eds. *Historia: Empiricism and Erudition in Early Modern Europe.* Cambridge, Mass.: MIT Press, 2005.

Poppi, Antonio. *Introduzione all'aristotelianismo padovano.* Padua: Antenore, 1970.

Prażmowska, Anita J. *A History of Poland.* Hampshire, UK: Palgrave Macmillan, 2004.

Principe, Lawrence M. "Alchemy Restored." *Isis* 102, no. 2 (2011): 305–312.

———. "Revealing Analogies: The Descriptive and Deceptive Roles of Sexuality and Gender in Latin Alchemy." In *Hidden Intercourse: Eros and Sexuality in the History of Western Esotericism,* edited by Wouter J. Hanegraaff and Jeffrey J. Kripal, 209–229. Leiden: Brill, 2008.

———. *The Secrets of Alchemy.* Chicago: University of Chicago Press, 2012.

Principe, Lawrence M., ed. *Chymists and Chymistry: Studies in the History of Alchemy and Early Modern Chemistry.* Sagamore Beach, Mass.: Watson Publishing International, 2007.

Principe Lawrence M., and William R. Newman. "Some Problems with the Historiography of Alchemy." In *Secrets of Nature: Astrology and Alchemy in Early Modern Europe,* edited by William R. Newman and Anthony Grafton, 385–432. Cambridge, Mass.: MIT Press, 2001.

Pullan, Brian S., ed. *Crisis and Change in the Venetian Economy of the Sixteenth and Seventeenth Centuries.* London: Methuen, 1968.

Pumfrey, Stephen, Paolo L. Rossi, and Maurice Slawinski, eds. *Science, Culture and Popular Belief in Renaissance Europe.* Manchester: Manchester University Press, 1994.

Quaintance, Courtney. *Le Feste, Written By Moderata Fonte.* In *Scenes from Italian Convent Life 1480–1680: An Anthology of Theatrical Texts and Contexts,* edited by Elissa B. Weaver, 193–231. Ravenna: Longo, 2009.

Quint, David. *Epic and Empire.* Princeton, N.J.: Princeton University Press, 1993.

Randall, John H. Jr. "Paduan Aristotelianism Reconsidered." In *Philosophy and Humanism: Renaissance Essays in Honor of Paul Oskar Kristeller,* edited by Edward Patrick Mahoney. Leiden: Brill, 1976.

Rankin, Alisha. "Becoming an Expert Practitioner: Court Experimentalism and the Medical Skills of Anna of Saxony (1532–1585)." *Isis* 98, no. 1 (March 2007): 23–53.

———. "Exotic Materials and Treasured Knowledge: The Valuable Legacy of Noblewomen's Remedies in Early Modern Germany." *Renaissance Studies* 28, no. 4 (2014): 533–555.

———. *Panaceia's Daughters: Noblewomen as Healers in Early Modern Germany.*
Chicago: University of Chicago Press, 2013.

Rattansi, Piyo, and Antonio Clericuzio, eds. *Alchemy and Chemistry in the 16th and 17th Centuries.* Dordrecht: Kluwer Academic. 1994.

Ray, Meredith K. "Experiments With Alchemy: Caterina Sforza in Early Modern Scientific Culture." In *Gender and Scientific Discourse in Early Modern Culture,* edited by Kathleen P. Long, 139–164. Aldershot: Ashgate, 2010.

———. "Impotence and Corruption: Sexual Function and Dysfunction in Early Modern Italian Books of Secrets." In *Cuckoldry, Impotence and Adultery in Europe (15th–17th Century),* edited by Sara Matthews-Grieco, 125–146. Aldershot: Ashgate, 2014.

———. "La castità conquistata: The Function of the Satyr in Pastoral Drama." *Romance Languages Annual* 9 (1996): 312–321.

———. "Letters and Lace: Arcangela Tarabotti and Convent Culture in Seicento Venice." In *Early Modern Women and Transnational Communities of Letters,* edited by J. Campbell and A. Larsen, 45–72. Aldershot: Ashgate, 2009.

———. "Prescriptions for Women: Alchemy, Medicine and the Renaissance *Querelle des Femmes.*" In *Women Writing Back/Writing Women Back: Transnational Perspectives from the Late Middle Ages to the Dawn of the Modern Era,* edited by Anke Gilleir, Alicia C. Montoya, and Suzan van Dijk, 135–162. Leiden: Brill, 2010.

———. "Textual Collaboration and Spiritual Partnership in Sixteenth-Century Italy: The Case of Ortensio Lando and Lucrezia Gonzaga." *Renaissance Quarterly* 62, no. 3 (Fall 2009): 694–747.

———. "Un'officina di lettere: le *Lettere di molte valorose donne* e la fonte della 'dottrina femminile.'" *Esperienze letterarie* 3 (2001): 69–91.

———. *Writing Gender in Women's Letter Collections of the Italian Renaissance.* Toronto: University of Toronto Press, 2009.

Redondi, Pietro. *Galileo eretico.* Torino: Einaudi, 1983.

Reeves, Eileen. *Galileo's Glassworks: The Telescope and the Mirror.* Cambridge, Mass.: Harvard University Press, 2008.

Renzetti, Emmanuela, and Rodolfo Taini. "Le cure dell'amore: Desiderio e passione in alcuni libri di segreti." *Sanità scienza e storia: Rivista del Centro italiano de storia sanitaria e ospitaliera* (1986): 33–86.

Righini, Guglielmo. "L'oroscopo galileaiano di Cosimo II de' Medici." *Annali dell'Istituto e Museo di Storia della Scienza di Firenze* I (1976): 29–36.

Riskin, Jessica, ed. *Genesis Redux: Essays in the History and Philosophy of Artificial Life.* Chicago: University of Chicago Press, 2007.

Rizzardini, Massimo. "Lo strano caso della Signora Isabella Cortese, professoressa di *secreti.*" *Philosophia* 2, no. 1 (2010): 45–84.

————. *Secretum. Alchimia, medicina e politica del corpo nel Rinascimento.* Milan: Bevivino, 2009.

Roberts, Gareth. *The Mirror of Alchemy: Alchemical Ideas and Images in Manuscripts and Books from Antiquity to the Seventeenth Century.* London: British Library, 1994.

Robin, Diana. *Publishing Women: Salons, The Presses, and the Counter-Reformation in Sixteenth-Century Italy.* Chicago: University of Chicago Press, 2007.

————. "A Renaissance Feminist Translation of Xenophon's *Oeconomicus.*" In *Roman Literature, Gender, and Reception: Domina Illustris. Essays in honor of Judith Peller Hallett. Routledge monographs in classical studies, 13,* edited by Donald Lateiner, Barbara K. Gold, and Judith Perkins, 207–221. London and New York: Routledge, 2013.

————. "Women on the Move: Trends in Anglophone Studies of Women in the Italian Renaissance." *I Tatti Studies in the Italian Renaissance* 16, nos. 1/2 (2013): 13–25.

Römer, Zdenka Janeković. "Marija Gondola Gozze. *La querelle des femmes* u renesansnom Dubrovniku." In *Žene u Hrvatskoj: Ženska I kulturna povijest,* edited by Andrea Feldman, 105–123. Zagreb: Ženska infoteka, 2004.

Rosenthal, Margaret. "Epilogue." In *Aretino's Dialogues,* edited and translated by Raymond Rosenthal, 387–402. New York: Marsilio, 1994.

————. *The Honest Courtesan: Veronica Franco, Citizen and Writer in Sixteenth-Century Venice.* Chicago: University of Chicago Press, 1993.

Ross, Sarah Gwyneth. *The Birth of Feminism: Woman as Intellect in Renaissance Italy and England.* Cambridge, Mass.: Harvard University Press, 2009.

Ruggiero, Guido. *Binding Passions: Tales of Magic, Marriage and Power at the End of the Renaissance.* New York: Oxford University Press, 1993.

————. "Witchcraft and Magic." In *A Companion To the Worlds of the Renaissance,* edited by Guido Ruggiero, 475–490. Oxford: Wiley-Blackwell, 2007.

Ryan, W. F., and Charles B. Schmitt, eds. *Pseudo-Aristotle: The Secret of Secrets. Sources and Influences.* London: Warburg Institute, 1982.

Sampson, Lisa. *Pastoral Drama in Early Modern Italy: The Making of a New Genre.* London: Legenda, 2006.

Sanesi, Ireneo. *Il cinquecentista Ortensio Lando.* Pistoia: Fratelli Bracali, 1893.

————. "Tre epistolari del Cinquecento." *Giornale storico della letteratura italiana* 24 (1894): 1–32.

Sarasohn, Lisa T. *The Natural Philosophy of Margaret Cavendish: Reason and Fancy during the Scientific Revolution.* Baltimore: Johns Hopkins University Press, 2010.

Schiebinger, Londa. "Maria Winkelman at the Berlin Academy: A Turning Point for Women in Science." *Isis* 78 (1987): 174–200.

————. *The Mind Has No Sex? Women in the Origins of Modern Science.* Cambridge, Mass.: Harvard University Press, 1989.

———. *Nature's Body: Gender in the Making of Modern Science*. New Brunswick, N.J.: Rutgers University Press, 2004.

———. "Women of Natural Knowledge." *Cambridge Histories Online*. Available at http:dx.doi.org/10.1017/CHL9780521572446.008/.

Schmitt, Charles. "Experience and Experiment: A Comparison of Zabarella's View with Galileo's in *De Motu*." *Studies in the Renaissance* 16 (1969): 80–138.

Schutte, Anne Jacobson. *Aspiring Saints: Pretense of Holiness, Inquisition, and Gender in the Republic of Venice (1618–1750)*. Baltimore: Johns Hopkins University Press, 2001.

Segal, Harold. *Renaissance Culture in Poland: The Rise of Humanism 1470–1543*. Ithaca, N.Y.: Cornell University Press, 1989.

Seitz, Jonathan. *Witchcraft and Inquisition in Early Modern Venice*. New York: Cambridge University Press, 2011.

Servolini, K. "Le miracolose 'ricette' di Madonna Caterina da Forlí." *Romagna Eroica* 3 (1943): f. IX, 2–5.

Shaw, James, and Evelyn Welch. *Making and Marketing Medicine in Renaissance Florence*. *Clio Medica* 89/The Wellcome Series in the History of Medicine. Amsterdam: Rodopi, 2011.

Shea, William, and Mariano Artigas. *Galileo in Rome: The Rise and Fall of a Troublesome Genius*. Oxford: Oxford University Press, 2003.

Shemek, Deanna. "In Continuous Expectation: Isabella d'Este's Epistolary Desire." In *Phaethon's Children: The Este Court and Its Culture in Early Modern Ferrara*, edited by Dennis Looney and Deanna Shemek, 269–300. Tempe: Arizona Center for Medieval and Renaissance Studies, 2005.

Singer, Charles. *The Earliest Chemical Industry: An Essay in the Historical Relations of Economics and Technology Illustrated from the Alum Trade*. London: Folio Society, 1948.

Siraisi, Nancy. *Medieval and Early Renaissance Medicine: An Introduction to Knowledge and Practice*. Chicago: University of Chicago Press, 1990.

Slaski, Jan. "La letteratura italiana nella Polonia fra il Rinascimento e il Barrocco." In *Cultura e nazione in Italia e Polonia dal Rinascimento all'Illuminismo*, edited by Vittore Branca and Sante Graciotti, 219–252. Florence: Olschki, 1986.

Smarr, Janet Levarie. *Joining the Conversation: Dialogues by Renaissance Women*. Ann Arbor: University of Michigan Press, 2005.

Smith, Pamela H. *The Business of Alchemy: Science and Culture in the Holy Roman Empire*. Princeton, N.J.: Princeton University Press, 1994.

———. "Science on the Move: Recent Trends in the History of Early Modern Science." *Renaissance Quarterly* 62 (2009): 345–375.

Sobel, Dava. *Galileo's Daughter: A Historical Memoir of Science, Faith, and Love*. New York: Walker, 1999.

———. *Letters to Father: Suor Maria Celeste to Galileo, 1623–1633*. New York: Walker, 2001.

Sperling, Jutta. *Convents and the Body Politic in Late Renaissance Venice.* Chicago: University of Chicago Press, 1999.

Spiller, Elizabeth J. "Introduction." *Seventeenth-Century English Recipe Books: Cooking, Physic and Chirurgery in the Works of Elizabeth Talbot Grey and Aletheia Talbot Howard,* ix–li. Aldershot: Ashgate, 2008.

Spratt, George. *Flora medica.* 2 vols. London: Callow and Wilson, 1830.

Stone, Daniel. *The Polish-Lithuanian State, 1386–1795.* Seattle: University of Washington Press, 2001.

Storey, Tessa. "Face Waters, Oils, Love Magic and Poison: Making and Selling Secrets in Early Modern Rome." In *Secrets and Knowledge in Medicine and Science, 1500–1800,* edited by Elaine Leong and Alisha Rankin, 143–166. Aldershot: Ashgate, 2011.

Strocchia, Sharon T. *Nuns and Nunneries in Renaissance Florence.* Baltimore: Johns Hopkins University Press, 2009.

———. "The Nun Apothecaries of Renaissance Florence: Marketing Medicines in the Convent." *Reniassiance Studies* 25 (2011): 627–647.

———. "Women and Health Care in Early Modern Europe." *Renaissance Studies* 28, no. 4 (2014): 496–514.

Stuard, Susan Moshard. *A State of Deference: Ragusa/Dubrovnik in the Medieval Centuries.* Philadelphia: University of Pennsylvania Press, 1992.

Tabanelli, Mario. *Il biscione e la rosa: Caterina Sforza, Girolamo Riario e i loro primi discendenti.* Faenza, 1972.

———. "Ricette di medicina dal libro *Degli experimenti* di Caterina Sforza." *La Piê* 43 (1970): 79–82, 144–149, 195–198, 248–250.

Targioni Tozzetti, Giovanni, and Adolfo Targioni-Tozzetti, eds. *Notizie della vita e delle opere di Pier'Antonio Micheli.* Florence: Le Monnier, 1858.

Tazbir, Janusz. *A State without Stakes. Polish Religious Toleration in the Sixteenth and Seventeenth Centuries.* New York: The Kosciuszko Foundation Twayne Publishers, 1973.

Tebeaux, Elizabeth. "Women and Technical Writing, 1475–1700: Technology, Literacy, and Development of a Genre." In Hunter and Hutton, *Women, Science and Medicine,* 29–62.

Tessicini, Dario. "The Comet of 1577 in Italy: Astrological Prognostications and Cometary Theory at the End of the Sixteenth Century." In *Celestial Novelties on the Eve of the Scientific Revolution, 1540–1630,* edited by Patrick J. Boner and Dario Tessicini, 57–84. Florence: Leo Olschki, 2013.

Thorndike, Lynn. *A History of Magic and Experimental Science.* 8 vols. New York: Macmillan, 1923–58.

Tilche, Giovanna. *Maria Gaetana Agnesi: La scienziata santa del '700.* Milan: Rizzoli, 1984.

Torbarina, Josip. *Tassovi soneti i madrigali u čast Cvijete Zuzorić Dubrovkinje. Hrvatsko kolo,* no. 21. Zagreb, 1940.

Tosi, Luisa. "Marie Meurdrac: Paracelsian Chemist and Feminist." *Ambix* 48, no. 2 (2002): 69–82.

Trollope, T. Adolphus. *A Decade of Italian Women.* 2 vols. London: Chapman and Hall, 1859.

Turpin, Adriana. "The New World Collections of Duke Cosimo I de' Medici and their role in the creation of a *Kunst-* and *Wunderkammer* in the Palazzo Vecchio." In *Curiosity and Wonder from the Renaissance to the Enlightenment,* edited by R. J. W. Evans and Alexander Marr, 63–86. Aldershot: Ashgate, 2006.

Ulewicz, Tadeusz. "Polish Humanism and its Italian Sources: Beginnings and Historical Development." In *The Polish Renaissance in Its European Context,* edited by Samuel Fizman, 215–235. Bloomington: Indiana University Press, 1988.

Vaccalluzzo, Nunzio. *Galileo Galilei nella poesia del suo secolo.* Milan: Remo Sandron, 1920.

———. *Galileo letterato e poeta.* Catania: Giannotta, 1896.

Verdile, Nadia. "Contributi alla biografia di Margherita Sarrocchi." *Rendiconti dell'Accademia di Archeologia, Lettere e Belle Arti di Napoli* 61 (1989–1990): 165–206.

Veress, Endre. *Berzeviczy Márton, 1538–1596.* Budapest: A Magyar Történelm. Társulat Kiadása, 1911.

Vidovich, Renzo de'. *Albo d'Oro delle famiglie nobili patrizie e illustri nel Regno di Dalmazia.* Trieste: Fondazione Scientifico Culturale Rustia Traine, 2004.

Weaver, Elissa, ed. *Arcangela Tarabotti. A Literary Nun in Baroque Venice.* Ravenna: Longo, 2006.

Webster, Charles. *From Paracelsus to Newton: Magic and the Making of Modern Science.* Cambridge: Cambridge University Press, 1982.

———. *Paracelsus: Medicine, Magic and Mission at the End of Time.* New Haven, Conn.: Yale University Press, 2008.

Weeks, Andrew. *Paracelsus: Speculative Theory and the Crisis of the Early Reformation.* Albany: State University of New York Press, 1997.

Welch, Evelyn S. *Shopping in the Renaissance: Consumer Cultures in Italy 1400–1600.* New Haven, Conn.: Yale University Press, 2005.

Westfall, Richard S. "Science and Patronage: Galileo and the Telescope." *Isis* 76 (1986): 11–30.

Westwater, Lynn Lara. "A Cloistered Nun Abroad: Arcangela Tarabotti's International Writing Career." In *Women Writing Back/Writing Women Back: Transnational Perspectives from the Late Middle Ages to the Dawn of the Modern Era,* edited by Anke Gilleir, Alicia C. Montoya, and Suzan van Dijk, 283–307. Leiden: Brill, 2010.

———. "The Disquieting Voice: Women's Writing and Antifeminism in Seventeenth-Century Venice." Ph.D. diss. University of Chicago, 2003.

————. "'Le false obiezioni de' nostri calunniatori.' Lucrezia Marinella Responds to the Misogynist Tradition." *Bruniana e campanelliana* 12, no. 1 (2006): 1–15.

————. "Lucrezia Marinella (1571–1653)." In *Encyclopedia of Women in the Renaissance: Italy, France, and England,* edited by Diana Maury Robin, Anne Larson, and Carol Levin, 234–237. Santa Barbara: ABC-Clio, 2007.

————. "A Rediscovered Friendship in the Republic of Letters: The Unpublished Correspondence of Arcangela Tarabotti and Ismael Boulliau." *Renaissance Quarterly* 65 (2012): 67–134.

Whaley, Leigh Ann. *Women and the Practice of Medical Care in Early Modern Europe 1400–1800.* Basingstoke: Palgrave Macmillan, 2011.

Wheeler, Jo. "The Ragusan Connection." Available at http://renaissancesecrets. blogspot.com/2013/04/the-ragusan-connection.html (accessed September 16, 2013).

————. *Renaissance Secrets: Recipes and Formulas.* London: V and A Publishing, 2009.

Wlassics, Tibor. *Galilei critico letterario.* Ravenna: Longo, 1974.

Wootton, David. *Galileo: Watcher of the Skies.* New Haven, Conn.: Yale University Press, 2010.

Yates, Frances. *Giordano Bruno and the Hermetic Tradition.* Chicago: University of Chicago Press, 1991.

Zambelli, Paola. "Fine del mondo o inizio della propaganda?" In *Scienze credenze occulte livelli di cultura,* 291–368. Firenze: Olschki, 1982.

Zanette, Emilio. "Bianca Cappella e la sua poetessa." *Nuova antologia* 88 (1953): 455–468.

————. *Suor Arcangela monaca del Seicento.* Rome-Venice: Istituto per la collaborazione culturale, 1960.

Zatti, Sergio. *Il modo epico.* Rome: Laterza, 2000.

Zupko, Ronald. *Italian Weights and Measures from the Middle Ages to the Nineteenth Century.* Philadelphia: American Philosophical Society, 1981.

Acknowledgments

This book began as a mission to learn more about the enigmatic Caterina Sforza and her alchemical manuscript. That quest soon turned into a much larger—and more daunting—project. I am grateful to the many people who generously shared their expertise with me along the way, and to the institutions and organizations that made it possible for me to devote myself to research and writing. The Penn Humanities Forum at the University of Pennsylvania provided support and an intellectual community in 2008–2009, and again in 2010–2011. In 2010–2011 a fellowship from the National Endowment for the Humanities supported my archival research in Italy; a 2011–2012 University of Delaware General University Research Grant and a 2013 research award from the University of Delaware's Institute for Global and Area Studies helped fund return trips to study the Sarrocchi–Galileo correspondence. I am grateful to the many librarians and archivists who helped me navigate the labyrinths of the Biblioteca Nazionale Centrale, Museo Galileo, and Archivio di Stato in Florence, and the Biblioteca Universitaria Alessandrina, Biblioteca dell'Accademia Nazionale dei Lincei e Corsiniana, and Archivio di Stato in Rome. I am especially indebted to the owners of the sixteenth-century manuscript of Sforza's *Experimenti,* who were so very kind as to allow me access to their library: exploring its contents was truly the experience of a lifetime.

Back in the United States, I would like to thank Christine Poggi of the University of Pennsylvania's Alice Paul Center for Research on Gender, Sexuality and Women for appointing me as a Research Fellow in 2013–2015, and Fabio Finotti for welcoming me to Penn's Italian Department as a visiting scholar. John Pollack of Penn's Kislak Center for Special Collections, Rare Books, and Manuscripts helped me locate materials and obtain images for reproduction. At the University of Delaware, Gary Ferguson and Richard Zipser were wonderfully supportive department chairs who encouraged this project at every stage; my thanks also go to Laura Salsini and my colleagues in the Italian Program. For their support and feedback at various stages of this project, I am grateful to Virginia Cox, Paula Findlen, Valeria Finucci, Diana Robin, and Elissa Weaver. I also wish to thank my anonymous readers at Harvard University Press for their insightful suggestions on the manuscript. Kate Lowe was enthusiastic about this project from the beginning; C. Ian Stevenson and Andrew Kinney shepherded it through the editorial process with care. Many thanks to Kimberly Giambattisto and Therese Malhame for their attentiveness and expertise in the preparation of the book.

Themes presented in Chapter 1 were developed in my earlier article, "Experiments with Alchemy: Caterina Sforza in Early Modern Scientific Culture," in *Gender and Scientific Discourse in Early Modern Scientific Culture,* ed. Kathleen P. Long, 139–164 (Aldershot: Ashgate, 2010). Sections of Chapter 3 are informed by my previous article, "Prescriptions for Women: Alchemy, Medicine, and the Renaissance *Querelle des Femmes,*" published in *Women Writing Back/Writing Women Back: Transnational Perspectives from the Late Middle Ages to the Dawn of the Modern Era,* ed. Anke Gillier, Alicia C. Montoya, and Suzan Van Dijk, 135–162 (Leiden: Brill, 2010).

Writing a book can be a solitary undertaking, and I am grateful for the friendship of those who have kept me company along the way: Arcana Albright, Susan Barr-Toman, Meridith Greenbaum, Ash Hanson, Mark Jurdjevic, Dana Katz, Lia Markey, and Lynn Westwater. Micah Kleit offered endless support and a sharp editorial eye. Courtney Quaintance gave me a home base in Rome; in Philadelphia, Lillyrose Veneziano Broccia generously shared her office with me. My family encouraged me at every stage: Lauren, John, Miranda, and Grace Pollard, David Ray, and, especially, Crennan Ray, who went above and beyond and read every word.

Most of all, I want to thank my son, Owen Johnson. He is my biggest supporter and my greatest inspiration, and he knows more about Caterina Sforza than any other ten-year-old I know.

Index